Hazards and Safety in Process Industries

Hazards and Safety in Process Industries

Case Studies

Mihir Kumar Purkait, Piyal Mondal,
Murchana Changmai, Vikranth Volli
and Chi-Min Shu

CRC Press
Taylor & Francis Group
Boca Raton London New York

CRC Press is an imprint of the
Taylor & Francis Group, an **informa** business

First edition published 2021
by CRC Press
6000 Broken Sound Parkway NW, Suite 300, Boca Raton, FL 33487-2742

and by CRC Press
2 Park Square, Milton Park, Abingdon, Oxon, OX14 4RN

CRC Press is an imprint of Taylor & Francis Group, LLC

ISBN: 978-0-367-51651-2 (hbk)
ISBN: 978-0-367-51652-9 (pbk)
ISBN: 978-1-003-05476-4 (ebk)

Typeset in Times
by SPi Global, India

This book is dedicated to our parents and families, for their endless support and motivation...

Contents

Preface

Throughout human history, our errors, indecorous disposal, storage, manufacturing, transportation, and usage of hazardous materials from process industries has resulted in serious destruction to property, people, and environment compromising the safety, and hindering the developing technological world. Several years of research, mistakes from the past, and the recorded history of tragedies, incidents, and accidents have illustrated the importance of robust safety systems, design improvements, error diagnostic procedures and policies for a safer society.

This book is designed to give a practical look at specific hazards, case studies and real-life experiences, the root cause of accidents, empirical and theoretical studies, methods and models to predict their consequences, and design procedures needed to prevent them from happening again. Many of the accidents described could occur in the different type of plants, attracting widespread interest. Some of them illustrate the hazard associated with storage tanks (Chapter 2), accidents caused by static electricity, and the hazards of flammable liquid mixtures while stirring. Theoretical investigations include reconstructing a fire accident scenario in a hydroelectric power plant, parametric studies involving an accident of methyl ethyl ketone peroxide and tert-butylperoxy-3,5,5-trimethyl hexanoate using a calorimetric approach.

Many of these incidents were not only near-misses, but also resulted in fatalities. Few of these incidents were simple; they needed hazard prevention measures, appropriate training, and design alternatives. Our focus was to look most critically on mitigation techniques, the analysis of off-site emergency procedure and the evaluation of human risk factors for a proper safety management systems. Each of the case studies has a lot of lessons learned and techniques to provide in-depth analysis in mitigation of these accidents resulted in national and international policy-making. This book is neither a guide of regulations for handling hazardous materials nor a collection of case studies. It is a contribution toward providing key references from an industrial perspective in filling the gap between the understanding of the root cause of the accident/incident by academicians to develop, combat, prevent, and predict the emergency scenarios. This book is not only beneficial for undergraduate students, but also assists the safety professionals in the design of resolute safety systems.

In today's world, statutory bodies cannot write regulations quickly enough to cover all the hazard-making, where information should be shared unrestricted through cooperative safety assessment programs. We are very grateful to all the contributors for sharing with us their technical knowledge and expertise. We are also grateful to all the authors and owners of copyright who have kindly allowed us to reproduce diagrams and tables from their publications (Elsevier, Wiley & Sons, American Chemical Society, etc.).

We thank many of our friends and colleagues who continue to teach us the fundamentals of chemical process safety, and are indebted to the internet, individuals, and organizations for their kind assistance in providing historical incident/accident data.

Finally, we continue to acknowledge our families, who provided patience, under-standing, and encouragement throughout. We believe that the textbook is the right blend of both industry experience and theoretical studies, providing tremendous potential for learning, and helps chemical industries and academicians to contribute to a safer future and community.

Prof. M.K. Purkait, Dr. P. Mondal, Dr. M Changmai,
Dr. V Volli and Prof. Chi Min Shu

Authors

Prof. Chi-Min is energetic, assiduous and an outstanding researcher dedicated himself in safety science and technology. His contribution in this field covers both fundamental and applied research.

Dr. Chi-Min Shu is the Academic Vice President and Distinguished Chair Professor in the Department of Safety, Health and Environment, National Yunlin University of Science and Technology (YunTech), Douliou, Taiwan. He was successful in establishing a state-of-the-art laboratory facility exclusively for the process safety and disaster prevention (PS&DPL), which was the first of its kind in Taiwan. He was awarded his PhD and MS Degree from Department of Chemical Engineering, University of Missouri-Rolla (UMR), Rolla, Missouri, USA after completing his BS from Department of Chemical Engineering, Tunghai University, Taichung, Taiwan. He became a Fellow of North American Thermal Analysis Society (NATAS) in 2011, of the American Institute of Chemical Engineers (AIChE) in 2016, and of the Institution of Engineering and Technology (IET) and the Royal Society of Chemistry (FRSC) in 2018. He is the Chairman and Technical Advisor for Pressure Vessel Association (PAV), the President of the Institute of Industrial Safety and Disaster Prevention, Advisory Board Member of Environment Protection Administration (EPA), and the National Fire Administration; Ministry of Interior, Executive Yuan, Taiwan. He is presently serving on the Editorial Board of the *Journal of Thermal Analysis and Calorimetry* (JTAC), *Process Safety Progress* (PSP), the *Journal of Loss Prevention in the Process Industries* (JLPPI), and the *Journal of Safety Research* (JSR). He has received several awards, including: Most Cited Author 2006–2009, The Institution of Chemical Engineers (IChemE), UK; National Distinguished Academician 2011, Ministry of Education, Executive Yuan, Taiwan; Outstanding Engineering Professor Award, 2014, Chinese Institute of Engineers, Taiwan; Exceptional Academia-Industry Cooperation Award 2014 at YunTech; Academic Contribution Award 2016, Disaster Management Society of Taiwan; and METTLER Award 2017, North American Thermal Analysis Society (NATAS), USA. He has more than 25 years of experience in industry, academia and research, published more than 329 peer-reviewed articles in different reputed journals of importance. Prof. Chi-Min has guided approximately 30 PhD students and currently 8 students are pursuing their PhD degree under his supervision. Prof. Chi-Min is energetic and assiduous and an outstanding researcher, who has dedicated himself to the areas of safety science and technology. His contributions in this field cover both fundamental and applied research.

Dr. Mihir Kumar Purkait is a Professor in the Department of Chemical Engineering and Head Centre for the Environment at Indian Institute of Technology Guwahati (IITG). Prior to joining as faculty in IITG (2004), he had received his PhD and M.Tech. in Chemical Engineering from the Indian Institute of Technology, Kharagpur

(IITKGP) after completing his B.Tech and B.Sc. (Hons) in Chemistry from the University of Calcutta. He has received several awards, including: Dr. A.V. Rama Rao Foundation's Best PhD Thesis and Research Award in Chemical Engineering from IIChE (2007), the BOYSCAST Fellow Award (2009-10) from the DST; the Young Engineers Award in the field of Chemical Engineering from the Institute of Engineers (India, 2009); and the Young Scientist Medal award from the Indian National Science Academy (INSA, 2009). Prof. Purkait is a Fellow of Royal Society of Chemistry (FRSC) UK, and a Fellow of Institute of Engineers (FIE) India. He is the director of two incubated companies (viz. RD Grow Green India Pvt. Ltd. and Vixudha Bio Products Ltd.). He is also technical advisor of Gammon India Ltd and Indian Oil Corporation, Bethkuchi for their treatment plant. His current research activities are focused in four distinct areas: (i) advanced separation technologies; (ii) waste to energy; (iii) smart materials for various applications; (iv) process intensification; and (v) hazard safety analysis of process. In each of the areas, his goal is to synthesize stimuli-responsive materials and to develop a more fundamental understanding of the factors governing the performance of the chemical and biochemical processes. He has more than 20 years of experience in academics and research and published more than 200 papers in different reputed journals (Citation: >9500, h-index = 55, 10 index = 116). He currently has 8 patents and completed 24 sponsored and consultancy projects from various funding agencies. Prof. Purkait has guided the studies of some 18 PhD students and is the author of 6 books.

Dr. Murchana Changmai currently works as Assistant Professor at BITS Pilani, Dubai Campus was formerly engaged with Hindustan Petroleum Corporation Limited R&D Centre (HPGRDC) as an active researcher. She completed her PhD (2019) and Master's (2014) in Chemical Engineering from the Indian Institute of Technology Guwahati under the guidance of Prof. Mihir K. Purkait, where she worked extensively in the field of water treatment utilizing various methods such as membrane separation, electrocoagulation and adsorption. She also has experience in the fabrication of inorganic membranes and adsorbents (nanoparticles, organic adsorbents and composite adsorbents). She has also done internships at Gifu University, Japan and SeoulTech, South Korea during her tenure as a PhD scholar. She has more than 10 peer-reviewed international publications in reputed journals, and has also presented more than 10 papers in international and national conferences.

Dr. Piyal Mondal received his BTech in Chemical Engineering from the National Institute of Technology Durgapur, West Bengal (INDIA) during 2012. He completed his Master's and PhD in Chemical Engineering from the Indian Institute of Technology Guwahati (INDIA) in Chemical Engineering. His research work is dedicated to preparing various surface-engineered polymers for specific environmental applications. The synthesis of polymeric membranes, green synthesized nanomaterials, and hybrid techniques to combat wastewater treatment as well as dealing with their safety hazards during operation is his research focus. He has fabricated different prototypes for environmental separation applications. Currently, he is the author of three reference books, namely: *Stimuli-responsive polymeric membrane* (Elsevier, 2018), *Treatment of industrial effluents* (CRC Press, 2019), *Thermal Induced Membrane Separation Processes* (Elsevier, 2020), and he has a few more in progress.

Moreover, his publication consists of 11 peer-reviewed articles in reputed international journals, with several more under review. He has presented more than 10 papers and received several awards in poster and paper presentations in his field at international and national conferences.

Dr. Vikranth Volli is an Assistant Professor in the Bachelor Program in Interdisciplinary Studies, Department of Safety, Health and Environment, at National Yunlin University of Science and Technology (YunTech), Douliou, Taiwan. Before joining YunTech, he worked as Assistant Professor in VIT University, Tamil Nadu, India. He has received his PhD in Chemical Engineering from the Indian Institute of Technology, Guwahati and MTech from National Institute of Technology, Rourkela after completing his BTech from Jawaharlal Nehru Technological University, Hyderabad, India. His research work is dedicated to studies relating the thermal behavior, oxidative stability, and hazard analysis of ionic liquids, ester-based lube, and commercial oils. His past research involved the production of liquid biofuels from lignocellulosic biomass via pyrolysis. He has also studied the synthesis of catalysts and value-added products from industrial wastes (fly ash and red mud) and their application in biodiesel production. He has published research articles in reputed journals like the *Chemical Engineering Journal*, the *Journal of Hazardous Materials*, the *Science of the Total Environment*, *Fuel*, etc.

1 Introduction to Industrial Safety and Hazard

1.1 BACKGROUND

Over decades, mankind has paid focus on the treatment of hazardous contaminants, and effluents (liquids and gases) from various industrial sectors for providing safety to humans and the environment (Volli and Purkait, 2015; Purkait et al., 2005; Mondal and Purkait, 2017). For safety reasons, attention was given to wastewater treatment (Mondal et al., 2020; Sriharsha et al., 2014), the treatment of underground water containing metals (Changmai et al. 2020a, 2020b) and also the treatment of hazardous contaminants (Mondal and Purkait, 2018, 2019). But it was analyzed after various serious incidents occurring in different industries that, apart from such attention, an increased should also be devoted towards working safety and following precautionary steps and measures in process units while working. A chemical manufacturing process is described as being inherently safer if it reduces or eliminates hazards associated with materials and operations used in the process, and this reduction or elimination is permanent and an inseparable part of the process technology. In these circumstances, a hazard is defined as a physical or chemical characteristic that has the potential for causing harm to people, the environment or property. The key to this definition is that the hazard is intrinsic to the material or to its conditions of storage or use. For example, chlorine is toxic by inhalation; gasoline is flammable and steam at 600 psig contains significant potential energy. These hazards are the basic properties of the materials and their conditions of usage and there is no way in which they can be changed. An inherently safer process reduces or eliminates the hazard by reducing the quantity of hazardous material or energy, or by completely eliminating the hazardous agent. A traditional approach to managing the risk associated with a chemical process is by providing layers of protection between the hazardous agent and the people, environment or property, which is potentially impacted.

1.2 WORK DOMAIN AND SAFETY INSTRUCTIONS

Recently in all process manufacturing industries safety-related measures has been widely accepted and implemented. Each and every industry has come up with the proposal of safety protocols and measures to implement among its different process units. Chemical plants where high temperature, pressure and hazardous chemicals are involved should follow a policy on strict adherence to such safety protocols and the workers should follow such measures for their own benefit and also to reduce life risk and other environmental hazards that could affect the entire society (CCPS, 2003). Plant operations verifying that the licensee has instituted effective management measures to provide for the safe operation of the facility during both routine and abnormal conditions, to recognize non-routine events affecting safety, to utilize an internal reporting system and to identify and execute corrective actions to return the plant to a safe and secure condition after possible upsets. A few points for safe plant operations are outlined below:

- Chemical process safety: Verifying that the licensee has implemented adequate measures for the protection of their workers, the public and the environment from hazardous chemicals that could adversely affect radiological safety or could be released from the processing of licensed radioactive material.
- Criticality safety: Verifying that the licensee has implemented adequate controls to prevent accidents, fire and explosions with reference to the Factory Act.
- Fire Protection: Verifying that the licensee has implemented controls to ensure that fires would not occur or would be limited through the safe handling and storage of hazardous materials.
- Safeguards material: The control, accounting and physical protection of special hazardous materials, classified material and information security.
- Radiological controls: Radiation protection, environmental protection, waste management and transportation.
- Environmental protection: Verifying that the licensee has established and implemented a program that effectively protects the environment by measuring and controlling releases.
- Waste management: Verifying that the licensee has established and implemented an effective program for hazardous waste management.
- Maintenance and surveillance: Verifying that the licensee has established and implemented effective programs for both corrective and preventive maintenance, configuration management and surveillance testing activities, that cover all items relied on for safety and safeguards.
- Training: Verifying the qualification and training of personnel relied on to perform functions necessary for adequate safety and safeguards.
- Emergency preparedness: Verifying that the licensee has established and implemented an effective emergency management program to protect the workers, public and the environment in the event of reasonably postulated events that could threaten the facility.

1.3 RECORDS AND REPORTS OF PREVALENT INDUSTRIAL ACCIDENTS

Several casualties have been reported over years because of industrial accidents, where sufficient precautious steps were not taken and implemented. There are several industries which have succumbed to such disasters, but, principally based on their high social impact, the energy, food and manufacturing industries have been discussed below, since these have faced several safety hazard issues over recent years.

1.3.1 ENERGY INDUSTRY

- **October 1957:** The Windscale fire, the worst nuclear accident in the UK's history, released substantial amounts of radioactive contamination into the surrounding area at Windscale, Cumberland (now Sellafield, Cumbria).
- **May 1962:** The Centralia mine fire began, forcing the gradual evacuation of the Centralia borough. This fire continues to burn in the abandoned borough.
- **March 4, 1965:** The Natchitoches explosion: A 32-inch gas transmission pipeline, north of Natchitoches, Louisiana, and belonging to the Tennessee Gas Pipeline, exploded and burned from stress corrosion cracking on March 4, killing 17 people. At least 9 others were injured, and 7 homes up to 450 feet from the rupture were destroyed. The same pipeline also had an explosion on May 9, 1955, just 930 feet (280 m) from the 1965 failure.
- **March 28, 1979:** Three Mile Island accident. Partial nuclear meltdown. Mechanical failures in the non-nuclear secondary system, followed by a stuck-open pilot-operated relief valve in the primary system, allowed large amounts of reactor coolant to escape. Plant operators initially failed to recognize the loss of coolant, resulting in a partial meltdown. The reactor was brought under control, but not before up to 481 PBq (13 million curies) of radioactive gases were released into the atmosphere (Walker, 2004).
- **June 3, 1979:** Ixtoc oil spill. The Ixtoc I exploratory oil well suffered a blowout, resulting in the third-largest oil spill and the second-largest accidental spill in history.
- **November 20, 1980:** A Texaco oil rig drilled into a salt mine transforming Lake Peigneur, which had been a freshwater lake before the accident, into a saltwater lake.
- **January 7, 1983:** An explosion in Newark, New Jersey was experience about 100–130 miles from the epicenter, but it only claimed 1 life, and only injured some 22–24 people.
- **July 23, 1984:** Romeoville, Illinois. The Union Oil refinery explosion killed 19 people.
- **August 17, 2009:** Sayano–Shushenskaya power station accident. Seventy-five people were killed at a hydroelectric power station when a turbine failed. The failed turbine had been vibrating for a considerable time. Emergency doors to stop the incoming water took a long time to close; a self-closing lock would have stopped the water in minutes.

- **February 7, 2010:** 2010 Connecticut power plant explosion. A large explosion occurred at a Kleen Energy Systems 620-megawatt, Siemens combined cycle gas- and oil-fired power plant in Middletown, Connecticut, United States. Preliminary reports attributed the cause of the explosion to a test being carried out on the plant's energy systems. The plant was still under construction and scheduled to start supplying energy in June 2010. The number of injuries was eventually established to be 27. Five people died in the explosion (The Telegraph, 2010).
- **June 21, 2019:** Philadelphia Refinery Explosion. An explosion at the Philadelphia Energy Solutions' refinery destroyed the alkylation unit, where crude oil is converted to high octane gas, and led to the planned closure of the financially troubled plant. While the explosion and fire only resulted in a few minor injuries, it was catastrophic for the business (CNN Business, 2019).

1.3.2 FOOD INDUSTRY

- **May 2, 1878:** The Washburn "A" Mill in Minneapolis was destroyed by a flour dust explosion, which killed 18. The mill was rebuilt with updated technology. The explosion led to new safety standards in the milling industry (Minnesota Historical Society, 1956).
- **January 15, 1919:** Great Molasses Flood. A large molasses tank in Boston, Massachusetts burst and a wave of molasses rushed through the streets at an estimated 35 mph (56 km/h), killing 21 and injuring 150. The event has entered local folklore; residents claim that on a hot summer day, the area still smells of molasses.
- **February 6, 1979:** The Roland Mill, located in Bremen, Germany, was destroyed by a flour dust explosion, killing 14 and injuring 17.
- **September 3, 1991:** Hamlet chicken processing plant fire in Hamlet, North Carolina, where locked doors trapped workers in a burning processing plant, causing 25 deaths.
- **September 3, 1998:** Grain elevator explosion in Haysville, Kansas. A series of dust explosions in a large grain storage facility resulted in the deaths of seven people (Kansas, 1998).
- **February 7, 2008:** The 2008 Georgia sugar refinery explosion in Port Wentworth, Georgia, United States. Thirteen people were killed and 42 injured when a dust explosion occurred at a sugar refinery owned by Imperial Sugar.

1.3.3 MANUFACTURING INDUSTRY

- **May 4, 1988:** PEPCON disaster, Henderson, Nevada. A massive fire and explosions at a chemical plant killed two people and injured over 300.
- **May 10, 1993:** Kader Toy Factory fire. A fire started in a poorly built factory in Thailand. Exit doors were locked and the stairwell collapsed. 188 workers were killed, mostly young women.

- **May 13, 2000:** Enschede fireworks disaster. A fire and explosion at a fireworks depot in Enschede, Netherlands resulted in 24 deaths, and another 947 injured. About 1,500 homes were damaged or destroyed. The damage was estimated to be over US$300 million in insured losses.
- **November 3, 2004:** Seest fireworks disaster. N. P. Johnsens Fyrværkerifabrik fireworks factory exploded in Seest, a suburb of Kolding, Denmark. One firefighter died; seven from the rescue team as well as 17 locals were injured. In total, 2,107 buildings were damaged by the explosion, with the cost of the damage estimated at €100 million.
- **December 6, 2006:** Falk Corporation Explosion. A gas leak triggered a large explosion and ensuing fire at a gear manufacturing facility in Milwaukee, Wisconsin. Three were killed and 47 injured, with several of the buildings at the facility being leveled.
- **April 18, 2007:** Qinghe Special Steel Corporation disaster. A ladle holding molten steel separated from the overhead iron rail, fell, tipped, and killed 32 workers, injuring another 6.
- **February 1, 2008:** Istanbul fireworks explosion. An unlicensed fireworks factory exploded accidentally, leaving, according to some reports, at least 22 people dead and at least 100 injured.
- **September 11, 2012:** Karachi, Pakistan, 289 people died in a fire at the Ali Enterprises garment factory, which made ready-to-wear clothing for export to the West.
- **November 24, 2012:** Dhaka Tasreen Fashions fire. A seven-story factory fire outside of Dhaka, the capital of Bangladesh, killed at least 112 people, 12 of whom had jumped out of windows to escape the blaze.

1.4 HAZARDS IN PROCESS UNITS

The general process hazards that are associated with any chemical plant are basically due to the various reaction processes occurring within a process unit (Kharabanda and Stallworthy, 1988; Alaimo 2001) . They are mainly divided by exothermic and endothermic reactions. Other process parameters during a reaction also pose a severe threat and should be taken care intensively. Such various hazards associated within a manufacturing plant are discussed in detail below:

1.4.1 GENERAL PROCESS HAZARDS

1.4.1.1 Exothermic Reactions with Mild Risk

Hydrogenation: The addition of hydrogenation to both sides of a double or triple bond is the use of hydrogen under pressure and at a relatively high temperature.

Hydrolysis: The reaction of a compound with water, such as the manufacture of sulfuric or phosphoric acids from oxides.

Alkylation: Addition of an alkyl group to a compound to form various organic compounds.

Isomerization: Rearrangement of the atoms in an organic molecule, such as a change from a straight chain to a branched molecule or displacement of a double bond. Hazards are dependent on the stability and the reactivity of the chemicals involved and may, in some cases, require a penalty of 0.50.

Sulfonation: Introduction of an SO_3H radical into an organic molecule through reaction with H_2SO_4.

Neutralization: Reaction between an acid and a base, to produce salt and water.

1.4.1.2 Exothermic Reactions with Sufficient Risk

Esterification: Reaction between an acid and an alcohol or unsaturated hydrocarbon moderate hazard, except in cases where the acid is highly reactive or where the reacting substances are unstable.

Oxidation: It's a combination of oxygen with some substances in which the reaction is controlled and does not go to CO_2 and H_2O as in the case of combustion, where vigorous oxidizing agents such as chlorates, nitric acid, hypochloric acids and salts are used.

Polymerization: It occurs by joining together of molecules to form chains or other linkages. Heat must be dissipated to keep the reaction under control.

Condensation: It takes place by joining together of two or more organic molecules with the splitting of H_2O, HCl or other compound.

1.4.1.3 Exothermic Reactions with High Risk

Halogenation: Introduction of halogen atoms (fluorine, chlorine, bromine, or iodine into an organic molecule. This is both a strongly exothermal and a corrosive process.

Nitration: It involves the replacement of a hydrogen atom in a compound with a nitro group, very strong exothermal reaction possibly with explosive by-products. Temperature controls must be good, impurities can act as catalysts for further oxidation or nitration, and rapid decomposition can occur.

1.4.1.4 Endothermic Reactions with Low Risk

Calcinations: Heating of a material to remove moisture or other volatile material.

Electrolysis: Separation of ions by means of electric current; there are hazards because of the presence of flammable or highly reactive products.

Pyrolysis or cracking: Thermal decomposition of large molecules by temperatures, pressures and a catalyst regeneration of the catalyst by a separate combustion process can be dangerous. If a combustion process is used as a source of energy for calcinations, pyrolysis or cracking, the risk is very high.

1.4.2 SPECIAL PROCESS HAZARDS

Process Temperature: Materials such as hexane and carbon disulfide have low auto-ignition temperatures and can be ignited on hot steam lines. Special attention should be given to operations above the flash point and above atmospheric boiling point.

Low Pressure: Air leakage into a system could create hazard when dealing with pyrophoric materials, diolefins with hazard of peroxide formation and catalyzed polymerization. Hydrogen collection system, and any vacuum distillation system operating at less than 0.57 bar, needs extreme attention.

Operation near flammable range: For processes that operate close to the flammable limits or where it is necessary to use instrumentation and or nitrogen or air purging to stay outside the explosion limits, such process requires carefulness. Examples include oxidation of toluene to benzoic acid, dissolving of rubber direct oxidation in the ethylene oxide process.

Operating pressure: The safety precautions increases for operations involving higher pressure reactions. Moderate risk is associated with operations involving viscous materials such as tars, bitumen, heavy lubricating oil and asphalts. Whereas high risks are involved for operations related with compressed gas and pressurized liquefied flammable gases.

Low temperature: Moderate and considerable safety precautions should be required for reactions occurring within 0°C to -30°C and below that. The purpose is to make allowance for presumed brittleness. Moreover, in the case of leakage, cold liquid will come into contact with the relatively hot environment, which can cause considerable evaporation.

Quantity of flammable materials: Flammable materials poses higher risk of fire and explosion hazards, and thus preventive and safety measures should be implemented when operations are related with flammable materials.

In storage/transportation: During storage or transportation of volatile and flammable materials huge precautions should be maintained since such materials due to shaking and movement can generate heat and lead to hazardous situation.

Loss of material through corrosion and erosion: This hazard should be assessed for both internal and external corrosion. Some areas to consider are:

- The influence of minor impurities in the process fluid on corrosion.
- External corrosion from breakdown in paint and coatings.
- Resistant linings (plastics, brick, etc.) liable to damage at seams, joints or pinholes.

Leakage of joints and packing: Gaskets, sealing of joints or shafts and packing can be sources of leaks, particularly where the thermal of pressure cycling occurs.

Toxic release: Various hazardous and toxic substances are utilized in industries today. People working in the industry should be made well aware of the health hazards caused by such chemicals when they are directly exposed to them.

1.4.3 CONFINED SPACE HAZARDS

The hazards encountered and associated with entering and working in confined spaces is capable of causing bodily injury, illness, and death to the worker. Accidents occur among workers because of the failure to recognize that a confined space is a potential hazard. It should therefore be considered that the most unfavorable situation exists in every case and that the danger of explosion, poisoning, and asphyxiation will be present at the onset of entry.

Confined spaces can be categorized generally as those with open tops and with a depth that will restrict the natural movement of air, and enclosed spaces with very limited openings for entry. In either of these cases, the space may contain mechanical equipment with moving parts. Any combination of these parameters will change the nature of the hazards encountered. Degreasers, pits, and certain types of storage tanks may be classified as open-topped confined spaces that usually contain no moving parts. However, gases that are heavier than air (butane, propane, and other hydrocarbons) remain in depressions and will flow to low points where they are difficult to remove. Open-topped water tanks that appear harmless may develop toxic atmospheres such as hydrogen sulfide from the vaporization of contaminated water. Therefore, these (heavier than air) gases are a primary concern when entry into such a confined space is being planned. Other hazards may develop due to the work performed in the confined space or because of corrosive residues that accelerate the decomposition of scaffolding supports and electrical components.

Confined spaces such as sewers, casings, tanks, silos, vaults, and compartments of ships usually have limited access. The problems arising in these areas are similar to those that occur in open-topped confined spaces. However, the limited access increases the risk of injury. Gases which are heavier than air such as carbon dioxide and propane may lie in a tank or vault for hours or even days after the containers have been opened. Because some gases are odorless, the hazard may be overlooked with fatal results. Gases those are lighter than air may also be trapped within an enclosed-type confined space, especially those with access from the bottom or the side.

Hazards specific to a confined space are dictated by: (1) the material stored or used in the confined space; as an example, damp-activated carbon in a filtration tank will absorb oxygen, thus creating an oxygen-deficient atmosphere; (2) the activity carried out, such as the fermentation of molasses that creates ethyl alcohol vapors and decreases the oxygen content of the atmosphere; or (3) the external environment, as in the case of sewer systems that may be affected by high tides, heavier-than-air gases, or flash floods.

The most hazardous kind of confined space is the type that combines limited access and mechanical devices. All the hazards of open-topped and limited access confined spaces may be present together with the additional hazard of moving parts. Digesters and boilers usually contain power-driven equipment which, unless properly isolated, may be inadvertently activated after entry. Such equipment may also contain physical hazards that further complicate the work environment and the entry and exit process.

1.4.4 ELECTRICAL HAZARDS

Electrical hazards in industrial space are quite common and sometimes can take a disastrous shape if not handled with extensive precautionary measures. Industrial electrical hazard can be broadly divided into two categories, namely human hazard and property hazard.

Flashover burns

- They are caused when the victim is in close proximity of a high current contact breaking, during which a severe arc is established in the air in an attempt to maintain the high load current.
- Such burns are also caused when the earth accidentally goes close to the high voltage point, setting up an arc due to the ionization of the air between the body and the live parts. The result of such an accident is the severe burning of the body. The currents are too high to cause ventricular fibrillation.
- The victim generally does not actually touch the live point and hence falls away from the conductor. As he falls away, the arc is extinguished and hence the time of passage is also quite brief.
- Nevertheless, the accident causes quite severe burns due both to the direct passage of current as well as flashover.

Another category of electric hazards is the severe destruction of the property due to fires or explosions. If the heat generated due to electrical process is not dissipated properly or allowed to generate uncontrollably (by allowing I to increase), it can cause the arc heating of the medium, surrounding and thereby result in fire.

The fire hazard is also eminent when sparking takes place. This can be due to:

- Loose contact.
- Phase to phase, phase to earth faults.
- Lightning strike.
- Static electricity.
- Sudden breaking of loading circuit, causing an arc. A tiny spark in a highly flammable atmosphere like diesel, petrol or LPG installations may result in an explosion.

1.4.5 FIRE AND EXPLOSION HAZARD

Industrial fires and explosions cause considerable damage to lives and property in addition to impeding productivity. The key to prevention is a management plan with which fire hazards can be identified and controlled effectively. A good plan should identify fire hazards in buildings, plants, processes, machinery and operating procedures before outlining measures to minimize the outbreak of fires. Every plant should also be equipped with adequate hardware and trained personnel to detect fires quickly and to limit their spread. The plan should also incorporate procedures to contain major emergencies and to bring back production to normal with minimum delay.

- Accidental fires and explosions in industry inflict the greatest damage to lives and property. Annually, about 25,000 persons die in India due to domestic and industrial fires which are also a major cause of fatal accidents in factories. The insurance industry paid about Rs.900 crores on account of fire claims in the year 2002–2003.

- Dust explosions are possible whenever the process produces combustible dusts. To produce a conflagration, the dust must have a sufficient surface area ratio to weight to sustain rapid oxidation to create and sustain an explosion. When a dust can sustain an explosion, the dust concentration must be within the explosive limits.
- Ignition of the dust depends on several factors:
  o Chemical composition.
  o Shape and fineness.
  o Dust distribution in the gas stream or atmosphere.
  o Concentration of oxygen in the gas stream.
  o Initial temperature and pressure of the gas.
  o Energy level available to detonate the explosion.

1.5 ACCIDENT INVESTIGATIONS AND ANALYSIS

1.5.1 ACCIDENT INVESTIGATION

It is always helpful to have an initial knowledge of the accident which could further help for investigation. Figure 1.1 provides a schematic diagram of a chain to be followed for investigation. Although accident investigation is a complicated process, it is necessary to pay particular attention to the following principles (Haddon 1972; Harms, 2004):

- **Basic assumption:** An investigation should be a fact-finding activity to learn from the experience of the accident, not an exercise designed to allocate blame or liability. The emphasis in conducting investigations should be on identifying the underlying causes in a chain of events leading to an accident, the lessons to be learned, and ways to prevent and mitigate similar accidents in the future.

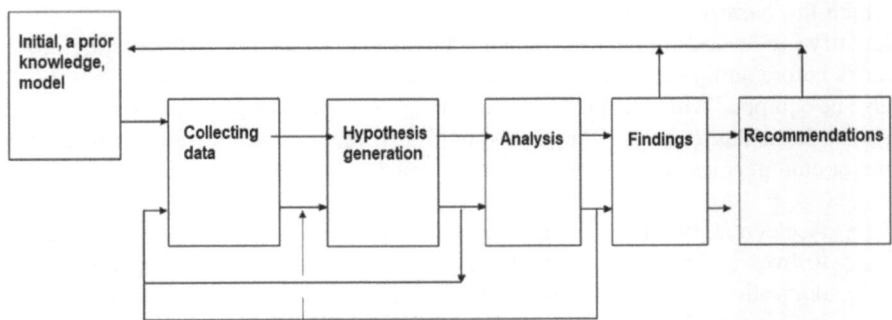

FIGURE 1.1   Initial Knowledge and accident investigation (Reproduced © ESReDA, 2009).

- **Protocols:** Protocols should be established for conducting investigations. These protocols should identify the roles and responsibilities of the individuals involved in the investigation, specify the steps to be taken in the investigative process and establish a common terminology to be used in preparing investigation reports in order to facilitate sharing information related to investigations. Caution should be taken regarding "anonymity" (e.g. persons interviewed, victims, organizations). The decision regarding this issue has to be made at a very early stage of the investigation process, and this decision has to be communicated to the participants and the stakeholders.
- **Coordination:** As there can be more than one body with the authority to investigate an accident, efforts should be made to co-ordinate the investigations to avoid duplication, improve effectiveness and help ensure access to all relevant evidence.
- **Competence:** A team should be established and should consist of participants from different disciplines, with different skills, including those with knowledge of the specific installation and work practices (operators, engineers, managers) subject to the investigation. All members of the investigation team should have the appropriate knowledge, competency and experience to carry out investigations. They should comply with the professional criteria of independence and objectivity.
- **Data and evidence:** Investigations should take account of the various types of information/evidence that might be available, including testimony from people (e.g. witnesses, experts) collected by face-to-face interviews or by hearings, relevant data, documentation and physical evidence. Evidence should be protected in order to facilitate the investigation process. There should be clear identification of who has responsibility for evidence and who can release evidence. Caution should be taken to ensure that all involved parties agree about the correct procedure for handling all collected material.
- **Reporting:** Investigation reports should include a factual chronology of the events leading up to the accident/near-miss, a statement of the underlying causes and contributing causes, and recommendations for follow-up actions. The recommendations should be specific, so that they can lead to adaptations (expected improvements) of technology and management systems. The objective should be to seek optimum solutions under the given circumstances, recognising that it might not be possible to achieve perfect solutions.
- **Follow-up of investigations:** When following up an investigation, there should be a review of the investigative process to help ensure that it has been effective, that there has been appropriate communication of its findings and to learn for future investigations. Efforts should be made to improve sharing experience related to the methodologies and approaches used in investigations of incidents.
- **Communication:** All communications concerning the investigation should be as transparent as possible without compromising the investigative process.

1.5.2 PHASES OF ACCIDENT INVESTIGATION AND BACKGROUND KNOWLEDGE

Phase I - Data collection: Data to be collected could be objective (chronological record of events, parameters and/or values, status of systems involved, written reports), subjective (feelings about a situation, explanation about relationships with other people) or mixed, i.e. "objective phenomena" described/rationalized by a person (such as an explanation of actions, description of situations).

Every fact that seems relevant to the analyst(s) for explanation and/or understanding of the event has to be collected. Data to be collected is not only "linked" with field personnel and/or line operators in the workplace, or only related to the direct (immediate) cause of an event. This means that, in particular, history-related data and managers' actions (e.g. decision-making) also have to be taken into account.

Phase II - Hypotheses generation: First set of data collected allows to defining assumptions concerning causes of the accident. Hypotheses can reflect several standpoints: for example, technical, human, organizational and cultural causes. Assumptions shape analysis and lead to other data to be collected so that they may be challenged. At the end of the investigation, hypotheses can be either confirmed or denied.

Phase III - Analysis: Analysis is the stage during which assumptions can be challenged. This means that they can be proven – either as relevant or as non-pertinent – thus requiring some new hypotheses to be defined (and processed in the same manner as were earlier hypotheses).

Phase IV – Findings: At the end of the analysis phase, the analyst is left with a set of proven hypotheses. They represent causes (direct and root) of the accident. Findings are a synthesis of accident explanation, i.e. they mainly deal with the causes that led to the accident. Findings also deal with phenomena that did not contribute to the accident itself but are discovered during the process of the investigation.

Phase V – Recommendations: Once these causes (technical, human, organizational, societal and cultural) have been established, corrective measures must be defined, tested, implemented and validated in operation in order to ensure that this type of accident does not recur.

1.5.3 ANALYSIS OF DIRECT AND ROOT CAUSES OF ACCIDENT

For a good understanding of the investigative process, several phases and steps are delineated. Such phrasing (phases and steps) might suggest a linear process, but essentially the fact-finding and analysis phases are interconnected by the iterative processing of facts, findings and analysis. Figure 1.2 is a schematic diagram that locates the analysis phase throughout the overall investigative process.

There are two major goals that drive the steps to be taken in the analysis:

- To validate WHAT happened and HOW it happened: implies an assessment of the plausibility (proving or invalidating) of hypotheses generated based upon the sequence of events, to challenge the various scenarios with available evidence, to validate the most probable scenario taken from observed consequences and traced back to its direct causes;

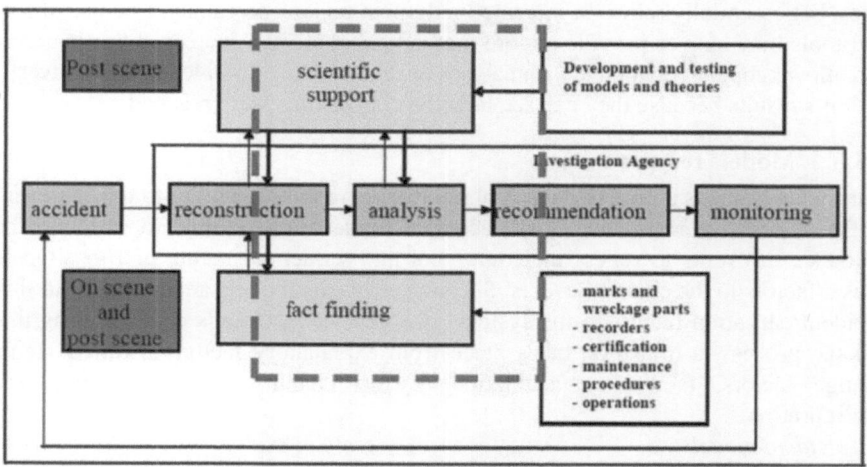

FIGURE 1.2 Analytical phase positioning during investigation (Reproduced © ESReDA, 2009).

- To answer WHY the accident could occur: requires identifying root causes, and asking WHY it was not prevented.

The question of WHAT happened should be answered in a structured manner and carefully guided in order to achieve credible and "objective" conclusions as a basis for consensus about an evidence-based explanation of the event.

Preliminary outcomes of this process could be. A decision:

- To collect more factual information;
- To generate additional hypotheses;
- To render the investigation as "inconclusive" due to the lack of a satisfactory explanation.

In general, the intermediate product of this investigative phase (challenging the hypotheses generated) can be one, or more, accident scenarios in which a consensus may be reached as an acceptable explanation of the event under investigation. As soon as the most probable scenario is identified, the analysis of root causes can start, on the basis of direct causes and a search for safety measures that could have prevented the accident.

The analysis lies at the heart of the investigative process: between the fact-finding phase and drawing up recommendations. Analysis is an iterative process, clarifying needs for collecting additional information as well as changing the content of the recommendations. Analysis has two aims: structuring what we know and structuring what we do not know. Analysis occurs throughout the investigative tasks and forms the basis for the investigation's management decisions on performance efficiency and resource allocation (Stoop, 2007; Gibson, 1961). Analysis has no prescriptive rules, but essentially relies on informed judgment under uncertainty. The use of formal tools may help to provide a more methodical approach, increased transparency and allow people to challenge the analyses or to have more confidence in the investigation's results because they can see how the conclusions were reached.

1.5.3.1 Models required

During the analysis phase, two types of models are required in order to link the event to the systems' performance. First, accident models are required to structure the sequence of events to reflect their temporal and sequential nature and to allocate causal factors to the chain of events. Secondly, systems models are necessary to link accident causation factors to the systems in which the accidents occur. During this linkage process, a transition takes place from explanatory factors toward systems change factors, facilitating adaptation of the system to its new state and configuration.

Accident models:

- Provide structure and transparency in the dynamics and complexity of the event;
- Allocate factors and actors to the sequence of events;
- Clarify relations and interactions between factors, actions and decisions.

These models, though, may contain generic pitfalls:

- They represent metaphors that should not be interpreted as depicting models of an accident (such as Heinrich, 'Iceberg' and Reason, 'Swiss Cheese' model);
- Only a very small number of models can be considered as systems-oriented (such as AcciMap or STEP).

Systems models:

- Should cover the overall systems architecture: its structure, culture, and context and including the life cycles for the design and operation of the primary systems;
- Should incorporate systems complexity and dynamics;
- Should facilitate identification of systems and knowledge deficiencies;
- Should facilitate the transition from explanation to systems change.

These models also may contain pitfalls, as they:

- May take a static, prescriptive form;
- May adopt a perspective from a specific discipline (such as technical, behavioral, cognitive, organizational or institutional);
- May be overly simplistic, focusing only on accident causation, explanatory variables, and not on systemic deficiencies and control variables.

Systems models should take into account the various dimensions that are characteristic of a systems approach:

- The various life cycle phases (such as design, development, construction and operations);
- The various systems levels (such as practice, management, policy-making and governance);
- The various design levels (such as the conceptual, functional and physical form levels).
- These types of systems models can be seen in Figure 1.3 of the Design, Control and Practice diagram. The diagram shows how the systems models facilitate the representation of possible accident scenarios and system adaptations.

These systems models can be depicted in the following Design, Control and Practice diagram (Figure 1.3) through which accident scenarios and system adaptations can be related:

1.5.3.2 Pitfalls in analysis

1.5.3.2.1 Pitfalls in Systems Modelling

Several pitfalls exist in applying systems models for representing complex and dynamic socio-technical environments. Such systems may be decomposed in a structural manner and take static, prescriptive form (such as the ICAO Annex 13 investigation protocol from 1951, with several updated editions since then). Such systems modelling may adopt the perspective of a specific discipline (technical, behavioral, cognitive, organizational or institutional). The modelling may be overly simplistic (such as the SHEL model – Software, Hardware, Environment, and Liveware), focusing only on accident causation and explanatory variables – not on systemic deficiencies and control variables. Thus, it is important to be aware of the perspective from which the modelling is being carried out and the assumptions made in order to

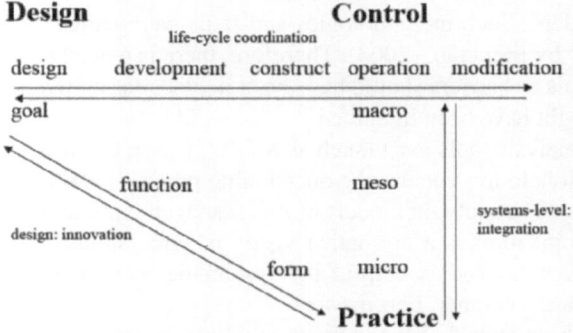

FIGURE 1.3 Design, Control and Practice diagram (Reproduced © ESReDA, 2009).

incorporate the desired aspects in the communication of results that lead to the decision-making process.

1.5.3.2.2 Fallacies in Analytic Reasoning

Analytical reasoning may contain several fallacies that may hamper the quality of the conclusions. The level of analysis may restrict itself to either technical failure or individual actions, thereby excluding higher systems levels. The arguments may consequently be based on assumptions instead of evidence, creating uncertainty in the likelihood of findings. The reasoning may contain fallacies of a suggestive, restrictive nature and may be based on ignorance of significant factors. The reasoning may not be representative, and rather based on exclusion and a false analogy, or may focus on correlation instead of cause. The reasoning may be ambiguous and appeal to popularity and focus on affirmation without denial (or false presumption) as an option.

Finally, biases may exist due to the manner in which groups process information (such as confirmation, groupthink, risky shift, tunnel vision, hindsight and pigeon holing).

1.5.3.3 Specificity of Root Causes Analysis

The problems of identifying root causes pose additional challenges to investigators. The first is to identify those remote causal factors and the second is to assess their causal influence to the event generation. The aim is to link general factors (such as human, organizational, cultural) to specific conditions that directly influenced decisions, actions and event sequence.

It must be acknowledged that identifying and qualifying root causes requires additional competencies from the human and social sciences. These last types of competencies are traditionally very rare in a world of technicians and engineers and even amongst managers of those socio-technical systems. Recent major accident investigations have explicitly involved researchers from the human and social sciences. This posture was then used as a reference by the US CSB when conducting the Texas City 2005 accident investigation. One way to identify and link the root causes to direct causes is to look for safety controls and barriers that have not or could have prevented the event. This implies that investigators should look for standards that are often applied in working procedures, but that may have not been met within the context of the accident. Such methodologies and tools were called "Norms, Novelties and Deviations" by Frei et al. (2003). Therefore, there is a need to question whether or not the controls or barriers should have been in place (as an industry standard), or perhaps they might have been imagined.

Root cause analysis tools exist (such as MORT, Cause Control Change Analysis, Tripod, etc.) that help to structure the questioning process (WHY did it happen?) in a systematic way. They rely on models of risk management that have their own limitations. This point implies a normative vision of what should have been the risk management practices and is helpful for systematic recommendations. But, as a reminder, tools are 'servants' not 'masters.'

In addition, these tools do have limits in highlighting the rationale behind actions, decisions, beliefs, and strategies of actions. Comprehensive approaches are therefore required to address these particular dimensions of human and social systems.

Descriptive approaches (based on social sciences models and theories) provide alternative perspectives (but are also complementary to normative models) given the complexity of systems involved.

1.5.3.4 Various Other Analysis Techniques

1.5.3.4.1 Gross Hazard Analysis

Perform a gross hazard analysis (GHA) to get a rough assessment of the risks involved in performing a task. It is "gross" because it requires further study. It is particularly useful in the early stages of an accident investigation in developing hypotheses. A GHA will usually take the form of a logic diagram or table. In either case, it will contain a brief description of the problem or accident and a list of the situations that can lead to the problem. In some cases, analysis goes a step further to determine how the problem could occur. A GHA diagram or table thus shows at a glance the potential causes of an accident. One of the following analysis techniques can then expand upon a GHA.

1.5.3.4.2 Job Safety Analysis

Job safety analysis (JSA) is part of many existing accident prevention programs. In general, JSA breaks a job into basic steps, and identifies the hazards associated with each step. The JSA also prescribes controls for each hazard. A JSA is a chart listing these steps, hazards, and controls. Review the JSA during the investigation if a JSA has been conducted for the job involved in an accident. Perform a JSA if one is not available. Perform a JSA as a part of the investigation to determine the events and conditions that led to the accident.

1.5.3.4.3 Failure Mode and Effect Analysis

Failure mode and effect analysis (FMEA) determines where failures occurred. Consider all items used in the task involved in the accident. These items include people, equipment, machine parts, materials, etc. In the usual procedure, FMEA lists each item on a chart. The chart lists the manner or mode in which each item can fail and determines the effects of each failure. Included in the analysis are the effects on other items and on overall task performance. In addition, make evaluations about the risks associated with each failure. That is, project the chance of each failure and the severity of its effects. Determine the most likely failures that led to the accident. This is done by comparing these projected effects and risks with actual accident results.

1.5.3.4.4 Fault Tree Analysis

Fault tree analysis (FTA) is a logic diagram. It shows all the potential causes of an accident or other undesired event. The undesired event is at the top of a "tree." Reasoning backward from this event, determine the circumstances that can lead to the problem. These circumstances are then broken down into the events that can lead to them, and so on. Continue the process until the identification of all events can produce the undesired event. Use a logic tree to describe each of these events and the manner in which they combine. This information determines the most probable sequence of events that led to the accident.

1.6 HAZARD CONTROL FOR SAFETY PROVISIONS

Hazardous control can be thought of in three ways. Each describes how and where the controls are placed on the path between the worker and the hazard (Gordon, 1949; Groeneweg 1998). Such controls are discussed below:

Control at the Source: The best way to control a hazard is to eliminate it. If this is not possible, the next step is the substitution of a non-hazardous or less-hazardous material or process. If there is no acceptable substitution, then the hazard is enclosed or isolated from workers. An example of this may be enclosing a high-voltage electrical panel and sealing it off from workers in an office. This would be controlling the hazard 'at the source.'

Control Along the Path: Some hazards, and the work processes that they are part of, cannot be enclosed or isolated. Placing a control 'along the path' means different protective measures are put in place between the hazard and workers. In the electrical panel example, office workers have been sealed off from the hazard but electricians will still have to be able to safely work on the panel. To protect the electricians, controls 'along the path' would probably include using energy lockout procedures and devices and non-conductive tools.

Control at the Worker: If controls 'at the source' and 'along the path' may not be enough to prevent injury, then placing controls 'at the worker' will be necessary. Control at the worker often consists of personal protective clothing and equipment that must be worn while performing certain tasks. Common types of this control are wearing gloves to protect the hands, hearing protection, or masks or respirators to protect airways. 'At the worker' is often the first type of hazard control that businesses put into place. Employers also always need to consider controlling hazards 'at the source' and 'along the path'.

1.7 SUMMARY

- A chemical manufacturing process is described as inherently safer if it reduces or eliminates hazards associated with materials and operations used in the process.
- A permit-to-work system is a formal written system used to control certain types of work that are potentially hazardous.
- The permit-to-work form must help communication between the parties involved. The company issuing the permit, taking into account dividable site conditions and requirements, should design it.
- Risk analysis in chemical process industries is an elaborate exercise involving several steps from preliminary hazard identification to the development of credible accident scenarios, to preparation of strategies for prevention or control of damage.
- Operating pressures above atmospheric pressure requires extensive carefulness and expert operation which would rather create serious risk.
- Gaskets, sealing of joints or shafts and packing can be sources of leaks, particularly where thermal of pressure cycling occurs.

- When toxic chemicals are present in the workplace, an individual exposure can be determined by measuring the concentration of a given chemical in the air and the duration of exposure.
- Investigation and analysis of a specific industrial hazard should be followed for tackling future accidents. Various investigation process and analysis techniques are discussed in this chapter which would help any process unit to obtain safety measure to prevent any disastrous accident.

REFERENCES

Alaimo, R.J. (2001). *Handbook of Chemical Health and Safety*, Washington: An American Chemical Society Publication.

Allen, Nick (7 February, 2010). *"Connecticut gas explosion at power plant 'leaves up to 50 dead'"*. London: Telegraph Media Group Limited.

Center for Chemical Process Safety (CCPS). (2003). *Guidelines for Investigating Chemical Process Incidents*, 2nd ed., New York: Wiley-AIChE.

Changmai, M., Das, P. P., Mondal, P., Paswan, M., Sinha, A., Biswas, P., Sarkar, S., Purkait, M.K. (2020b). Hybrid electrocoagulation-microfiltration technique for treatment of nanofiltration rejected steel industry effluent. *Int. J. Environ. Anal. Chem.* doi: 10.1080/03067319.2020.1715381.

Changmai, M., Mondal, P., Sinha, A., Biswas, P., Sarkar, S., Purkait, M.K. (2020a). Metal removal efficiency of novel LD slag incorporated ceramic membrane from steel plant wastewater. *Int. J. Environ. Anal. Chem.* doi:10.1080/03067319.2020.1734198.

Egan, M. (22 July, 2019). "Philadelphia refinery goes bankrupt after fire". *CNN Business*. 22 July 2019.

Frei, R., Kingston, J., Koornneef, F., Schallier, P. (2003). Investigation Tools in Context, Proceedings of 24th ESReDA Seminar, Safety Investigation of Accidents, Petten, May 12–13.

Gibson, J.J. (1961). The Contribution of Experimental Psychology to the Formulation of the Problem of Safety – A Brief for Basic Research, In: Haddon, W., Suchman E.A., and Klein D. Edits, *Accident Research: Methods and Approaches*. New York: Harper and Row.

Gordon, J.E. (1949). The Epidemiology of Accidents, *Am. J. Pub. Hea.*, Vol. 39 pp. 504–515.

Groeneweg, J. (1998). *Controlling the Controllable. The Management of Safety*. Netherlands: DSWO Press, Leiden University.

Guidelines for safety investigation of Accidents (2009). European Safety Reliability and Data Association (ESReDA). ISBN 978-82-51-50309-9.

Haddon, W. (1972), A Logical Framework for Categorizing Highway Safety Phenomena and Activity, *J Trauma*, Vol. 12, pp 193–207.

Harms, R. L. (2004). Relationships between Accident Investigations, Risk Analysis and Safety Management, *Journal. Hazard. Mater.*, Vol. 111, pp. 13–19.

Kansas, H. (1998). Fire Investigation Summary, grain Elevator Explosion. National Fire Protection Association (NFPA), Fire Investigations Department, 1999.

Kharabanda, O.P. and Stallworthy, E.A. (1988). *Safety in the Chemical Industry: Lessons from Major Disasters*, London: Butterworth-Heinemann Ltd.

Mondal, P., Purkait, M.K. (2017). Green synthesized Iron nanoparticle embedded pH-responsive PVDF-co-HFP membranes: Optimization study for NPs preparation and Nitrobenzene reduction. *Sep. Sci. Technol.* 52 (14), 2338-2355.

Mondal, P., Purkait, M.K. (2018). Green synthesized Iron nanoparticles supported on pH-responsive polymeric membrane for Nitrobenzene reduction and fluoride rejection study: Optimization approach. *J. Cleaner Prod.* 170, 1111-1123.

Mondal, P., Purkait, M.K. (2019). Preparation and characterization of novel green synthesized iron–aluminum nanocomposite and studying its efficiency in fluoride removal. *Chemosphere* 235, 391–402.

Mondal, P., Samanta, N., Kumar, A., Purkait, M.K. (2020). Recovery of H_2SO_4 from wastewater in presence of NaCl and $KHCO_3$ through pH responsive polysulfone membrane: Optimization approach. *Pol. Test.*, 86, 106463.

Purkait, M. K., Bhattacharya, P. K., De, S. (2005). Membrane filtration of leather plant effluent: Flux decline mechanism, *J. Membr. Sci.* 258, 85-96.

Sriharsha, E., Uppaluri, R., Purkait, M. K. (2014). Microfiltration of oil-water emulsions using low cost ceramic membranes prepared with uniaxial dry compaction method. *Ceramic Int.* 40, 1155–1164.

Stoop, J. (2007). Are Safety Investigations Proactive? Proceedings of the 33rd ESReDA Seminar, Future Challenges of Accident Investigations, Ispra, November 13–14.

Volli, V., Purkait, M.K. (2015). Selective preparation of zeolite X and A from flyash and its use as catalyst for biodiesel production. *J. Hazard. Mat.* 297, 101-111.

Walker, J. Samuel (2004). *Three Mile Island: A Nuclear Crisis in Historical Perspective.* Berkeley: University of California Press. ISBN 0-520-23940-7.

Washburn A Mill Explosion. (April, 1956). The Great Mill Explosion and Fire of 1878. Minnesota Historical Society. In Hennepin County History, vol. 16-2, no. 62: pp. 9-10.

2 A Study of a Caprolactam Storage Tank Accident through Root Cause Analysis with a Computational Approach

2.1 INTRODUCTION TO CAPROLACTUM STORAGE IN INDUSTRIES

Over the past four decades, the reason behind the progressive development of industries and global economic prosperity, especially in Taiwan, is mainly due to the advances in the emerging petrochemical industry (Chen et al., 2014). A variety of chemicals is conserved in storage tanks after the crude oil-refining process, for further commercial usage in refineries. Studies shows that such chemicals are usually hazardous and flammable (Bai and Liu, 1995; Chang and Lin, 2006). If any accidental release of such hazardous chemicals occurs, it can have devastating effects such as the loss of life, injuries to workers, property damage, economic downturns and social protest. Petroleum refineries are found to be among the most accident-prone locations, according to many investigations (Chang and Lin, 2006; Wang et al., 2013). Reports (De La Fuente et al., 2014; Jonsson et al., 2015; Marucci-Wellman et al., 2015) reveal that property losses can be as high as USD1,000,000 if such a small accident occurs, which can be further accompanied by interruptions to production and employee injuries. Various problems, such as company bankruptcy, stock devaluation and lawsuits, can arise due to a large accident in an industry (Chang and Lin, 2006). Examples from the previous decade include the Buncefield Oil Storage Depots disaster of December 11, 2005 (MIIB, 2008); a massive tank fire at Caribbean Petroleum Refining on October 23, 2009 (U.S. CSB, 2009); a devastating vapor cloud explosion that occurred in a large fuel storage area at the Indian Oil Corporation Depot in Jaipur, India on October 29, 2009 (Sharma et al., 2013); and an explosion that occurred at Xingang Port in Dalian, China on July 16, 2010 (Zhang et al., 2013). In the petrochemical industry, storage tanks play an important critical role behind the continuous and steady growth of oil which is maintained by the strategic petroleum reserves (Bai and Liu, 1995; Shi et al., 2014).

The atmospheric storage tank is the most common for on-site oil storage, since it is a convenient and lockable form (Bai and Liu, 1995). However, the relatively highly flammable products of crude oil after fractional distillation, which are in danger of exploding are conserved in storage tanks and floating roof storage tanks. In order to analyze the failure probability values and the potential risks involved during the oil-conserving process, the fault tree analysis (FTA) and why tree analysis (WTA) are therefore the tools most often used (Chi et al., 2014; Dong and Yu, 2005; Ejlali and Miremadi, 2014; Hu et al., 2014).

The deductive principles of FTA and WTA were employed broadly during the investigation of an accident's root causes in graphical form (Shi et al., 2014; Ejlali and Miremadi, 2014) to discover the logical functional relationships among the components and subsystems of a system (Baysari et al., 2008; Chauvin et al., 2013; Li et al., 2008; Naderpour et al., 2014; Olsen and Shorrock, 2010; Patterson and Shappell, 2010; Schroder-Hinrichs et al., 2011).

The direct and indirect causes of a systems failure are generally clarified by this format and FTA evaluates an event occurring probability (Ejlali and Miremadi, 2014; Zemva and Zajc, 2005). FTA and WTA provides a wide range of applications, which includes oil and gas transmission (Dong and Yu, 2005), railway systems (Svedung et al., 2008), electric power (Volkanovski et al., 2009), bio-energy production (Hu et al., 2014), nuclear power (Purba et al., 2011), machine robot systems (Lin and Wang, 1997), the aerospace industry (Ale, 2006), the petrochemical processing industry (Wang et al., 2013; Lavasani et al., 2015), and the construction industry (Dong and Yu, 2005).

The possible array of events which leads to an oil storage tank explosion can be demonstrated through such forms of analysis. However, factors such as the distribution of stress during the explosion process is not taken into account during FTA or WTA. In order to simulate the particular explosion process, ANSYS computational fluid dynamics software is used. Among the numerous researches which investigated about chemical explosions, a few was found to focus on the causes of physical explosions. This chapter combines the analysis by WTA and ANSYS to investigate the physical causes of accidents related to caprolactum (CPL) involved in oil storage tanks. The case study in this chapter helps in understanding and analyzing the physical aspects of explosions in oil storage tanks. It also sets a reference related to standard operating procedures (SOPs) for the storage of oil derivatives, which will help towards providing a safer handling protocols and measures.

2.2 CASE STUDY OF A CAPROLACTUM ACCIDENT

2.2.1 Personnel Interview Record

This study involved a storage tank located at a plant in Taiwan. During the explosion, the top cover collapsed, which deformed it completely and also destroyed the feed piping. Approximately 164.72 tons of CPL were stored in the tank. This had a melting point of 69.0 °C, a specific gravity of 1.05, vapor pressure of 0.1 mmHg, an explosion limit of 1.4–8.0 vol% (LEL-UEL), vapor density of 3.9, and a vapor flash point of approximately 100.0 °C.

SUS304 stainless steel was utilized for preparing the storage tank. In June 2000, a floating top design was constructed, and in 2003 a modification with fixed top design to the storage tank was made. The storage tank was maintained at 80.0 °C by enclosing it with thermal insulating material embedded with hot water piping and a zinc-coated metal plate. The approximate volume was estimated to be of 2500.0 kL. There were connected auxiliary features such as a CPL feeding and discharging pipe, a backflow gas pipe, a hot water pipe, a low-pressure vapor pipe, a trace oxygen analyzer, a storage tank, a nitrogen purging system, a thermocouple, a level meter, a pressure transmitter, a flow meter, a loading system, a hot water pump, a filter system, a hot water tank, and a hot water recycling system. In addition, there was a fire hydrant system, a sprinkler system, a cooling sprinkler, a rainwater discharge valve, a vapor cooling system, a bridge pipe support, a measuring port, a sampling port, an oil fence and stairs, a ground wire, and a purging nitrogen supply, as shown in Figure 2.1a and b.

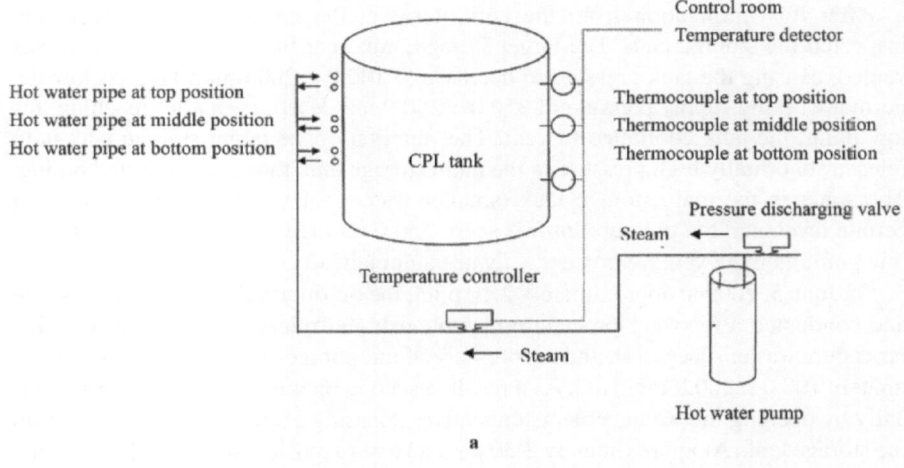

a

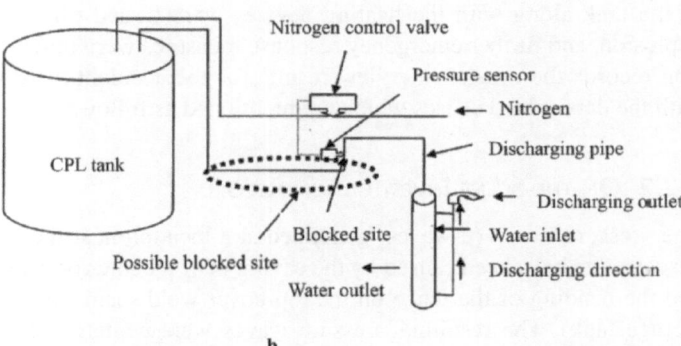

b

FIGURE 2.1 (a) The sketch of tank pressure control system. (b) The sketch of tank undergoes heating process (Reproduced with permission from Liu et al. 2017, Copyright © Elsevier).

The design and construction of the storage tank was produced following the API Standard 650 (API, 2013). The storage tank was equipped with three 4-inch pipelines on the external wall: One for unloading material from a ship, another for releasing pressure through vents connected to the ship, and another for nitrogen blanketing. After the storage tank was modified in 2003, the venting pipe leading toward the ship was sealed, and a new ¾-inch pipe was constructed on the ground surface. Nitrogen was introduced through a 4-inch pipeline into the storage tank. In the new design, the venting pipe was connected to the storage tank. CPL tank operations personnel in charge of the operation at the time of the explosion were interviewed about two weeks later, at 10:40 a.m. on Friday, June 5, 2009. The storage tank explosion occurred on May 15, 2009 at approximately 2:00 p.m. The head of operations stationed in the control room informed on-duty staff that a pressure alarm had sounded for the storage tank (the alarm pressure value was set at 600.0 mm H_2O, normal pressure value was 500.0 mm H_2O, oxygen concentration was 10.0 ppm, and the system was purged with nitrogen).

After this notification from the control room, the on-duty staff immediately inspected the storage tank. The target storage tank near the main storage tank was vented, causing the tank pressure to decrease to 100.0 ± 200.0 mm H_2O (below the normal working supply pressure of 350.0 ± 250.0 mm H_2O); even after reaching this low point, the tank continued to vent. The function of the target storage tank is to release abnormally high pressure in the main storage tank through connected piping. The water in the main storage tank (smaller water seal vessel) is maintained at a certain level, and the pressure limit is set at 550.0 mm H_2O. If the pressure exceeds this limit, then the venting process activates automatically.

On June 5, 2009 at approximately 2:10 p.m., the on-duty staff arrived on the scene and conducted a recovery procedure to replenish and release water. However, this procedure was unsuccessful, and the pressure of the storage tank fluctuated at approximately 100.0 ± 200.0 mm H_2O. As a result, a second operation was performed manually by opening the water-replenishing valve, releasing a small amount of gas from the storage tank. At approximately 2:30 p.m. a large explosion occurred. The thermal insulating materials, such as calcium silicate boards and insulation wool, were scattered near the tank due to the sudden rise in pressure in the storage tank. The power to the tank along with the heating devices were turned off immediately after the explosion, and further emergency response measures were initiated. In the investigation records the on-site interview results for the accident were summarized along with the determined causes which are mentioned as follows.

2.2.2 On-the-Scene Investigation Record

The pressure waves (p-waves) generated at a location near the explosion center are suspected to have been caused by the caving in of the interior wall of the storage tank and the bending of the fence on the top-cover weld seam (the weakest point of the storage tank). The resulting pressure waves were centered at the explosion point inside the storage tank and expanded outward spherically. The wave surface area increased as the radius of the storage tank enlarged. In addition, air gas pressure changed substantially and rapidly over time. Because of the aforementioned factors, the top cover of the storage tank was unable to withstand the largest pressure wave during the sudden and continuous rise and fall in pressure (DP), leading to the explosion.

The fractured surface had a dull metal color that did not show any sign of corrosion along the weld seam between the body and top cover, as depicted in Figure 2.2a and b. Furthermore, the lifting hole installed on the tank body for the purposes of relocation also showed damage near the weld seam. Deformation of the weld seam was clearly observed on the fractured surface between the lifting hole and the tank body.

a

b

FIGURE 2.2   (a) Photograph of the fractured surface between the storage tank body and its top cover. (b) Photograph of the fracture surface between the lifting hole and tank body (Reproduced with permission from Liu et al. 2017, Copyright © Elsevier).

CPL crystals and CPL blockage inside the pipes were noted on the interfaces between the vertical and horizontal hose pipes and pressure releasing pipes. A false signal was sent to the nitrogen control valve because of the improper location and the low sensitivity of the pressure sensor installed on the pressure release pipe, as shown in Figure 2.3a and b. The sensor sent a signal to the nitrogen control valve that pressure was low, so the pressurized nitrogen flow continued, eventually resulting in an explosion.

2.3 COMPUTATIONAL APPROACH

The principle of WTA is as follows: A "why" question is proposed, an answer is developed, and another "why" question is crafted accordingly. This question-and-answer root cause analysis is similar to a game of tic-tac-toe, and the analysis can delineate the logical connectivity between events and all their causes from a wide range of possibilities. This form of analysis enables researchers to identify all the situations that could have been involved in a failure event (LRS, 2014) and is a method that enables investigative teams to remain focused. Most crucially, WTA is a means of showing others what evidence has been discovered using the rigorous methods of root cause analysis (Failsafe Network, 2014).

2.3.1 THREE-DIMENSIONAL FINITE ELEMENT SIMULATION ON STRESS ANALYSIS

The demand for performance prediction has increased alongside the development of high-level technology, attracting the attention of scientists, merchandisers, and munitions producers in a wide variety of fields, such as electronics, fluid dynamics, structural analysis, multi-physics, systems, and embedded software applications.

Following this trend, the ANSYS Corporation has developed a series of software that predicts the performance of a mechanical instrument before it is manufactured. The ANSYS Mechanical software is a comprehensive finite element analysis (FEA) tool for structural analysis, including linear, nonlinear, and dynamic studies (Introduction of ANSYS cooperation, 2016). A complete set of element behavior,

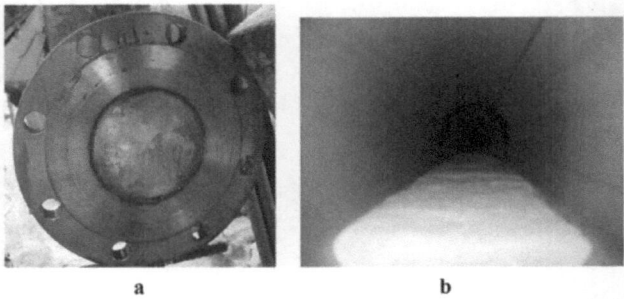

a   b

FIGURE 2.3   a and b Plots of the interfaces between the vertical and horizontal hose and pressure releasing pipes (Reproduced with permission from Liu et al. 2017, Copyright © Elsevier).

equation solvers, and material, and material models for a wide range of mechanical design problems can be applied using this engineering simulation product. Moreover, ANSYS Mechanical offers coupled-physics capabilities and piezoelectric, acoustic, thermo-electric, and thermal-structural analysis.

FEA is a computerized approach for anticipating how a product responds to real-world forces, vibration, fluid flow, heat, and other physical effects. FEA demonstrates whether a product will crack, rupture, wear out, or function the way it was originally designed. The software splits the real object in question into a large number (in some cases over 100,000) of finite elements, such as minute cubes that predict the behavior of each element according to summed by a computer to predict the behaviors of the actual object.

Furthermore, FEA facilitates the forecasting of the behaviors of products affected by physical processes, such as mechanical stress, mechanical vibration, motion, fatigue, heat transfer, fluid flow, electrostatics, and plastic injection molding (Introduction of Autodesk's Product, 2014).

2.4 INVESTIGATION AND ANALYSIS FOR THE CAUSE OF ACCIDENT

An explosion can be described as a process caused by a rapid change in physical and chemical energies under a certain volume of state or space. In this process, the internal energy of the system is rapidly transformed into mechanical energy, light, kinetic energy, and thermal radiation. Therefore, an explosion can also be described as a rapid process that releases physical or chemical energy. This type of process can be divided into three categories according to cause: Physical explosions, chemical explosions, and physicochemical explosions.

Swift changes in the composition of materials are defined as physical explosions. However, the molecular structure of a material does not change before and after it undergoes an explosion. However, a chemical explosion is caused by a rapid chemical reaction in the material itself. Chemical reactions often generate a combination of high temperature and pressure that triggers an explosion. Explosions involve intense physicochemical changes that generate light, heat, and pressure, consequently creating dissociation and an oxygen-deficient environment within the explosion area. The damage created by an explosion depends on the location, quantity, and properties of the exploding material, as well as the conditions under which the event occurs. The possible harms induced by an explosion may include shock waves, vibration effects, flying debris, gas leakage, and fire hazards.

Explosions cause changes in air gas pressure that lead to sudden increases and decreases in pressure among affected materials. The high temperature and pressure created during an explosion induces detrimental effects, referred to as explosion effects, such as deformation, compression, fracturing, and scattering, on surrounding objects. Figure 2.4 shows that explosions can be classified according to their speed, characteristics, and reaction phases.

In this case study, the explosion of the main storage tank generated destructive forces that caused the objects in close proximity to the explosion to rupture and vibrate. Because the shock waves created by the explosion were inversely

Pressure (bar)

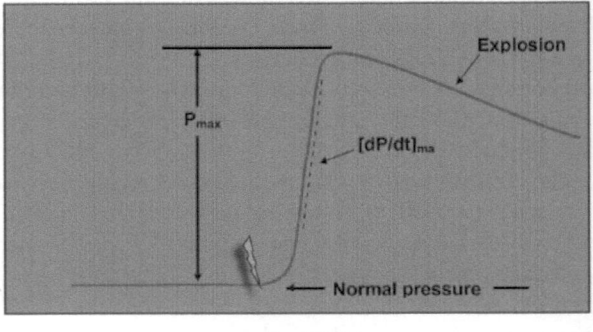

TIME

FIGURE 2.4 Correlation between the pressure released and the rate of pressure increase in an explosion (Reproduced with permission from Liu et al. 2017, Copyright © Elsevier).

proportional to the distance from the exploding object (shock waves are usually formed from the superimposed positioning of compressed waves), when the waves propagated throughout the storage tank, the medium underwent a skipping effect. Three-dimensional pressure waves were generated by the exploding gas, which then expanded outward spherically from the explosion point. The sudden and ongoing rises and drops in pressure (DP) eventually rendered the top cover of the storage tank unable to withstand the accumulated pressure drop, and an explosion ensued.

2.5 CLARIFICATION FOR THE CAUSE OF ACCIDENT

WTA was performed systematically to clarify the direct and indirect causes of the explosion. This investigation focused on performing a material and stress analysis for the storage tank, in order to further verify the appropriateness of the construction or design of the storage tank. The assessment results are elucidated by the WTA process flow diagram as listed in Figure 2.5. As shown in Table 2.1, the causes of the explosion accident were assessed according to the following scenarios: (1) improper design and construction of the storage tank; (2) corrosion of the weld seam of the storage tank; (3) blockage of the storage tank venting pipe; (4) nitrogen system malfunction; and (5) improper operation and management. The assessments are described as follows (Table 2.2).

2.5.1 IMPROPER DESIGN AND CONSTRUCTION OF THE STORAGE TANK

Regulation 3.10.6 of the API 650 Standard was used to conduct the assessment. According to this regulation, the minimum thickness for the design of the top cover is 3/16 in. (5 mm). The diameter of the tank body should be 15.29 m (50.164 ft) with R = 0.93 m.

Therefore, the thickness of the top cover could be estimated as 46.65/200 in. = 0.2335 in. = 5.924 mm ≥ 4.76 mm. The designated thickness for the storage tank top cover was deemed acceptable, because the actual stress was 169.38 kgf/m², which

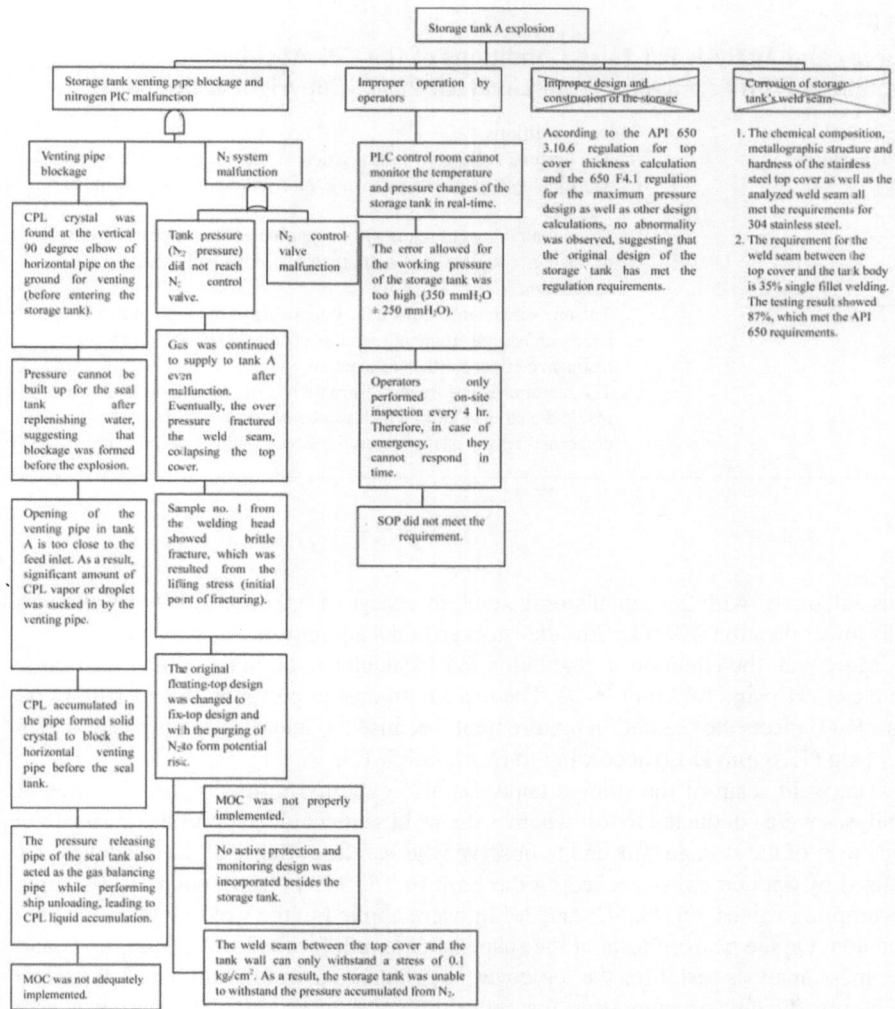

FIGURE 2.5 Results of WTA for the storage tank (Reproduced with permission from Liu et al. 2017, Copyright © Elsevier).

TABLE 2.1

Possible Causes of The CPL Explosion Accident as Determined Through WTA (Data Obtained With Permission From Liu Et Al. 2017, Copyright © Elsevier)

Accident	Possible Cause Scenario
CPL explosion	N_2 system malfunction
	Storage tank venting pipe blockage
	Incorrect operation by operators
	Improper design and construction of the storage tank
	Corrosion of the storage tank's weld seam

TABLE 2.2

Survey and Analysis For False Conditions of the CPL Accident (Data Obtained With Permission From Liu Et Al. 2017, Copyright © Elsevier)

Accident	False Conditions
CPL explosion	1. The original design and construction of the storage tank have no abnormality because of following the regulation requirements of API 650.
	2. The remaining weld seam will be insufficient to support the weight of the entire top cover, if more than 1/2 of the weld seam is in suspension.
	3. The pressure change before the collapsing of the top cover will not approach natural frequency, so this change is not able to create resonance effect for the storage tank.
	4. The major cause of fracture formation is the exceeding pressure inside the storage tank, which can be demonstrated from the combined results of the material analysis and stress analysis.

was calculated with the actual storage tank thickness of 6.0 mm. This level of stress was lower than the 219.6 kgf/m^2 that storage tanks are required to withstand.

Moreover, the equation in regulation F.4.1 calculates the maximum design pressure as 2.31 psig (1.62 mm H$_2$O). The maximum design pressure of 2.31 psig (1.62 mm H$_2$O) meets the regulation requirement, because the internal pressure is less than 2.5 psig (1.76 mm H$_2$O) according to regulation F.1.3.

The weld seam of the storage tank was also explored in this report. A series of analyses were conducted to test whether the weld seam could support the mass of the top cover of the storage tank and to observe whether the collapse of the top cover was caused by applied stress exceeding the limit of 155 MPa (15,810 mm H$_2$O). Three assumption values, of 1/4, 1/2, and 3/4 in, were applied to the welding length in suspension. On the basis of tests of the suspension length for the three assumptions and the mass analysis result for the top cover, the remaining weld seam was determined to be insufficient for supporting the entire top cover when the weld seam was suspended for more than 1/2 in, leading to the collapse of the top cover, as illustrated in Figures 2.6–2.8.

In addition, three-dimensional finite element modeling was employed to evaluate the natural frequency of the storage tank, because 10 min. fluctuations in pressure had occurred before the collapse of the top cover of the storage tank, as indicated in Figure 2.9. The possibility of resonance caused by pressure fluctuation (frequency) was confirmed through tests involving three-dimensional finite element modeling. The natural frequency was calculated as 8.548 (1/s) under the first modeling vibration conditions, where the resonance effect can be triggered by the lowest frequency value.

The pressure change before the collapsing of the top cover was insufficient to create a resonance effect in the storage tank, because the variation in pressure was too slow to generate a fluctuation of 8–9/s.

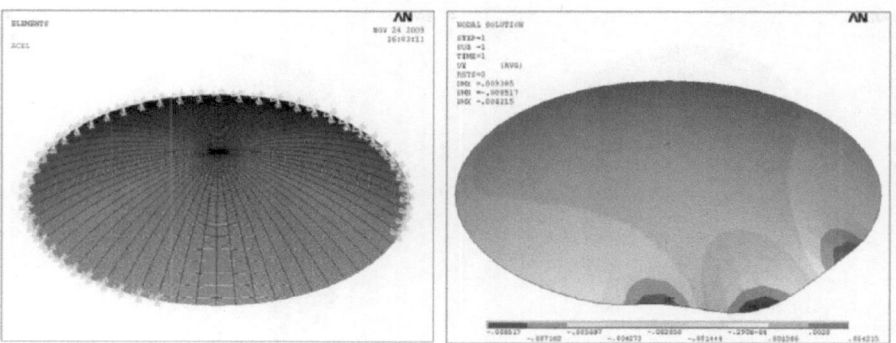

FIGURE 2.6 Three-dimensional finite element modeling results, assuming the suspension is limited to 1/4 of the weld seam length (Reproduced with permission from Liu et al. 2017, Copyright © Elsevier).

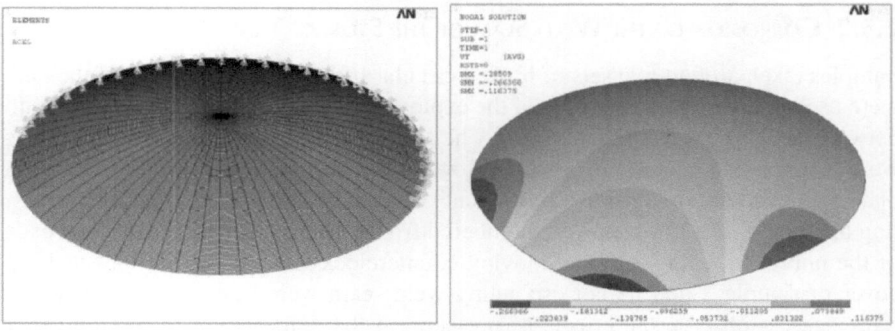

FIGURE 2.7 Three-dimensional finite element modeling results, assuming the suspension is limited to 1/2 of the weld seam length (Reproduced with permission from Liu et al. 2017, Copyright © Elsevier).

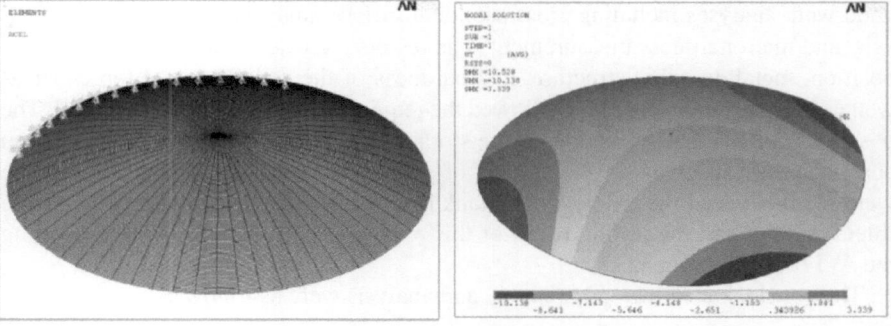

FIGURE 2.8 Three-dimensional finite element modeling results, assuming the suspension is limited to 3/4 of the weld seam length (Reproduced with permission from Liu et al. 2017, Copyright © Elsevier).

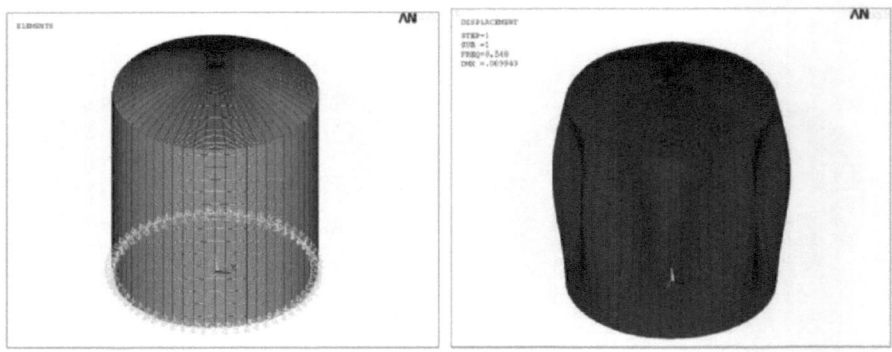

FIGURE 2.9   Three-dimensional finite element dynamic simulation results for the storage tank (Reproduced with permission from Liu et al. 2017, Copyright © Elsevier).

2.5.2 CORROSION OF THE WELD SEAM OF THE STORAGE TANK

Samples taken around five sets of broken circular lifting holes on the top of the tank were used to investigate the cause of the explosion, which occurred at the weld seam between the top cover and tank body. The five sets of broken circular lifting holes were taken as the center and extended toward the left and right for 20.0 cm; steel plate samples with dimensions of 40.0 cm² were then obtained through a laser-cutting technique. The samples were numbered from 1 to 5, starting from the two ends of the unbroken weld seam and moving counterclockwise. Moreover, the steel top cover of sample 3 and its corresponding weld seam were selected as comparison references to observe the effects of pressure on the deformation of the top cover before and after the explosion. A set of control samples was obtained from the unbroken weld seam for subsequent comparison and analysis to distinguish the materials on the broken and unbroken weld seam. Six controls were obtained for the storage tank body.

Part of the weld seam for each control was sampled, and the sample acquired then underwent analyses including cross-sectional surface analysis, metallographic analysis, and microhardness measurement. The results indicated that the chemical composition, metallographic structure, and hardness of the stainless steel top cover as well as the analyzed weld seam followed the requirements for 304 stainless steel. The possibility of fracture formation on the storage tank body was ruled out because no corrosion was observed on the surface of any of the weld seam. Furthermore, the weld seam between the top cover and tank body is required to contain 35.0% single fillet welding, and the testing result of this explosion analysis was 87.0%, meeting the API 650 requirement.

The conclusions drawn after survey and analysis were as follows:

(1) The original design and construction of the storage tank had no abnormalities because they fully complied with the regulation requirements of API 650.

(2) The remaining weld seam was insufficient to support the mass of the entire top cover, if more than half of the weld seam was in suspension.

(3) The pressure change before the collapse of the top cover did not approach natural frequency and, as a result, could not create a resonance effect in the storage tank.

(4) The main cause of fracture formation was the excessive pressure inside the storage tank, which was indicated by the combined results of the material and stress analyses.

2.5.3 BLOCKAGE OF THE STORAGE TANK VENTING PIPE

Near the vertical 90° elbow of the horizontal pipe used for venting, solid CPL crystals were identified in the pipe during an onsite inspection. The initial evidence did not confirm whether the crystallization of CPL or the blockage had appeared at the interfaces between the vertical and horizontal hose pipes and pressure releasing pipes before or after the explosion.

However, several material analysis and stress analysis tests determined that the risk of improper venting would still be not induced even if the water-replenishing valve was left completely open and the operator had improperly closed the pressure releasing pipe. The blockage was formed before the explosion, because pressure could not rise in the storage tank even after water was replenished.

In addition, a substantial amount of CPL vapor was sucked in by the venting pipe, because the opening of the venting pipe in the crystals accumulated in the horizontal venting pipe and blocked it. The gas-balancing pipe used for unloading CPL from ships had a high probability of cross-contamination with CPL in the original design. Instead of the gas-balancing pipe being returned to the ship, the pressure-releasing pipe of the storage tank was used as the gas-balancing pipe while the ship was being unloaded to forestall cross-contamination. Therefore, the root cause of the explosion was actions and assessments related to the management of change (MOC) when the gas-balancing pipe was altered during ship unloading.

2.5.4 NITROGEN SYSTEM MALFUNCTION

The pressure sensor of the storage tank was installed on the pressure release pipe in the original design. The fact that CPL crystals had blocked the interfaces between the vertical and horizontal hose pipes and pressure releasing pipes was only discovered after the explosion because of the improper location and low sensitivity of the sensor. This caused the pressure sensor of the tank to send false signals to the nitrogen control valve. Because the nitrogen control valve received an incorrect low pressure signal (low pressure), the storage tank built up the tank pressure by continuing to replenish the tank with nitrogen.

The normal pressure for the storage tank was designed to be only 1.623 kgf/cm^2, and the weld seam between the top cover and the tank wall could only withstand a stress of 0.1 kgf/cm^2. However, the working supply pressure for the nitrogen system was 7.0 kgf/cm^2. As a consequence, the storage tank was unable to withstand the pressure from the nitrogen, leading to fracturing at the weld seam.

Figure 2.10 indicates that the fractured portion of sample 1 at the welding toe displayed a cupped shape, and the welding head showed brittle fractures, differing from other samples. These results can be explained by the faster rate of deformation at this location than that at other sampling points. Therefore, strong evidence indicates that the weld seam near sample 1 was the point where the top cover of the storage tank began to crack. The internal pressure of the storage tank was higher than the yield strength of the material, damaging the top cover. In this case, part of the welding toe began to fracture and this continued to spread outward because of deformation of material. The high pressure inside the storage tank will cause more fractures along the weld seam once breakage is formed at the weakest point, eventually causing the top cover of the storage tank to crack.

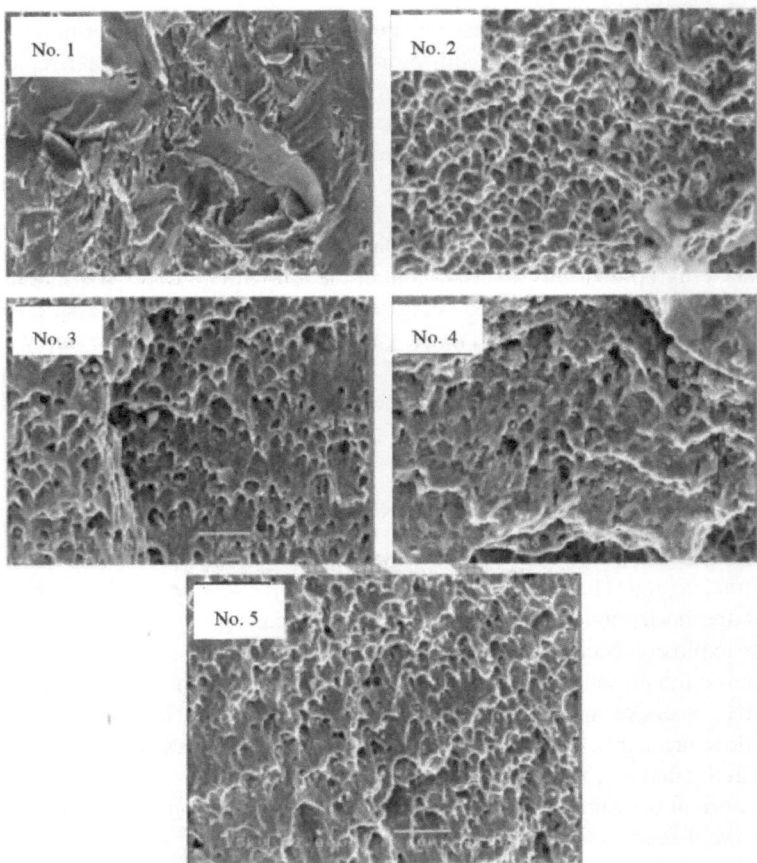

FIGURE 2.10 Evidence of the cup-shaped brittle fracture on the welding head (Reproduced with permission from Liu et al. 2017, Copyright © Elsevier).

The indirect cause of the top cover fracturing along the weld seam is the change in the original design of the storage tank top from the floating type (used for oil storage) to the fixed type (used for CPL storage). Accordingly, more risks were added to the calculations because an atmospheric pressure storage tank and pressurized storage tank were incorporated into the operation during the nitrogen purging process. The fundamental causes of the accident included a lack of active protection and monitoring of the storage tank, indicating improper MOC practices after the change in design, and an overload of pressure on the welding point of the storage tank. The explosion was ultimately caused by the continuing supply of nitrogen creating enormous pressure that exceeded the limit strength of the storage tank.

2.5.5 IMPROPER MANAGEMENT

The company provided SOP and daily CPL management checklist, the mentioned working supply pressure for the storage tank was 350.0 ± 250.0 mm H_2O. However, it was found that the pressure limit exceeded about more than 20 times the working supply pressure of nitrogen system > 7.0 kgf/cm^2 (equivalent to 70,000.0 mm H_2O), as per the designed limit for the storage tank. Furthermore, records showed some discrepancy in working supply pressure to the storage tank. On May 14, 2009 at 7:00 p.m. the working supply pressure was maintained at 263.0 mm H_2O, whereas on the following day it surprisingly increased to 460.0 mm H_2O at 11:00 a.m. An obvious increase of working supply pressure from the standard pressure of 350.0 mm H_2O approximately by 110.0 mm H_2O, should have raised an alarm, but nothing such happened since the on-duty staff was not tracking the monitor. According to his report since the pressure did not pass the working limit of ±250.0 mm H_2O at the moment he checked it, the pressure was not monitored again.

An alarm was noted in the storage tank pressure chamber, on the day of the explosion at approximately 2:00 p.m., since the exceeded pressure surpassed the alarm pressure value of 600.0 mm H_2O. The target storage tank near the main storage tank was vented to reduce the pressure to 100.0–200.0 mm H_2O, and the main tank still continued to vent. After the pressure of the main storage tank reached approximately 100.0–200.0 mm H_2O, the water replenishing valve was opened manually. As a result, only a small amount of gas was released from the storage tank. The explosion, in turn, occurred at approximately 2:30 p.m. Apart from the previously described false signals on venting pipelines, the control room could not monitor temperature and pressure changes in the storage tank, and the operators were not able to adequately conduct real-time, on-site inspections. Therefore, appropriate and timely responses to emergencies were impossible.

2.6 SUMMARY

The accident investigated in this chapter can be regarded as a physical explosion since no chemical reaction, fire hazard, or thermal radiation were involved. Four main causes that were attributed due to the excessive pressure which led to the explosion of the storage tank were: Improper operation of the storage tank nitrogen system, and venting pipe blockage along with its poor management, inefficient designing, and construction of storage tank, along with the management system. The pressure transmitter sensor didn't correctly detect the actual pressure signal, which resulted in the transmission pressure being too low to cause the false signal to occur. Such happenings led to the improper control of the nitrogen control switch valve. Moreover, the control valve maintained in the state of nitrogen added, and the tank to the water tank only occurred at the condition, where the pressure relief line has been blocked (certified by the pressure transmitter and CPL in the lowest point line). The pressure relief at the reservoir pipe, which was the only way to release the pressure, was blocked and this resulted in the tank being unable to exercise the normal pressure relief function. Such a series of events was found to be the main reason for this accident. Hence, there is a need for a safer design of storage tanks which should be followed with proper SOPs and procedures to avoid any abnormalities. Moreover, there is a need to revise the existing SOPs carefully in order to improve working operations which would further help the personnel to tackle the process upsets more readily and with better capability.

REFERENCES

Ale, B.J.M., 2006. Towards a causal model for air transport safety-an ongoing research project. *Safe. Sci.* 44, 657–673.

API, 2013. American Petroleum Institute. API Standard 650. *Welded tanks for oil storage.* http://www.api.org/.

Bai, M., Liu, Z.W., 1995. Economic benefit analysis of large-scale oil tank. *Pet. Eng. Constr.* 1, 8–10.

Baysari, M.T., McIntosh, A.S., Wilson, R., 2008. Understanding the human factors contribution to railway accidents and incidents in Australia. *Accid. Anal. Prev.* 40, 1750–1757.

Buncefield Major Incident Investigation Board, 2008. The Buncefield Incident. December 11, 2005, Final report, Hemel Hempstead, Hertfordshire, UK.

Chang, J.I., Lin, C.C., 2006. A study of storage tank accidents. *J. Loss Prev. Process Ind.* 19, 51–59.

Chauvin, C., Lardjane, S., Morel, G., Clostermann, J.P., Langard, B., 2013. Human and organisational factors in maritime accidents: analysis of collisions at sea using the HFACS. *Accid. Anal. Prev.* 59, 26–37.

Chen, W.T., Chen, W.C., Hsueh, K.H., Chiu, C.W., Shu, C.M., 2014. Thermokinetic parameters analysis for 1,1-bis-(tert-butylperoxy)-3,3,5-trimethylcyclohexane at isothermal conditions for safety assessment. *J. Therm. Anal. Calorim.* 118, 1085–1094.

Chi, C.F., Lin, S.Z., Dewi, R.S., 2014. Graphical fault tree analysis for fatal falls in the construction industry. *Accid. Anal. Prev.* 72, 359–369.

De La Fuente, V.S., López, M.A.C., González, I.F., Alcántara, O.J.G., Ritzel, D.O., 2014. The impact of the economic crisis on occupational injuries. *J. Saf. Res.* 48, 77–85.

Dong, Y.H., Yu, D.T., 2005. Estimation of failure probability of oil and gas pipelines by fuzzy fault tree analysis. *J. Loss Prev. Process Ind.* 18, 83–88.

Ejlali, A., Miremadi, S.G., 2014. FPGA-based Monte Carlo simulation for fault tree analysis. *Microelectron. Reliab* 44, 1017–1028.

Failsafe Network, 2014. *Four significant problems with root cause analysis* http://www.failsafe-network.com/root-cause-analysis-definitions/why-tree.

Hu, J., Chu, J.Y., Liu, J.H., Qin, D.Y., Cheng, S.K., Li, Z.F., Mang, H.P., Neupane, K., Wauthelet, M., Huba, E.M., 2014. Application of fault tree approach for technical assessment of small-sized biogas systems in Nepal. *Appl. Energ* 113, 1372–1381.

Introduction of ANSYS cooperation, 2016. http://www.ansys.com.

Introduction of Autodesk's product, 2014. *Finite element analysis*, http://usa.autodesk.com/adsk/servlet/item?siteID¼123112&id¼17670721.

Introduction of explosion protection, 2016. *Fike official website* http://www.fike.com/solutions/explosion-protection/.

Jonsson, A., Bergqvist, A., Andersson, R., 2015. Assessing the number of fire fatalities in a defined population. *J. Saf. Res.* 55, 99–103.

Lavasani, S.M., Zendegani, A., Celik, M., 2015. An extension to Fuzzy Fault Tree Analysis (FFTA) application in petrochemical process industry. *Process. Saf. Environ. Prot.* 93, 75–88.

Li, W.C., Harris, D., Yu, C.S., 2008. Routes to failure: analysis of 41 civil aviation accidents from the Republic of China using the human factors analysis and classification system. *Accid. Anal. Prev.* 40, 426–434.

Lifetime Reliability & Solutions, 2014. *Understanding How to Use the 5-whys for Root Cause Analysis*. Available in http://www.lifetime-reliability.com/index.html.

Lin, C.T., Wang, M.J.J., 1997. Hybrid fault tree analysis using fuzzy sets. *Reliab. Eng. Syst. Saf.* 58, 205–213.

Liu, W.Y., Chen, C.H., Chen, W.T., Shu, C.M., 2017. A study of caprolactam storage tank accident through root cause analysis with a computational approach. *J. Loss Prev. Proc.* 50, 80–90.

Marucci-Wellman, H.R., Courtney, T.K., Corns, H.L., Sorock, G.S., Webster, B.S., Wasiak, R.Y.N.I., Matz, S., Leamon, T.B., 2015. The direct cost burden of 13 years of disabling workplace injuries in the U.S. (1998–2010): findings from the Liberty Mutual Workplace Safety Index. *J. Saf. Res.* 55, 53–62.

Naderpour, M., Lu, J., Zhang, G., 2014. The explosion at institute: modeling and analyzing the situation awareness factor. *Accid. Anal. Prev.* 73, 209–224.

Olsen, N.S., Shorrock, S.T., 2010. Evaluation of the HFACS-ADF safety classification system: inter-coder consensus and intra-coder consistency. *Accid. Anal. Prev.* 42, 437–444.

Patterson, J.M., Shappell, S.A., 2010. Operator error and system deficiencies: analysis of 508 mining incidents and accidents from Queensland, Australia using HFACS. *Accid. Anal. Prev.* 42, 1379–1385.

Purba, J.H., Lu, J., Ruan, D., Zhang, G., 2011. Failure possibilities for nuclear safety assessment by fault tree analysis. *Int. J. Nucl. Knowl. Manag.* 5, 162–177.

Schroder-Hinrichs, J.U., Baldauf, M., Ghirxi, K.T., 2011. Accident investigation reporting deficiencies related to organizational factors in machinery space fires and explosions. *Accid. Anal. Prev.* 43, 1187–1196.

Sharma, R.K., Gurjar, B.R., Wate, S.R., Ghuge, S.P., Agrawal, R., 2013. Assessment of an accidental vapour cloud explosion: lessons from the Indian oil corporation ltd. accident at Jaipur, India. *J. Loss Prev. Process Ind.* 26, 82–90.

Shi, L., Shuai, J., Xu, K., 2014. Fuzzy fault tree assessment based on improved AHP for fire and explosion accidents for steel oil storage tanks. *J. Hazard. Mater* 278, 529–538.

Svedung, I., Andersson, R., Ra, H., 2008. Suicide prevention in railway systems: application of a barrier approach. *Safe. Sci.* 46, 729–737.

U.S. Chemical Safety Board, 2009. Cited at December, 05, 2011, http://www.csb.gov/caribbean-petroleum-investigative-photos-/.

Volkanovski, A., Cepin, M., Mavko, B., 2009. Application of the fault tree analysis for assessment of power system reliability. *Reliab. Eng. Syst. Safe* 94, 1116–1127.

Wang, D., Zhang, P., Chen, L., 2013. Fuzzy fault tree analysis for fire and explosion of crude oil tanks. *J. Loss Prev. Process Ind.* 26, 1390–1398.

Zemva, A., Zajc, B., 2005. Test generation for technology-specific multi-faults based on detectable perturbations. *Microelectron. Reliab* 45, 163–173.

Zhang, D., Ding, A., Cui, S., Hu, C., Thornton, F.S., Dou, J., Sun, Y., Huang, W.E., 2013. Whole cell bioreporter application for rapid detection and evaluation of crude oil spill in seawater caused by Dalian oil tank explosion. *Water Res.* 47, 1191–1200.

3 Fire Explosion Accident Caused by Static Electricity in a Propylene Plant

3.1 STATIC ELECTRICITY AND PROPYLENE PLANT INTEGRATED INDUSTRIES

Static electricity is the static charge developed between two surfaces under relative motion and has been for the cause of many industrial accidents. It has a potential ignition risk, and it is very challenging to identify the source of formation and discharge. For example, liquid/solid particles flowing along the pipeline walls, agitated liquids in tanks, a belt pulley assembly, gas particles rubbing against solid surface, dust clouds in air, vacuum devices, walking and getting up from a seat develop equal and opposite charges that flow rapidly when grounded. When two bodies of different materials are brought into physical contact with each other, when separated, creates a counterforce, making one body positively charged (surplus protons) and the other negatively charged (excess of electrons) which acts as a source of potential difference. Static charge is developed in contact with non-conductive materials (plastics) where most of the hydrocarbons containing oxygen are stored (non-grounded). A spark produced during high static discharge can ignite the flammable vapors (in liquids) or a dust cloud (in powders), resulting in fire and explosion.

In 1969, static charge developed by water jets during washing resulted in a major explosion in a three 20000-ton cargo oil tanker (Gibson, 1997). Netherlands organization for applied scientific research investigated major industrial accidents where static electricity was the ignition source. An explosion in a compressor/high-pressure air buffer system was instigated due to the long-term accumulation of lubricating oil and the residual acids. Fire was noticed in the oil demister and this was followed by a secondary explosion, resulting in severe injuries to the operator. An explosion in the buffer tank of an incinerator took place after 15 hours of regular operation due to a leak in the instrumental air system, which formed a natural gas (fuel)-air mixture ignited by static charge. An explosion took place in a waterways storage tank while benzene was transferred from an onshore storage tank. Traces of ammonia reacting

39

with the coating of metallic zinc layer resulted in the formation of flammable hydrogen gas could have ignited while loading at some point between tank wall and pump line (Logtenberg, 2001). The ignition risks associated with electrostatic phenomenon generated during various operations in petroleum, chemical, and powder handling industries encouraged both academia and industry to undertake research to help simulate and comprehend static electric discharge mitigation.

Polypropylene (PP) is a thermoplastic polymer (partially crystalline and non-polar) belonging to the group of poly-olefins, and is the second-most widely used commodity plastic, with improved hardness and heat resistance similar to polyethylene. The PP was first polymerized in 1951 and was commercialized in 1954 by the Italian chemist, Professor Giulio Natta, in Spain. Due to its high demand in the manufacture of piping systems, medical or laboratory items, and daily household items such as bags, containers, ropes and other wearable items, the global industrial demand for this product increased to 56 million tons in 2018, and is projected to reach 88 million tons in 2026 at a growth rate of 5.7%. In the international market over the past few years, the electronics, computer, and biotechnology sectors under domestic high-tech industries have gained attention. The risk of fire and explosion along with various poisonous gas releases have intensified because of the installation of various high-tech plants in Taiwan. Therefore, an increasing need for such disaster management has emerged which needs to be implemented in high-tech plants (Hsieh, 2014; Li, 2014). The present research explored the fire and explosions caused mainly due to the leakage of acetone involved in high-tech plants with polypropylene (PP) and copper-clad laminates (CCLs) units. The investigation was focused mainly on the space management of the high-tech plants at the fire hazard incident and through the analysis of similar spatial features determined the potential risk of plants (Suardin et al., 2009; Dana et al., 2014).

3.2 STRUCTURAL FEATURES OF A HIGH-TECH PLANT

1. *Solid framework structure*: Mainly steel and reinforced concrete are used for framing the high-tech plant to increase the load capacity. Such strong frameworks can resist strong winds as well as seismic shaking and provides a minimal chance of process failure. Due to such strong basement, the firefighters are relieved from the possibility of building collapse during a fire out-break operation.

2. *Spacious floor area*: In order to create convenient process operations within a high-tech plant, the building design focuses more on the opening up of broader spaces (office, clean room, and laboratory), where fire propagates readily.

3. *Interior complex decorating materials*: Within a high-tech plant building, complex materials are found associated with interior decorations, furnished items which exists in good amount. During a fire, such complex materials burn rapidly and produces a large amount of toxic gases, along with hot smoke, which are considered very harmful for survival.

4. *Heating process*: A slight error during the manufacturing operation can lead to the outbreak of a fire within any confined space that is operating in a high-temperature environment.

5. *Long-term operation of mechanical configurations*: An excessive quantity of equipment is being stacked in most high-tech plants in each unit. The bulky machines are hard to transport, such bulky commodities act as hindrance during disaster rescue operation, by blocking the site of ignition points.

6. *Complex arrangement of pipelines and dense covers*: Controlling the leakage during any continuous operation is not easily achieved, since there is a variety of equipment, and instruments interconnected by pipelines involving hazardous chemicals and gases, that might introduce fire and explosion gases.

7. *Special zoning design*: During the outbreak of a fire, the elimination of smoke and heat is challenging as most of the operations of a clean room are accompanied in the confined space and the destruction of zoning increases the possibility of a fire spreading rapidly.

3.3 FIRE PREVENTION FEATURES

1. *Dense smoke:* The combustible materials present in high-tech plants produce a significant quantity of smoke that extends via various openings, channels, stairs, and ducts obstructing vision and rescue procedures.

2 *High temperature*: The high-tech plants are secured structures, restricting the dissipation of accumulated heat generated from combustion.

3. *Fire outbreak and rapid spread*: The special zoning in high-tech plants results in the increase of rate of fire spread due to the use of interior decoration acting as a source of flammable materials.

4. *Prolonged firefighting*: The wide-ranging area and commodious arrangement of high-tech plant makes it highly difficult to identify the ignition source.

5. *Extensive water damage*: The fire-rescue process involves pumping of water into the broad spaces (since it is difficult to precisely determine the ignition source) to reduce the temperature and the substantial damage caused when compared to fire.

6. *Difficult to escape*: The fire outbreak is accompanied by the generation of dense smoke due to the multifaceted spatial design and the operation of high-tech plants, hindering the escape of employees and firefighting personnel.

8. *Toxic and hazardous products*: The use of lethal and hazardous materials as feedstock or product increase the risk of the personnel and firefighters' exposure to toxic fumes hindering the relief operations.

7. *Fire compartment destruction*: The expansion of the combustion area is facilitated by the destruction of the fire compartment (commissioning of machines and equipment) as unit operations and processes are controlled by automated systems aggravates the technical hitches of fire-rescue operations.

3.4 CASE REVIEW ON DISASTER

The present chapter investigates a fire accident at the PP and CCL high-tech plant on December 10, 2010, combining the plant space, and fire characteristics to prevent similar occurrences effectively. The plant site consists of the administration building, plants 1 and 2 (reinforced concrete structures), automatic/raw material warehouse and an outdoor storage tank area (reinforced steel structures) for the production of CCL and PP. The fire and explosion resulted in a property loss of US\$20 million, with one fatality and injuries to five other employees. The source of ignition was a static charge that came in contact with leaking flammable liquid acetone resulted in fire and explosion (Lin, 2014; Ye, 2014). The time of the fire outbreak was 4:00 p.m. on a sunny day. The fire rescue upon reaching the accident site determined that the power supply was disconnected, doors and windows were closed, and flames vented from the rare side of the plant (Figure 3.1). The fire has spread to an area was 1000 m², injuring five personnel. The fire location was the first floor gluing area (Figure 3.2), where the smell of coke was dominant and the explosion was inevitable. Acetone was delivered to Plant 1 or 2 from the six outdoor storage tanks connected by air tube via pump (Figure 3.3). Approximately two hours before the fire, a night-shift engineer observed acetone shortage in the storage tank of plant 1. At about 4.00 p.m., an alert in plant 2 about acetone leakage was signaled. Acetone seeped from the third floor of plant 2 through the vacant spaces between the gluing area and floors to the second and first floors. The plant requested the firefighting personnel to provide support, but the explosion was unavoidable. The fire caused severe damage to the plant, automatic/raw material warehouse and the administration building.

FIGURE 3.1 Appearance of the fire scene (Reproduced with permission from Chou et al. 2015, Copyright © Elsevier).

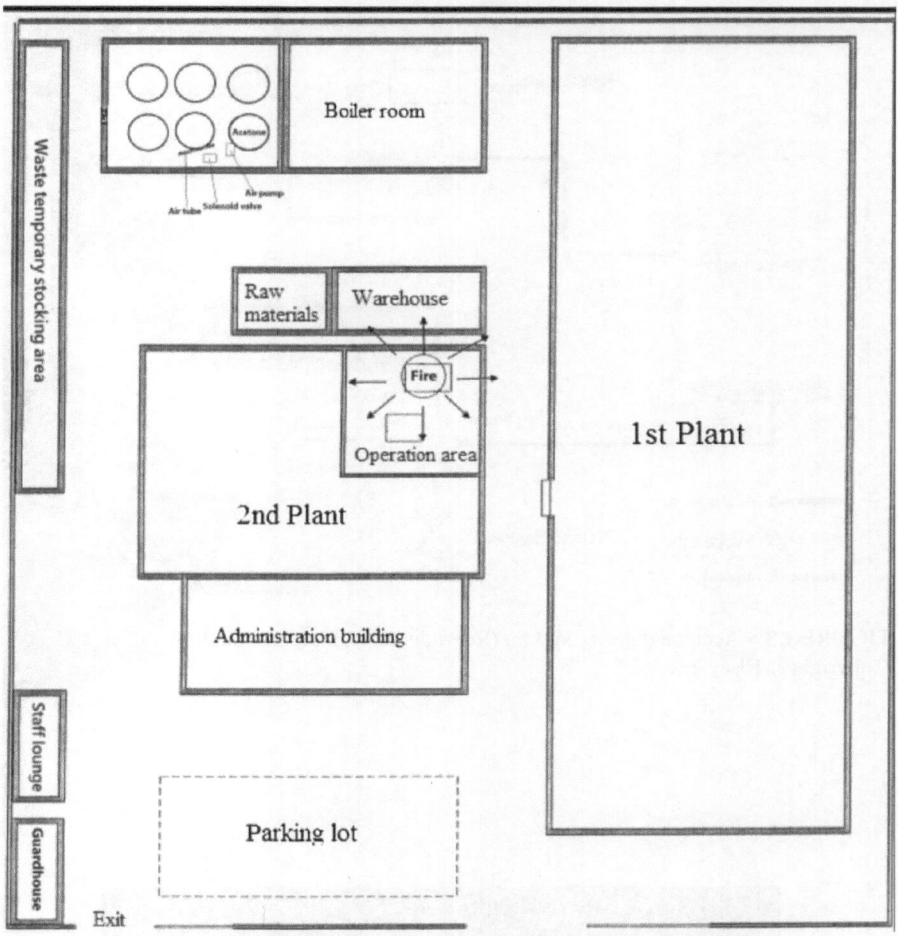

FIGURE 3.2 Plan view of the fire scene (Reproduced with permission from Chou et al. 2015, Copyright © Elsevier).

3.5 INVESTIGATION OF CAUSE FOR ACCIDENT

The standard operating procedures established by the National Fire Agency, Ministry of the Interior, Taiwan (NFA, 2009) was used for investigating fires and other major industrial mishaps (Chan, 2010). It was determined that the failure of a solenoid valve (air pump) to transfer acetone from the storage tank of plant 1 resulted in acetone shortage. Further, the force delivery of acetone from inaccurate locations of air pumps (removing the solenoid valve) in plants 1 and 2 by the night-shift engineer prompted the malfunction in automated control systems, resulting in delivery and overflow of high volumes of acetone in the intermediated tank located on the third floor of plant 2. The investigation team believed that the explosion has significantly affected the gluing area in the first floor (Figure 3.4 and 3.5).

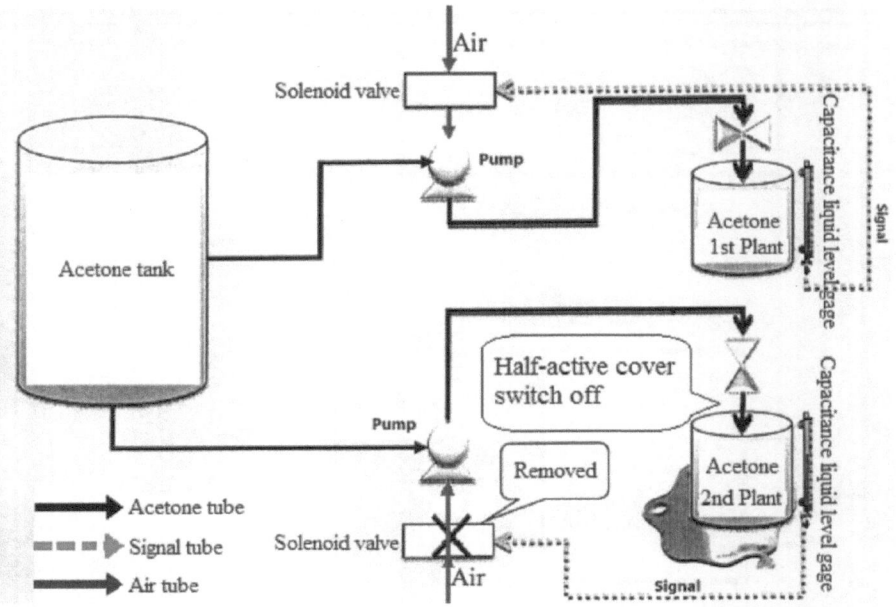

FIGURE 3.3 Acetone delivery system (Reproduced with permission from Chou et al. 2015, Copyright © Elsevier).

FIGURE 3.4 Disaster damage condition of the first floor of Plant 2 (Reproduced with permission from Chou et al. 2015, Copyright © Elsevier).

FIGURE 3.5 Disaster damage condition of the gluing area of the first floor of Plant 2 (Reproduced with permission from Chou et al. 2015, Copyright © Elsevier).

An alumina compartment around the gluing machine became the fire center, spreading outside burning all the interior decoration, walls, and ceilings. On the third floor of plant 2, the manual control valve of the acetone intermediate tank on the top was closed that should actually be open. The automatic control of the up-and-down motion of the half-activated cover ceased due to the closed manual control valve. The solenoid valve and air tube that deliver acetone from the outdoor storage tank were disconnected and the other end of the air tube was detached in plant 2. These two errors were attributed to human error and resulted in liquid acetone leakage.

The surveillance video revealed that one employee confirmed the acetone leakage and fled along with two others immediately from the gluing area covering their noses and mouths and returned to the scene after three minutes when the fire ignited. The overflown acetone from the intermediate tank flooded from the third floor into the second and first floor due to the tunnel design of the oven in the gluing area. The discharge of accumulated static charge from the regular process operations in PP production, the leakage of acetone from the third to the first floor, the vapors generated readily ignited and an explosion was inevitable. The deceased and the injured (5) were working in plant 2. The explosion resulted in loss to property and considerable injury to the personnel.

3.6 SUMMARY

Those process areas that are prone to generate static electricity, such as manufacturing, processing, operating, packing, and moving objects, should be grounded. Control measures in restricting process speeds, humidity, ionization, and the use of static eliminators are essential. Further, a plant should strengthen professional training, establish standard operating procedures (SOP), and organize frequent emergency response drills for the delivery of acetone. High-tech plants comprise precision equipment, congested pipelines, confined space, special zoning, clean room, and broad floor area when compared with conventional plants, which destroy fire compartments and can rapidly achieve smolderingly high temperatures. The high quantities of acids, alkaline, flammables, explosives, and toxic chemicals are stored for production that possesses a serious risk of fire and explosion. Proactive plant safety management and disaster prevention measures based on methodical planning should be implemented for loss prevention.

REFERENCES

Chan, J.H., 2010. The development of industrial fire investigation technology and cases study. National Chiao Tung University, Hsinchu, Taiwan, Master's Thesis.
Chou, H. C., Yeh, C. T., Shu, C. M., 2015. Fire accident investigation of an explosion caused by static electricity in a propylene plant. *Proc. Saf. Environ. Protect.* 97, 116–121.
Dana, S., Lee, C.J., Park, J., Shin, D., Yoon, E.S., 2014. Quantitative risk analysis of fire and explosion on the top-side LNG-liquefaction process of LNG-FPSO. *Proc. Safe. Environ. Prot.* 92 (5), 430–441.
Gibson N., 1997. Static electricity - An industrial hazard under control? *J. Electrostat.* 40–41, 21–30.
Hsieh, W.K., 2014. Performance evaluation of fire protection systems in high-tech facilities. Chang Jung Christian University, Tainan, Taiwan, Master's Thesis.
Li, M.C., 2014. Applied analysis for prevention of fire accident caused from outdoor petrochemical pipeline leakage. Wu Feng University of Science and Technology, Chiayi, Taiwan, Master's Thesis.
Lin, H.Y., 2014. Electrostatic hazard evaluation and control for traditional paint and ink process in Taiwan. Central Taiwan University of Science and Technology, Taichung, Taiwan, Master's Thesis.
Logtenberg, M. T., 2001. Four explosions: Four times static electricity was the most probable ignition source. *Pasman, H.J.:* 1459–1464. doi: 10.1016/B978-044450699-3/50056-2.
National Fire Agency (NFA), Ministry of the Interior, Executive Yuan, Taiwan, 2009. Identification of the Standard Operating Procedures of Fire Investigation, Reference to Standards of Taiwan.
Suardin, J.A., McPhate Jr., A.J., Sipkemab, A., Childsc, M., Mannan, M.S., 2009. Fire and explosion assessment on oil and gas floating production storage offloading (FPSO): an effective screening and comparison tool. *Proc. Saf. Environ. Prot.* 87(3), 147–160.
Ye, C.H., (2014. To investigation the electrostatic discharge hazard prevention for powder loading and unloading operation in Taiwan. Central Taiwan University of Science and Technology, Taichung, Taiwan, Master's Thesis.

4 Thermal Hazard Accident during Hydrogen Peroxide Mixing with Propanone

Case Study

4.1 OVERVIEW OF H₂O₂ USES IN VARIOUS INDUSTRIES

Organic peroxides, especially hydrogen peroxide (H_2O_2) is a multi-functional green oxidant commonly used as bleach in paper and textile; and a cross-linking agent in petrochemicals and disinfectant in consumer products. It is an environmentally friendly chemical that, due to its weak O–O covalent bond, produces water and oxygen as by-products (Baciocchi, Boni, & Aprile, 2004; Liou & Lub, 2008; Poulopoulos, Arvanitakis, & Philippopoulos, 2006). Discovered in 1818, H_2O_2 was used for municipal/industrial wastewater treatment (1970) and more recently (since 2000) for antimicrobial applications. It is a very selective industrial bleaching agent, highly environmentally friendly, colorless and non-corrosive as it causes less textile fibre damage compared to other bleaching agents. In practice, H_2O_2 (35 or 50 mass%) is stored in drums for bulk quantities and can be crystallized below –261 to 0 ºC. Hydrogen peroxide is well known for incompatible contaminants (metal ions, in-/ organic acids, ethnalo and petroleum ether) due to its hazardous structure (H-O-O-H), and is soluble in water. The major drawback of H_2O_2 is its inclination to decomposition in the presence of transition metal ions to form compounds capable of active oxygen, catalase, and other oxidizable substance on garments worn in contact with the skin (Barton & Nolan, 1989; Chen, Lin, Shu, & Kao, 2006; Duxbury, 1980; Fauske, 1984; Huff, 1982; Leung, 1986; Lu, Yang, & Lin, 2006; Schreck et al., 2004). Thermal explosions of H_2O_2 have transpired in its whole lifecycle (production, storage, transportation, and utilization) since the marketable products containing H_2O_2 are formulated at acidic pH. The vapor phase explosion, fire due to decomposition (oxygen release), rapid pressurization and heat release are the major thermal risks involved during the handling and storage of H_2O_2. It is imperative to evaluate the thermal hazardous characteristics of H_2O_2 and the factors influencing this process for safety concerns. Yeh et al., (2003) estimated the thermo-kinetic data of H_2O_2 using differential scanning calorimetry (DSC) and vent sizing package 2 (VSP2). This chapter investigates the reactive hazards, thermal decomposition and runaway behaviors of H_2O_2 with propanone (CH_3COCH_3).

4.2 HAZARDOUS AND TOXIC EFFECTS OF H_2O_2

Hydrogen peroxide is an adaptation of nature. Its production, utilization, degradation, and concentration depend on the dynamic equilibrium between photochemical and biological processes. The toxic properties and chemical reactivity of H_2O_2 can be used sometimes as a natural disinfectant and is dependent on the intoxication and detoxification of the organism and cells. However the release of oxygen during degradation/detoxification can also lead to hazardous effects. Three main mechanisms that cause the toxicity of H_2O_2 are lipid peroxidation, corrosive damage, and the formation of oxygen. Micro-organisms, especially heterotrophic bacteria, account for the decomposition of H_2O_2 by chemical reduction and enzymatic/catalytic peroxidation where the decomposition can vary from a few minutes to several weeks depending on physical, chemical and biological factors. The half-life of H_2O_2 in fresh water is typically between two and eight hours, but this can be prolonged to several days in water denuded of micro-organisms. The occupational and household exposure of H_2O_2 can result in gas embolization and brain infarction. The accidental ingestion of small quantities of 35% H_2O_2 in children can cause significant toxicity, including vomiting, irritation, and headache. Chronic exposure tp high concentrations of H_2O_2 results in severe irritation, coughing, and the inflammation of mucous membranes. Similarly, prolonged dermal exposure to H_2O_2 causes inflammation, blistering and severe tissue damage. In infants, H_2O_2 toxicity causes intestinal irritation, gangrene, and portal veins. Rapid pressurization, heat release, fire due to oxygen release, vapor phase explosion and cytotoxin effects are the major safety and health concerns of hydrogen peroxide exposure.

4.3 INVESTIGATING THERMAL EXPLOSION ACCIDENTS OF H_2O_2

Numerous accidents were reported in the literature, which clearly show the potentially hazardous behavior of H_2O_2 due to catalytic runaway reaction. In one incident, which took place at a warehouse in Beijing, a spontaneous combustion accident in an open-air storage tank of H_2O_2 occurred at 2°C. In Taiwan, five separate accidents in peroxide production industries resulted in 55 fatalities and injuries to 156 people. In 1996, the Yung-Hsin plant explosion in an H_2O_2 storage tank in Taoyuan county was a major industrial accident that killed 10 people and wounded 47 during the fire-fighting (Duh, Lee, Hsu, Hwang, & Kao, 1997; Wu, Su, & Shu, 2008). The accident investigation reported a series of events in which H_2O_2, methyl ethyl ketone and dimethyl phthalate were used as raw materials in the production of methyl ethyl ketone peroxide. An exothermic reaction in the reactor due to the failure of cooling systems and wrong dosage resulted in an explosion. The fragment of the broken hot concrete propelled into a 10-ton H_2O_2 storage tank, which experienced a secondary explosion (Chi, Wu, & Shu, 2009). In the USA and Japan in the period between 1996 and 2005, some 11 major accidents involving H_2O_2 were reported. In the year 1999, for example, a tanker carrying 500 L of waste H_2O_2 blew up on an expressway in

Tokyo, scattering the broken wreckage to a radius of 100 m and injuring 19 people. Recent accidents in China have been triggered by the thermal hazards of H_2O_2. At low temperatures, for example, the incompatibilities of H_2O_2 can result in exothermic runaway, followed by a thermal explosion. Considering the incompatibility of other materials with H_2O_2, it was observed that the exothermic onset temperatures (71–92 °C) was higher and the heat of decomposition (97.3–486.6 J g^{-1}) was lower than the parent species (Chen et al., 2006). The thermal hazard studies of H_2O_2 suggested that of any commercial package of organic peroxides, the determination, self-accelerating is crucial (Chen, Wu, Wang, & Shu, 2008; Wu, Shyu, I, Chi, & Shu, 2009). The pressure relief characteristics of OPs were studied by the Design institute for emergency relief systems. The determined values of the exothermic threshold temperature were 50–120 °C and the reactive/incompatible hazards of H_2O_2 with other substances should be clearly identified and needs detailed investigation. The runway reaction of OPs are often complemented by an excessive heat release and thermal explosions with a maximum self-heating rate $((dT\,dt^{-1})_{max})$ of > 100 °C min^{-1}, influenced by metal ions, temperature, Lewis acid/bases and other impurities (Chen, Wu, Lin, Hou, & Shu, 2008; Wang, Shu, Duh, & Kao, 2001; Weber, 2006; Wu, Wang, Wu, Hu, & Shu, 2008). Acetone is widely used as a cleaning agent in process industries and its mixing with H_2O_2 is prohibited (Wu et al., 2010). In this chapter, the reactive characteristics, inherent and relative instabilities of H_2O_2 with acetone under runaway conditions were evaluated to prevent chemical reaction and thermal explosion using DSC, and VSP2. The thermokinetic data, such as the self-heating rate, the adiabatic time to maximum rate (TMR_{ad}), the exothermic onset temperature (T_o), and the adiabatic temperature rise (ΔT_{ad}), were determined to significantly affect the reactive hazards of H_2O_2 when mixed with propanone.

4.3.1 SAMPLES

The hydrogen peroxide (31 mass%) and propanone (100 mass%) was purchased from the Aldrich Co. For DSC experiments, H_2O_2 with different concentrations such as 10, 20, 31, and 45 mass% was considered that are usually encountered either in process or storage areas.

4.3.2 DIFFERENTIAL SCANNING CALORIMETRY (DSC)

The DSC experiments were performed using a Mettler DSC821e/TA8000 system with temperatures between 30 and 300 °C at a heating rate of 4 °C min^{-1}. About 5 mg of 10, 20, 31, and 45 mass % H_2O_2 dripped with proportion of different, incompatible samples using a pipette, sealed manually, and was placed in the testcell that could withstand a pressure of 100 bar (Fessas et al. 2005; Hou et al. 2001).

4.3.3 VENT SIZING PACKAGE 2 (VSP2)

The variation of temperature and pressure with time was obtained using a PC-controlled adiabatic calorimeter (Fauske & Associates, Inc.). To measure the phenomenon of self-exothermicity/self-heating rate, the heat-wait-search (H-W-S) mode was adopted for VSP2 at <0.1 °C min^{-1} (Wang et al., 2001). The heat released during the reaction was contained within the tests cell due to its low heat capacity of volume 112 mL with low thermal inertia factor (F) (1.05-1.32).

4.3.4 THERMOKINETICS APPLICATION FOR ADIABATIC SYSTEM

The kinetic parameters of an adiabatic process for a single reaction was evaluated based on the equation derived by Townsend and Tou (1980)

$$\ln k = \ln A \frac{E_a}{RT} \tag{4.1}$$

$$= \ln \frac{\dfrac{dT}{dt}}{C_0^{n-1}\left(\dfrac{T_f - T}{T_f - T_0}\right)^n (T_f - T_0)} \tag{4.2}$$

$$= \ln \frac{\dfrac{dT}{dt}}{\left(\dfrac{T_f - T}{T_f - T_0}\right)(T_f - T_0)} \tag{4.3}$$

$$= \ln \frac{\dfrac{dT}{dt}}{T_f - T} \tag{4.4}$$

$$k = A e^{\frac{E_a}{RT}} \tag{4.5}$$

where the rate constant is represented by the Arrhenius law (Budzianowski, 2005). A plot of *ln(k) vs −1000/T* of the experimental data from VSP2 by assuming n=1, represents a straight line and the kinetic parameters, such as frequency factor (*A*) and activation energy (*E$_a$*), were estimated.

4.3.5 CRITICAL TEMPERATURE (T$_c$) OF H$_2$O$_2$ CALCULATION

The Franke-Kamenetskii parameter (δ_{cr}) and critical temperature (T$_c$) were determined as a function of thermokinetic parameters in various geometries using numerical analysis (Luo, Hu, and Lu, 1997)

$$\delta_{cr} = a_0 + a_1 \exp\left(\frac{\alpha E_a}{RT_{0,cr}}\right) + a_2 \exp\left(\frac{\gamma E_a}{RT_{0,cr}}\right) \tag{4.6}$$

Linear regression was employed to calculate the T_c as follows:

$$T_c = \frac{E_a}{R\left[c_1 + c_2 \ln\left(\dfrac{RQAr_0^2}{E_a\lambda}\right)\right]} \tag{4.7}$$

Table 4.1 displays the fundamental parameters of various tank types in Eq. (4.7). Thermokinetic parameters of 20 mass% H_2O_2 by VSP$_2$ were as shown in Table 4.2. For sake of loss prevention, T_c is a very important index of the manufacturing process. Eq. (4.7) compared with Tables 4.1 and 4.2 were employed to determine the T_c. This model can be used to calculate the T_c of various storage tanks depending on differential chemical for safety storage, transportation, and use.

4.4 THERMAL ANALYSIS BY DSC FOR H_2O_2 WITH PROPANONE

Table 4.3 shows the typical temperature and heat flow curves for the thermal decomposition of 10, 20, 31, and 45 mass% H2O2, respectively. Heat flow versus sample temperature of 10, 20, 31, and 45 mass% H2O2 by DSC was shown in Table 4.3. The thermal decomposition of H_2O_2 mixed with pure propanone signify two peaks where the decomposition of H_2O_2 was represented by the first peak (Table 4.4) and the second peak represents the thermal decomposition of acetone peroxide (Table 4.5) which is formed by heat or acid catalysis (Wu et al., 2010). The typical temperature and heat flow curve (Figure 4.1), thermokinetic parameters, and heat production rate vs time curves of 10, 20, 31, and 45 mass% H_2O_2 mixed with 100 mass% propanone was presented in Table 4.5.

TABLE 4.1
Fundamental Parameters of Various Tank Types for T_c Calculation

Type	a_0	a_1	a_2	c_1	c_2	A	γ
Flat	0.8865	0.0705	0.0508	5.59234	1.04562	0.1875	0.0360
Barrel	2.0189	0.1681	0.1201	4.72918	1.04565	0.1876	0.0359
Sphere	3.3539	0.2846	0.2037	4.19710	1.04567	0.1866	0.0359

(Data obtained with permission from Chi et al. 2012, Copyright © Elsevier)

TABLE 4.2
Thermokinetic Parameters of 20 Mass% H_2O_2 by VSP2

	Q	A	Ea	λ	ρ	M
Property	(kJ mol^{-1})	(s^{-1})	(kJmol^{-1})	(Jcm^{-1} s^{-1}°C^{-1})	(g cm^{-3})	(g mol-1)
	97.5	9.74×10^9	128.9	8.75×10^{-3}	1.458	34

(Data obtained with permission from Chi et al. 2012, Copyright © Elsevier)

TABLE 4.3
Thermokinetic Parameters of Various Concentrations H_2O_2 by DSC Under 4 $^oCmin^{-1}$

H_2O_2 (mass%)	m (mg)	T_0 (^oC)	T_{max} (^oC)	ΔH_d (J g^{-1})
10	3	70	75	270
20	4	72	77	400
31	3	60	65	880
45	3	43	70	975

(Data obtained with permission from Chi et al. 2012, Copyright © Elsevier)

TABLE 4.4
Thermokinetic Parameters for Various Concentrations of H_2O_2 Mixed with Propanone by DSC Under 4^oC min^{-1} of Heating Rate (1st peak)

H_2O_2(mass%)	Propanone (mass%)	m (mg)	T_0 (^oC)	T_{max} (^oC)	ΔH_d (J g^{-1})
10	100	2	63	88	570
20	100	3	63	90	572
31	100	2	60	100	645
45	100	3	50	105	752

(Data obtained with permission from Chi et al. 2012, Copyright © Elsevier)

TABLE 4.5
Thermokinetic Parameters for Various Concentrations of H_2O_2 Mixed with Propanone by DSC Under 4 $^oCmin^{-1}$ of Heating Rate (2nd Peak)

H_2O_2(mass%)	Propanone (mass%)	m (mg)	T_0 (^oC)	T_{max} (^oC)	ΔH_d (J g^{-1})
10	100	2	200	230	1209
20	100	3	202	240	1322
31	100	2	205	250	1491
45	100	3	205	250	1633

(Data obtained with permission from Chi et al. 2012, Copyright © Elsevier)

The mixture of H_2O_2 and propanone represented multiple peaks with different peak temperatures. However, pure H_2O_2 had a single peak. From Table 4.4, it can observed be that T_0 was in the range of 43-72 ^oC with peak temperatures about 65-99 ^oC and mixing with incompatible materials resulted in the decrease in heat of reaction. The transformation of the reaction region was demonstrated by a slight increase of initial temperatures when H_2O_2 mixed with propanone as shown in the DSC

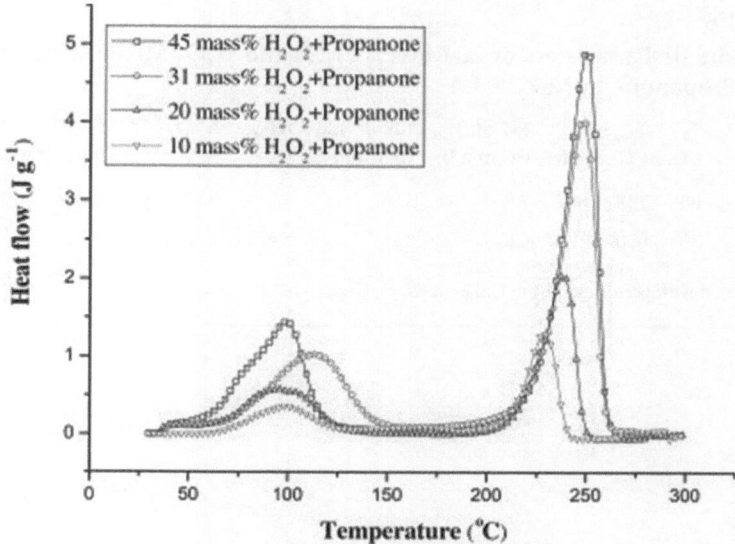

FIGURE 4.1 Thermal curves of 10, 20, 31, and 45 mass% H_2O_2 mixed with propanone by DSC (Reproduced with permission from Chi et al. 2012, Copyright © Elsevier).

curves. However, the mixture of while propanone to H_2O_2, significantly altered the curves, suggesting the complex reaction mechanism of propanone rather than the pure one. Therefore, the mixing of propanone with H_2O_2 is prohibited as it can easily form acetone peroxide. The thermal decomposition of H_2O_2/propanone is more dangerous than pure H_2O_2 and was identified as a physical hazard in the storage tank or reactor. Thermal decomposition peak of propanone mixed with H_2O_2 results in two peaks, where the second peak is more dangerous as the decomposition is accompanied with the release of large amounts of heat and pressure.

4.5 ADIABATIC KINETICS STUDY

The thermal hazard of H_2O_2 when mixed with propanone was analyzed by VSP2. The various thermokinetic parameters for 20 mass % H_2O_2 such as T_o, T_{max}, P_{max}, maximum of temperature and pressure rise rate $((dT\ dt^{-1})_{max}$ and $(dP\ dt^{-1}))$, was represented in Table 4.6.

The temperature vs time curve of VSP2 with 20 mass % H_2O_2 was represented in Figure 4.2. The H_2O_2 decomposed at 80 °C and increased to 158 °C with a change in phase and the P_{max} was determined from Figure 4.3. The E_a and A were evaluated from Eq. (4) using the slope of the curves in Figure 4.4. The adiabatic time to maximum rate (TMR_{ad}) was determined as 70 min with T_c ca. 116 °C. The E_a of H_2O_2 propanone mixture is higher than H_2O_2 (Figure 4.5). The reaction of H_2O_2/propanone mixture initiated at 70 °C with an increase in temperature of the reaction mixture to 650 °C. The $(dT\ dt^{-1})_{max}$ and $(dP\ dt^{-1})_{max}$ was calculated as 620 °C min^{-1} and 121 bar, respectively. The runaway reaction of H_2O_2 propanone mixture is much quicker than pure H_2O_2.

TABLE 4.6
Thermokinetic Parameters of 20 Mass% H_2O_2 and H_2O_2 Mixed with 100 mass% Propanone by VSP_2

Material	T_0 (°C)	T_{max} (°C)	P_{max} (bar)	$(dT\ dt^{-1})_{max}$ (°C min⁻¹)	$(dP\ dt^{-1})_{max}$ (bar min.₋₁)	ΔT_{ad} (°C)	A (s⁻¹)	E_a (kJmol⁻¹)	TMR_{ad} (min)
H_2O_2	80	158	54.2	7.8	100	78	9.13×10^{15}	128.9	70
H_2O_2/ propanone	70	650	91.6	620	121	580	1.60×10^{48}	316.3	27 (After T_0)

(Data obtained with permission from Chi et al. 2012, Copyright © Elsevier)

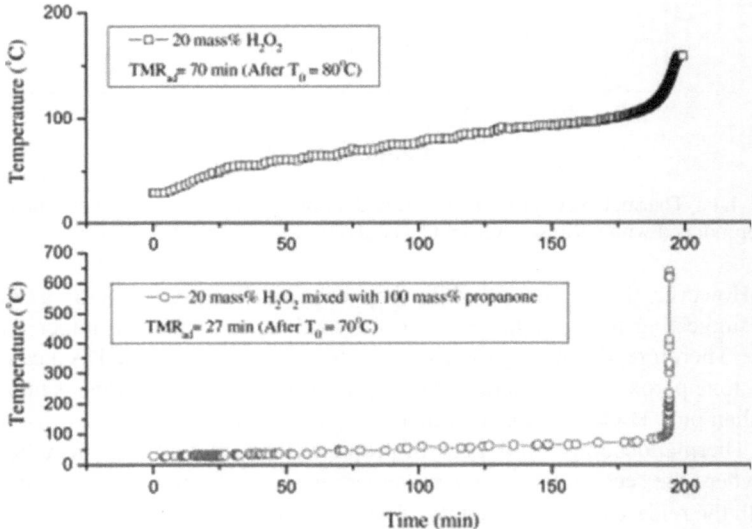

FIGURE 4.2 Temperature versus time for thermal decomposition of H_2O_2 and H_2O_2 mixed with propanone by VSP2 (Reproduced with permission from Chi et al. 2012, Copyright © Elsevier).

4.6 SUMMARY

The incompatible characteristics of H_2O_2 with propanone under runaway reactions were evaluated using DSC and VSP2. The determined values of T_0 provided the necessary basis in understanding the hazard characteristics of H_2O_2 and its incompatible mixing with other chemicals during transportation, handling, storage and even disposal. The thermal decomposition of propanone mixed with H_2O_2 represented two stages whereas pure H_2O_2 represented single stage. The adiabatic time to maximum

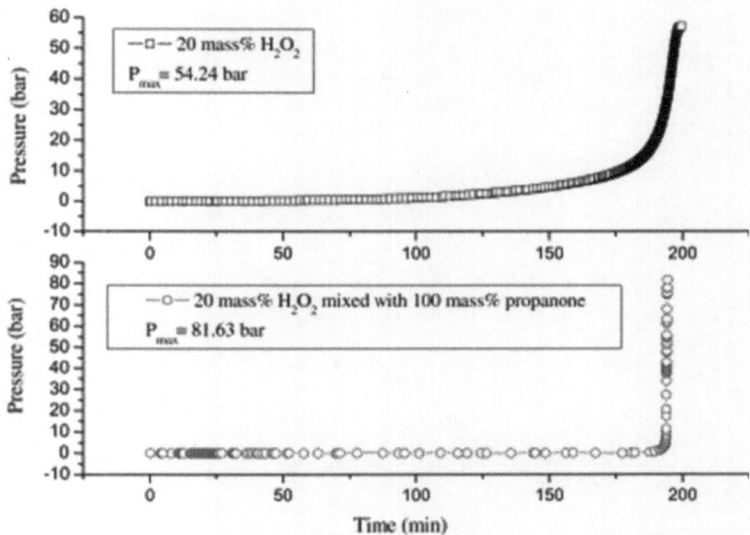

FIGURE 4.3 Pressure versus time for thermal decomposition of H_2O_2 and H_2O_2 mixed with propanone by VSP2 (Reproduced with permission from Chi et al. 2012, Copyright © Elsevier).

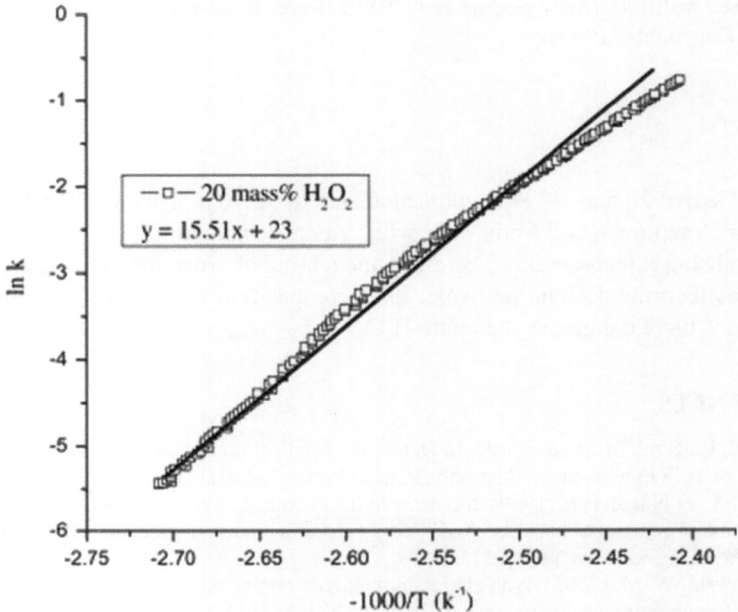

FIGURE 4.4 The correlation of overall rate constant (k) and temperature (T) for 20 mass% H_2O_2 by VSP2 (Reproduced with permission from Chi et al. 2012, Copyright © Elsevier).

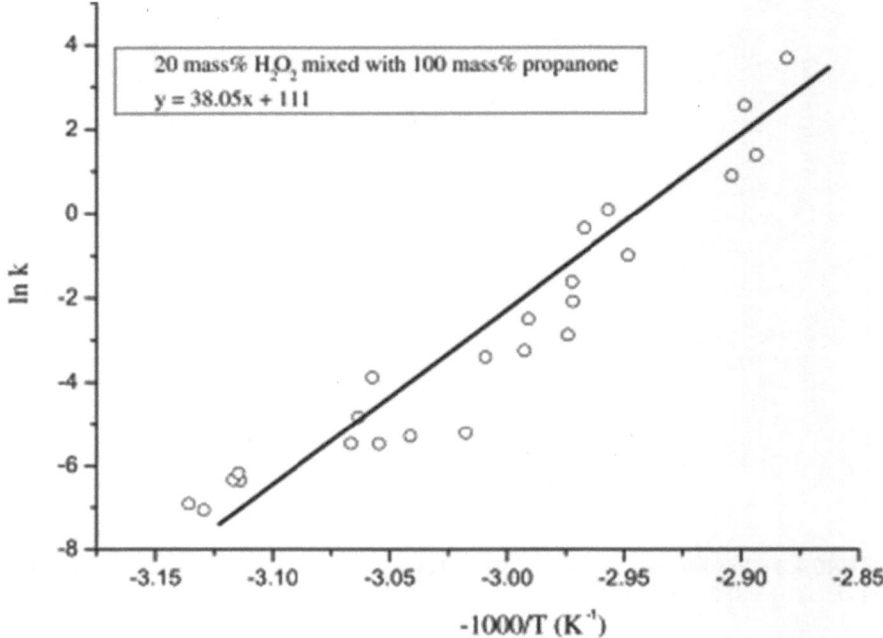

FIGURE 4.5   The correlation of overall rate constant (k) and temperature (T) for 20 mass% H_2O_2 mixed with 100 mass% propanone by VSP2 (Reproduced with permission from Chi et al. 2012, Copyright © Elsevier).

rate (TMR_{ad}) of 20 mass % H_2O_2 was calculated as ca. 70 min and the TMR of H_2O_2/ propanone mixture was 27 min only with critical temperature (T_c) at 116 °C after exothermic onset temperature. Therefore, the mixing of propanone with H_2O_2 is prohibited as it forms acetone peroxide, its decomposition releases large amounts of heat, and is more dangerous than pure H_2O_2.

REFERENCES

Baciocchi, R., Boni, M. R., & Aprile, L. D. (2004). Application of H2O2 lifetime as an indicator of TCE Fenton-like oxidation in soil. *J. Hazard. Mater.*, B107, 97–102.

Barton, J. A., & Nolan, P. F. (1989). Incidents in the chemical industry due to thermal runaway chemical reactions, hazards. X. Process safety in fine and specialty chemical plants. *IChemE Symposium Series*, 115, 3–18.

Budzianowski, W. M. (2005). Non-stationary catalytic combustion over a catalyst with internal temperature gradients. *Arch. Combust.*, 25(1-4),7–15.

Chen, K. Y., Lin, C. M., Shu, C. M., & Kao, C. S. (2006). An evaluation on thermokinetic parameter for hydrogen peroxide at various concentration by DSC. *J. Therm. Anal. Calorim.*, 85(1), 87–89.

Chen, J. R., Wu, S. H., Lin, S. Y., Hou, H. Y., & Shu, C. M. (2008). Utilization of micro-calorimetry for an assessment of the potential for a runaway decomposition of cumene hydroperoxide at low temperatures. *J. Therm. Anal. Calorim.*, 93(1), 127–133.

Chen, K. Y., Wu, S. H., Wang, Y. W., & Shu, C. M. (2008). Runaway reaction and thermal hazard simulation of cumene hydroperoxide by DSC. *J. Loss. Prev. Process. Ind.*, 21, 101–109.

Chi, J. H., Wu, S. H., Charpentier, J. C., Yet, P. I., Shu, C. M. (2012). Thermal hazard accident investigation of hydrogen peroxide mixing with propanone employing calorimetric approaches. *J. Loss Prev. Process Ind.* 25 (1), 142–147.

Chi, J. H., Wu, S. H., & Shu, C. M. (2009). Thermal explosion analysis of methyl ethyl ketone peroxide by non-isothermal and isothermal calorimetric applications. *J. Hazard. Mater.*, 171, 1145–1149.

Duh, Y. S., Lee, C., Hsu, C. C., Hwang, D. R., & Kao, C. S. (1997). Chemical incompatibility of nitro compounds. *J. Hazard. Mater.*, 53, 183–194.

Duxbury, H. A. (1980). Relief system sizing for polymerization reactors. *Chem. Eng.*, 31–38.

Fauske, H. K. (1984). Generalized vent sizing nomogram for runaway chemical reactions. *Plant/Operations Progress*, 3(4), 213–220.

Fessas, D., Signorelli, M., & Schiraldi, A. (2005). Polymorphous transitions in cocoa butter - a quantitative DSC study. *J. Therm. Anal. Calorim.*, 82(3), 691–702.

Hou, H. Y., Shu, C. M., & Duh, Y. S. (2001). Exothermic decomposition of cumene hydroperoxide at low temperature conditions. *Am. Inst. Chem. Eng. J.*, 47, 1893–1894.

Huff, J. E. (1982). Emergency venting requirements. *Plant/Operations Progress*, 1(4), 211–220.

Leung, J. C. (1986). A generalized correlation for one-component homogeneous equilibrium flashing choked flow. *AIChE Journal*, 32(10), 1643–1646.

Liou, M. J., & Lub, M. C. (2008). Catalytic degradation of explosives with goethite and hydrogen peroxide. *J. Hazard. Mater.*, 151, 540–546.

Lu, K. T., Yang, C. C., & Lin, P. C. (2006). The criteria of critical runaway and stable temperatures of catalytic decomposition of hydrogen peroxide in the presence of hydrochloric acid. *J. Hazard. Mater.*, B135, 319–327.

Luo, K. M., Hu, K. W., & Lu, K. T. (1997). The calculation of critical temperatures of thermal explosion for energetic materials. *J. Chin. Inst. Chem. Eng.*, 28(1), 21–28.

Poulopoulos, S. G., Arvanitakis, F., & Philippopoulos, C. J. (2006). Photochemical treatment of phenol aqueous solutions using ultraviolet radiation and hydrogen peroxide. *J. Hazard. Mater.*, 129, 64–68.

Schreck, A., Knorr, A., Wehrstedt, K. D., Wandrey, P. A., Gmeinwieser, T., & Steinbach, J. (2004). Investigation of the explosive hazard of mixtures containing hydrogen peroxide and different alcohols. *J. Hazard. Mater.*, A108, 1–7.

Townsend, D. I., & Tou, J. C. (1980). Thermal hazard evaluation by an accelerating rate calorimeter. *Thermochim. Acta*, 37, 1–30.

Wang, Y. W., Shu, C. M., Duh, Y. S., & Kao, C. S. (2001). Thermal runaway hazards of cumene hydroperoxide with contaminants. *Ind. Eng. Chem. Res.*, 40, 1125–1132.

Weber, M. (2006). Some safety aspects on the design of sparger systems for the oxidation of organic liquids. *Process. Saf. Prog.*, 25(4), 326–330.

Wu, S. H., Chi, J. H., Huang, C. C., Lin, N. K., Peng, J. J., & Shu, C. M. (2010). Thermal hazard analyses and incompatible reaction evaluation of hydrogen peroxide by DSC. *J. Therm. Anal. Calorim.*, 102, 563–568.

Wu, S. H., Shyu, M. L., I, Y. P., Chi, J. H., & Shu, C. M. (2009). Evaluation of runaway reaction for dicumyl peroxide in a batch reactor by DSC and VSP2. *J. Loss Prev. Process Ind.*, 22, 721–727.

Wu, S. H., Su, C. H., & Shu, C. M. (2008). Thermal accident investigation of methyl ethyl ketone peroxide by calorimetric technique. *Int. J. Chem. Sci.*, 6(2), 487–496.

Wu, S. H., Wang, Y. W., Wu, T. C., Hu, W. N., & Shu, C. M. (2008). Evaluation of thermal hazards for dicumyl peroxide by DSC and VSP2. *J. Therm. Anal. Calorim.*, 93(1), 189–194.

Yeh, P. Y., Shu, C. M., & Duh, Y. S. (2003). Thermal hazard analysis of methyl ethyl ketone peroxide. *Ind. Eng. Chem. Res.*, 42, 1–5.

5 Hydroelectric Power Plant Fire Accident
Fire Dynamics Simulator (FDS)

5.1 OVERVIEW OF A HYDROELECTRIC POWER PLANT IN DIFFERENT SECTORS

Hydroelectric power plants are one of the most important renewable energy sources in the world at present due to the low environmental and operational impacts of operating and maintenance costs. Hydraulic energy is an indispensable source of 71% of all renewable electricity generation worldwide. In 2016, it has reached 1.064 GW installed power, estimated at generating around 16.4% of the world's energy production. It has also been reported that there is much potential in the world to increase the number of hydroelectric power plants (World Energy Council, 2016). The availability of this hydraulic potential as much as possible will increase the energy supply security in the world as well as providing a significant reduction in energy production costs. Like wind energy, hydropower energy is used mostly for electricity generation as a contribution to the production of electricity around the world. Another major but mostly unknown use of hydro power is for storing energy. Using the existing dam infrastructure, utilities use hydro power to store energy in a process which is known as "pumped hydro storage". This use is becoming more important as there are limited options for cheap energy storage. In the past hydro power, like wind power, was used for agricultural use such as processing grain wherein the kinetic energy of the moving water was converted into mechanical energy. However, in recent years this use has almost completely disappeared.

Among the current uses of hydropower energy are:

(i) **Electricity** – Hydroelectricity is one of the world's most important sources of energy. It is one of the cheapest and non-polluting sources of power. Although it can cause ecological damage, initially it has better climate compatibility than other major forms of energy such as nuclear, coal, and gas. Many countries in the Nordic region and South America are almost completely dependent on hydro power for their energy needs. For some countries like China and India with massive energy needs, hydroelectricity is the only option currently amongst non-global warming energy choices to build in large capacities.

(ii) **Energy storage** – It is estimated that at present there is around 90 GW of global pumped hydro storage in the world and with increasing solar and wind energy production, this capacity is only going to grow. The main use of pumped hydro storage is for grid energy storage. Electric utilities are the main customers of this technology, using pumped hydro storage for:

    a. **Load balancing** – Storing power during low usage periods and generating power during high usage periods.

    b. **Accommodation of intermittent sources of energy** – In recent years, solar energy and wind energy have been growing at an extremely fast rate of 50% and 30% CAGR over the past few years. A larger share of these forms of renewable energy in the electricity mix is driving the growth in grid storage.

    c. **Reducing capital investments** as peak power plants like natural gas combined cycle plants are much more expensive to run than normal thermal and nuclear energy plants.

(iii) **Agriculture** – Hydro power was used in ancient times for producing flour from grain and was also used for sawing timber and stone, raised water into irrigation canals.

(iv) **Industry** – Hydro power was used earlier for some industrial applications such as driving the bellows in small blast furnaces and for extraction of metal ores in a method known as hushing.

(v) Owing to global energy shortages, the development of environmentally friendly and sustainable substitute energy is actively proposed by a number of nations. Hydroelectric power has the benefit of there being little detrimental impact on the environment, and it also helps in flood prevention and provides reservoirs for a variety of benefits. However, hydroelectric power plants are commonly constructed in underground locations, which can greatly increase the potential fire risks and serve as a detriment to escaping from a fire.

To improve the fire safety of existing hydroelectric power plants and set a reference point for future ones, the study in this chapter analyzed the fire at the Taiwan Dajia-River hydroelectric power plant, which involved six deaths and 26 injuries. Using a computer simulation, the characteristics of fire hazard, especially at hydroelectric power plants, were discussed. The information of this study can serve as a reference for fire escape and personnel protection at hydroelectric power plants in order to reduce the fire casualties.

5.2 SELECTION OF FIRE SCENE

The worst fire case recorded at a hydroelectric power plant in Taiwan was selected for our case study. From the investigation of site, this study gleaned the locale information to complete a sectional drawing of the case, as shown in Figure 5.1. The research team also conducted interviews with survivors of the fire. After all the collected information was processed, the fire dynamics simulator (FDS) version 4.05 computer-simulated program was used to reconstruct the fire scene. The result of computerized fire simulation was compared with the real fire scene to study the main cause of the deaths and injuries from this particular fire. The aforementioned fire took place at the Taiwan Dajia-River hydroelectric power plant at around 4 p.m. on October 28, 1993. When it occurred, the hydroelectric power plant was running under testing mode before the completion of its construction. It is suspected that improper operation procedures caused the oil inside the transformer to start the fire. The fire rapidly gutted the transformer room before spreading to the surrounding areas. The fire spread so fast and furiously that it destroyed any escape path, resulting in serious casualties.

5.3 FIRE SPACE SIMULATION

After on-site investigations and interviews, it was discovered that the fire started in the transformer room. There was a transformer set which was constructed of flammable polyurethane and contained 38.1 m^3 of transformer oil. Next to the transformer room was a machine installation platform and the main egress lane. The machine installation platform was used to maintain and repair the generator set, while the main egress lane was accessed by employees and repair vehicles. The main egress lane connected with a 300-m-long cable tunnel to reach outside, as shown in Figure 5.1. When the fire

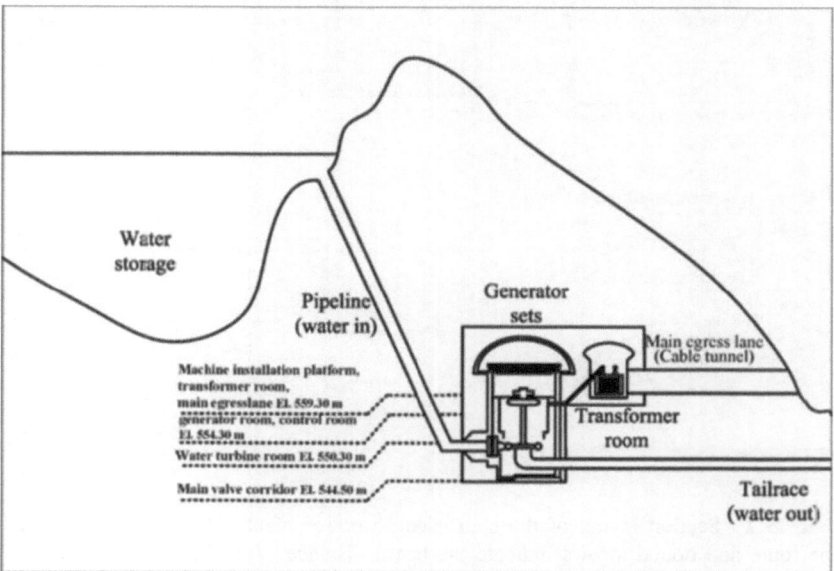

FIGURE 5.1 Sectional drawing of Taiwan Dajia-River hydroelectric power plant (Reproduced with permission from Chi et al. (2011) Copyright © Wiley & Sons).

started, the only escape route was through the machine installation platform to the main egress lane and cable tunnel.

The ignition transformer room was not airtight, and there was a rolling steel door about 8 m wide, 5 m tall, and 1.2 mm thick between the transformer room and the main egress lane, and two aluminum blinds each about 2.5 m wide and 1.5 m tall between the transformer room and machine installation platform, as demonstrated in Figure 5.2d. The rest of the transformer room surroundings were concrete walls about 20 cm in depth.

For computer simulation analysis, the heat release rate (HRR) calculation of combustible materials at a fire scene plays a very important role as it affects the results of the overall simulation (Christensen & Icove 2004). For the case study, the fire initiation area was the transformer room and the combustible material was the transformer oil. The volume of transformer oil was 38.1 m^3, multiplied by its density (889.9 kg/m^3) (Ko, 1993), and the weight of transformer oil was almost 3.39×10^4 kg. Based on the on-site investigation, two data were collected, the transformer area A = 6.4 m × 3.0 m = 19.2 m^2 and burning time t = 7200 sec. With the burning heat of transformer oil ΔH = 3920 kJ/mol and its molecular weight MW = 84 (Ko, 1993), this study adopted

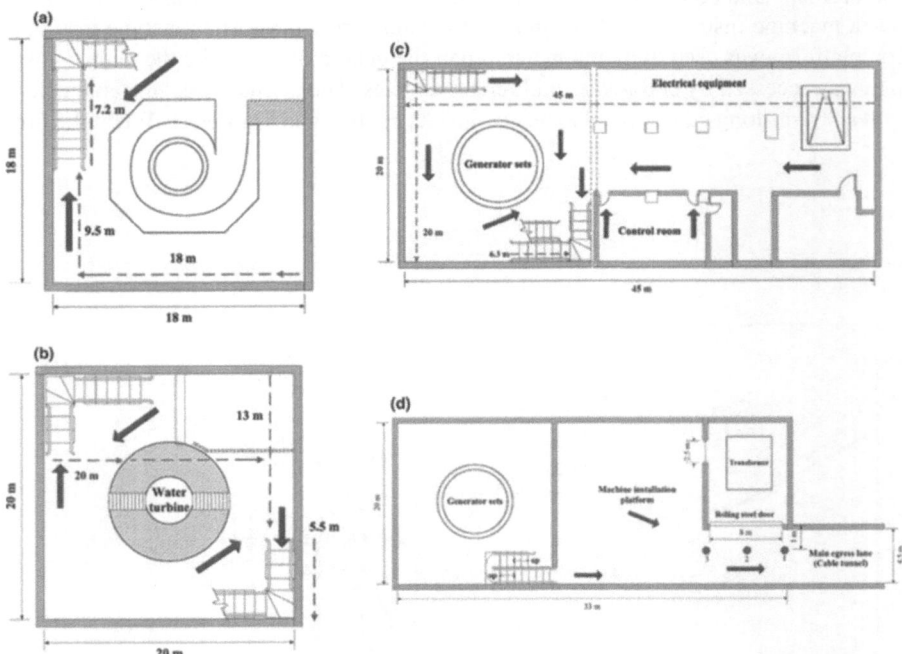

FIGURE 5.2 Section layout of the hydroelectric power plant. (Bold arrows indicate the escape route and dotted arrows indicate the travel distance.) (a) Main valve corridor (EL. 544.50   m); (b) water turbine room (EL. 550.30   m); (c) generator room and control room (EL. 554.30   m); and (d) machine installation platform, transformer room, and main egress lane, spots indicate the sensor location (EL. 559.30   m) (Reproduced with permission from Chi et al. (2011) Copyright © Wiley & Sons).

the method used to calculate the HRR of gasoline in the research of Shen et al. (2008), and the HRR is calculated as below:

$$Q(kW/m^2) - \Delta H(kJ/mol) \times W / MW(mol) \times$$
$$1 / A(1/m^2) \times 1 / t(1/sec) - 3920(kW \, sec /mol) \times \qquad (5.1)$$
$$3.39 \times 104/84(mol) \times 1 / 19.2(1/m^2) \times 1 / 7200(1/sec)$$

$$\sim 11,500(kW / m^2) - 11.5(MW / m^2)$$

5.4 RESULTS OF FIRE SPACE SIMULATION

The FDS computer simulation results have shown that the fire quickly spread after the ignition in the transformer room, as delineated in Figure 5.3a. As the room temperature increased, the aluminum blind fractured about 15 sec after the ignition of the fire, as shown in Figure 5.3b. The flame and smoke then spread rapidly to the

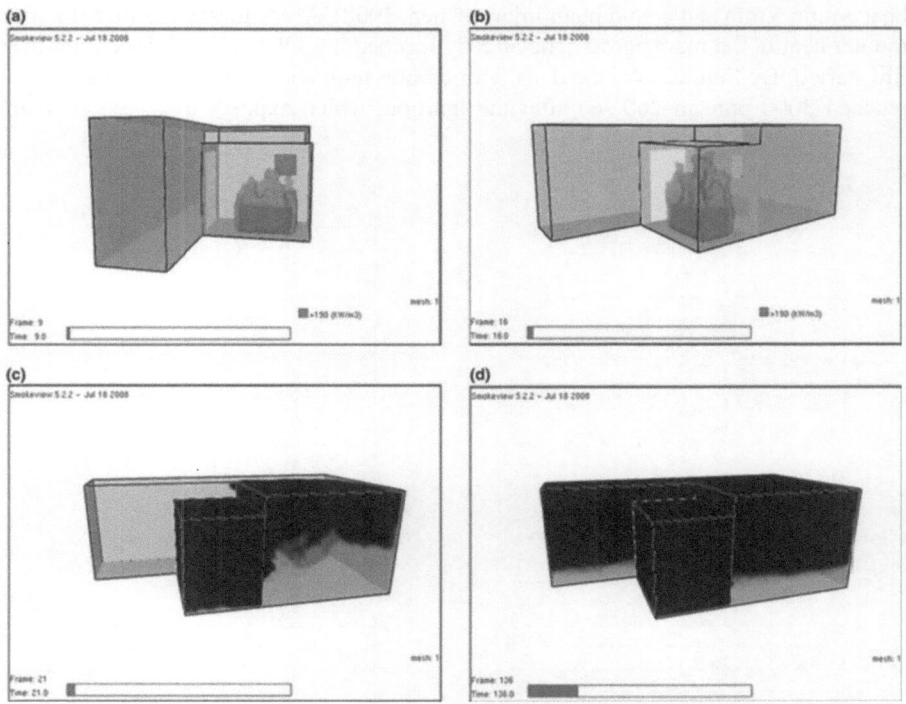

FIGURE 5.3 Fire and smoke spread profile of simulation; (a) 9 sec after fire ignited; (b) 15 sec after fire ignited, the aluminum blind fractured; (c) 21 sec after fire ignited; and (d) 136 sec after fire ignited (Reproduced with permission from Chi et al. (2011) Copyright © Wiley & Sons).

machine installation platform outside and extended to the main egress lane used for fire escape, as shown in Figure 5.3c, d. Not only did the broken window provide a path for the fire to reach outside the transformer room, but it also brought in additional air to fuel the fire. Figures 5.4a–d show that the curves move dramatically from 180 to 240 sec after the fire began, which is believed to be related to the broken window.

To examine the escape risks in the main egress lane during the fire, this study set up three sensors: two at each side and the third one at the midpoint of the steel rolling door of transformer room at 1 m ahead of the door along the main egress lane, as depicted in Figure 5.2d. The sensors were installed 1.8 m above the ground, which is the average height of a walking evacuee's nose and mouth according to the information by the Architecture and Building Research Institute (ABRI), Ministry of the Interior, Executive Yuan, Taiwan (Chen & Chien 2008). The changes over time of temperature, radiant heat, carbon monoxide, and oxygen at each detection point were collected by the sensors (Figure 5.4a–d). From Figure 5.4a, it is shown that the main egress lane temperature reached 60°C in 180 sec after the ignition and rose to 120°C in 236 sec. The increasing temperature would, of course, lower the evacuees' chance of escaping the fire safely (Ke, 2003). The research has showed that the radiant heat should not exceed 2.5 kW/m²; otherwise, humans would not be able to tolerant the heat within 5 min and would incur injury (Chen, 2008). According to Figure 5.4b, the radiant heat in the main egress lane already reached 2.5 kW/ m² in 155 sec after the fire started. By Figure 5.4c, the density of carbon monoxide at the main egress lane reached 3000 ppm in 260 sec after the ignition, which exposes the possibility of

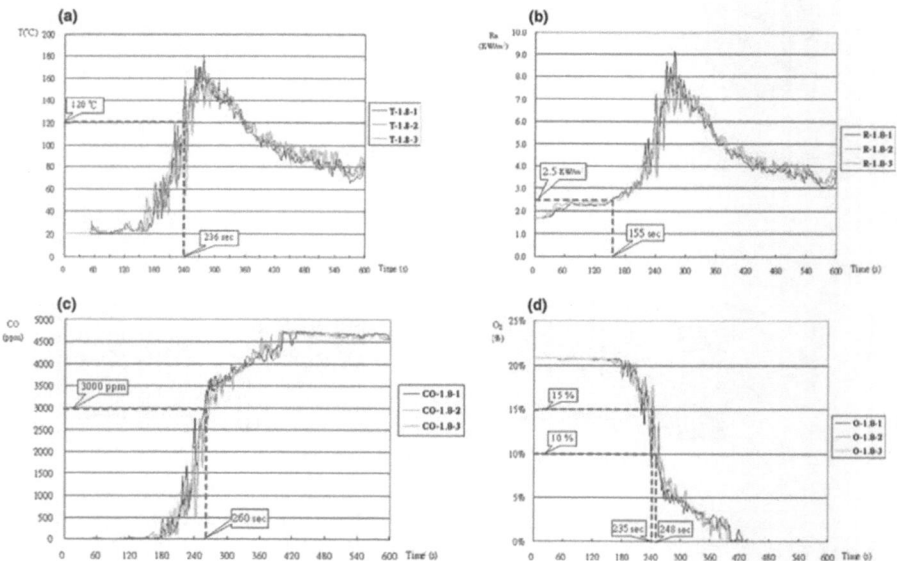

FIGURE 5.4 Variation of the fire hazard factors of simulation; (a) temperature; (b) radiant heat (c); carbon monoxide; and (d) oxygen (Reproduced with permission from Chi et al. (2011) Copyright © Wiley & Sons).

choking and death from breathing in too much smoke during the fire escape (Ke, 2003). Moreover, evacuees are likely to make mistakes owing to fatigue when oxygen levels go from 14% down to 10%. Figure 5.4d shows that the oxygen level at the main egress lane was 15% in 235 sec after the fire started and then down to 10% in 248 sec after the fire began.

5.5 ACTUAL ESCAPE TIME CALCULATION

To distinguish the difficulty of fire escape for evacuees from the fire, the formulae developed by ABRI (4) were adapted to calculate the actual escape time of this fire. By Figure 5.1, inside the underground hydroelectric power plant, the main valve corridor was located at the deepest section, an elevation of 544.50 m above sea level. The water turbine room was located at the second deep section, an elevation of 550.30 m above sea level. Upper section included the generator room and the control room, an elevation of 554.30 m above sea level. The nearest section to the ground was the machine installation platform, the transformer room, and the main egress lane, an elevation of 559.30 m above sea level. The floor layouts of each section are shown in Figure 5.2.

In calculating the actual fire escape time, it was assumed that the evacuees stayed farther away from the main egress lane at the main valve corridor and had to go through all the aforementioned floors to escape. Based on related formulae (Chen & Chien 2008) by ABRI, the actual escape time that the personnel spent is calculated as following:

$$t_{escape} - t_{start} + t_{travel} + t_{queue} \qquad (5.2)$$

where t_{escape} is escape finish time (min), made up of three durations: t_{start}, the escape start time, the time between fire started and evacuees perception; t_{travel}, the escape travel time, the traveling time for evacuees from any location to exit point; and t_{queue}, the queue time, the time for all evacuees pass through the exit.

Calculation of escape start time (t_{start})

$$t_{start} - \frac{\sqrt{\Sigma A_{area}}}{30} + \frac{\sqrt{A_1 + A_2 + A_3 + A_4}}{30} \qquad (5.3)$$

Where, A_{area} is the total floor area (m²), A_1 to A_4 is the floor area of each section, as depicted in Figure 5.2a–d, the calculation is $A_1 = 324$ (m²), $A_2 = 400$ (m²), $A_3 = 900$ (m²), and $A_4 = 660$ (m²). Thus:

$$t_{start} - \frac{\sqrt{\Sigma A_{area}}}{30} + \frac{\sqrt{324 + 400 + 900 + 660}}{30} = 1.59 (\text{min.})$$

Calculation of escape travel time (t_{travel})

$$t_{travel} - \max\left(\Sigma \frac{l_i}{\upsilon}\right) - \sum_{j-1}^{4} t_{ti} - t_{t1} + t_{t2} + t_{t3} + t_{t4} \qquad (5.4)$$

Where, l_i is the travel distance from any location to exit point (m), v is travel speed (m/min), which v is 35 m/min on stairs and 78 m/min not on the stairs (Chen & Chien 2008). Moreover, t_{t1} to t_{t4} is the travel time to exit point on each section. The calculation is detailed in Table 5.1.

Calculation of queue time (t_{queue})

$$t_{queue} = \frac{\Sigma p A_{area}}{\Sigma N_{eff} B_{eff}} = \sum_{i-1}^{4} t_{qi} = t_{t1} + t_{t2} + t_{t3} + t_{t4} \qquad (5.5)$$

Where, p is occupant density (p/m^2), acquired from the on-site investigation as 0.04 (p/m^2). A_{area} is total floor area (m^2), N_{eff} is effective flowing coefficient (p/min/m), acquired from the reference as 90 (p/min/m) (4), and B_{eff} is effective exit width (m). The calculation is given in Table 5.2.

Actual escape time (t_{escape})

Using Equation 5.2, $t_{escape} = t_{start} + t_{travel} + t_{queue} = 1.59 + 2.783 + 0.654 = 5.027$ (min) = 302 (sec)

5.6 ANALYSIS AND OBSERVATION IN REAL FIRE SCENE CASE

As mentioned previously, the actual escape time of an evacuee is calculated to be about 5.03 min (302 sec). However, the computer simulation results from the detection point on the main egress lane reveal that the simulated safe escaping time is about 236 sec, taking into account changes in temperature, of 155 sec, changes in radiant heat, of 260 sec, changes in carbon monoxide, and about 235–248 sec for

TABLE 5.1
Calculations of Escape Travel Time (t_{travel}) for the Investigated Case

Space	Situation	Travel Distance (l_i) (m)	Travel Speed (v) (m min^{-1})*	Travel Time (l_i/v) (min)	Subtotal t_{ti} (min)
Main valve	Stairs	7.2	35	0.206	0.559
corridor	Nonstairs	18	78	0.231	
	Nonstairs	9.5	78	0.122	
Water turbine	Stairs	5.5	35	0.157	0.58
room	Nonstairs	20	78	0.256	
	Nonstairs	13	78	0.167	
Generator room	Stairs	6.3	35	0.18	1.013
	Nonstairs	45	78	0.577	
	Nonstairs	20	78	0.256	
Machine	Stairs	5.5	35	0.157	0.631
installation	Nonstairs	20	78	0.256	
platform	Nonstairs	17	78	0.218	

(Reproduced with permission from Chi et al. (2011) Copyright © Wiley & Sons)
* v refers to the set travel speed of personnel at workplace by ABRI

TABLE 5.2

Calculations of Queue Time (t_{queue}) for the Investigated Case

Space	Occupant Density (p m^{-2})	Floor Area A_{area} (m^2)	Effective Flowing Coefficient N_{eff} (p min^{-1} m^{-1})*	Effective Exit Width B_{eff} (m)	Subtotal t_{ti} (min)
Main valve corridor	0.04	324	90	1	0.144
Water turbine room	0.04	400	90	1	0.178
Generator room	0.04	900	90	1.5	0.267
Machine installation platform	0.04	660	90	4.5	0.065

(Reproduced with permission from Chi et al. (2011) Copyright © Wiley & Sons)

* N_{eff} is the set value considering the capacity of main egress lane is enough for all evacuees

changes in oxygen level. These escape times are well below the actual escape time spent, 5.03 min (302 sec); thus, it was difficult for evacuees to escape after the fire began, which resulted in the tragic deaths of six and 26 injuries.

According to the fire investigation report, the fire area was located in a cave that is tens of meters underground. The changing curve in Figure 5.4d reveals the fire circumstance mentioned earlier.

After the fire occurred, oxygen levels dropped quickly and the fire area went into an incomplete burning stage. Then, smoke rapidly spread over the fire area, as diagrammed in Figure 5.3d. Based on Figure 5.4a–d, among all the fire hazards, radiant heat caused the most damage to fire escape. Figure 5.4b shows that less than 155 sec after fire started, radiant heat levels detected by the sensor at main egress lane would severely harm evacuees. The study results match what was found in the fire investigation that of 26 injuries, several people injured in the fire later died of severe burns. At 400 sec after the fire occurred, oxygen levels dropped to almost zero, as shown in Figure 5.4d. This matches the results from the fire investigation that five of six fire victims died from lack of oxygen.

5.7 SUMMARY

The HRR of transformer oil used in hydroelectric power plants was calculated in this study to be as high as 11.5 (MW/m^2). Compared to a burning car, this rate is between a compact car and a van (PIARC, 2007), which reveals how high the fire intensity was. Based on the computer simulation results, there were only 180–240 sec for evacuees to escape safely. However, to generate power effectively, the hydroelectric power plant generator is located as deep as tens or hundreds of meters underground. Therefore, when a fire occurs, especially in a transformer room next to the cable tunnel that is also used as main egress lane, the extremely high fire intensity would make it very difficult for evacuees to escape safely (Hu et al., 2006; Duarte, 2004). From the simulation results, the fractured aluminum blinds of the transformer room played a critical role in spread of flame and smoke. To avoid future similar fire trag-edies, the openings in the transformer room should sustain 1+ hour of fire resistance and automatic sprinkler and fire smoke systems should be equipped as they are in the long tunnel or even provide more than two escape routes (Ingason & Wickstrom 2006). Additionally, in order to protect the trapped evacuees, personnel at the hydro-electric power plant should carry a breathing apparatus that provides oxygen effec-tively up to 60 min., and several emergency shelters should also be provided (Ingason, 2006; Kang, 2007). It is expected that these suggestions will enhance the fire preven-tion and safety of both existing and future hydroelectric power plants to minimize the fire casualties.

REFERENCES

Chen, H.I. 2008. Smoke and heat. In: Chen HI, editor. *Fire science*. Taipei: Taiwan Tingmao Publishing. 9–15.
Chi, J.-H., Shu, C.-M., Wu, S.-H. 2011. Using Fire Dynamics Simulator to Reconstruct a Hydroelectric Power Plant Fire Accident. *J. For. Sci.* 56(6): 1639–1644.
Christensen, A.M., & Icove, D.J. 2004. The application of NIST's Fire Dynamics Simulator to the investigation of carbon monoxide exposure in the deaths of three Pittsburgh fire fighters. *J Forensic Sci.*, 49(1): 1–4.
Duarte, D.. 2004. A performance overview about fire risk management in the Brazilian hydro-electric generating plants and transmission network. *J Loss Prevent Proc. Indus.*, 17: 65–75.
Hu, L.H., Huo, R., Peng, W., Chow, W.K., Yang, R.X. & 2006. On the maximum smoke tem-perature under the ceiling in tunnel fires. *Tunn Undergr. Sp. Tech.*, 21: 650–655.
Ingason, H. 2006. Large fires in tunnels. *Fire Technol.*, 42(4): 271–272.
Ingason, H., & Wickstrom, U. 2006. The international FORUM of fire research directors: a position paper on future actions for improving road tunnel fire safety. *Fire Safety J.*, 41: 111–114.
Kang, K. 2007. Application of code approach for emergency evacuation in a rail station. *Fire Technol.*, 43(4): 331–346.
Ke, J.M. 2003. Computer simulation and design analysis of smoke management system in large stations [Dissertation]. Kaohsiung (Taiwan): University of Sun Yat-sen .
Ko, C.S. 1993. *Cheng-Wen the contemporary dictionary of chemistry and chemical engineer-ing*, 1st ed. Taipei, Taiwan: Cheng-Wen Publishing, 286.
PIARC. 2007. *Committee on Road Tunnels Operation (C3.3). Systems and equipment for fire and smoke control in road tunnels*. Paris, France: PIARC.
Shen, T.S., & Huang, Y.H., Chien, S.W. 2008. Using Fire Dynamic Simulation (FDS) to recon-struct an arson fire scene. *Build Environ.*, 43: 1036–1045.
World Energy Council, London, UK, 2016

6 Thermal Accident of Methyl Ethyl Ketone Peroxide Plant
Calorimetric Analysis

6.1 OVERVIEW ABOUT METHYL ETHYL KETONE PEROXIDE (MEKP) IN CHEMICAL PLANTS

Methyl Ethyl Ketone Peroxide (MEKP) is used as a hardener in the manufacture of resins, synthetic rubber and other petrochemical plastics. It is an ingredient of paints, varnishes and paint removers. MEKP is also used in the fiber glass and plastics industry as a curing agent. It is an organic peroxide, which is explosive in its pure form. Hence it is commercially available as a 40–60% solution with stabilizing agents such as dimethyl phthalate, cyclohexane peroxide, or diallyl phthalate (Gooch, 2011; Barbalace, 2009).

MEKP is listed as a highly toxic substance and is categorized into United Nations Hazard class 5.2 (Eller and Cassinelli, 1994). Its colorless nature and minimal odor has led to accidental ingestion among both adults and children (Prez-Martinez et al. 1997, Bates et al. 2001). Several cases of intentional ingestion for self-harm or suicide have been reported (Mittleman et al. 1986). In addition, poisoning by inhalation and spillage to eyes leading to corrosive damage has also been reported. Unprotected workers are victims of chronic exposure and toxicity (Brigham and Landrigan, 1985).

Commercially available preparations of MEKP are strong oxidizing substances. They are known to produce alkylperoxyl radicals upon contact with metal ions, a process accelerated by the presence of iron in the heme molecule in biological systems. Tissue damage is believed to be caused by these free radicals, which denature organic molecules, including the peroxidation of lipids. Further toxicity is caused by the acidity of the chemicals produced (Akaike et al. 1992).

The behavior of thermal explosions or runaway reactions has been widely studied for many years. In a reactor with an exothermic reaction, it is very easy to accumulate energy and temperature, when the heat generation rate exceeds the heat removal rate by Semenov theory (Semenov, 1984). Methyl ethyl ketone peroxide (MEKPO) is usually applied as initiators and cross-linking agents for polymerization reactions.

One reason for accidents involves the peroxy group (–O–O–) of organic peroxides, due to its instability and high sensitivity for thermal sources. Many thermal explosions and runaway reactions have been caused globally by MEKPO resulting in a large number of injuries and even death, as shown in Table 6.1 (Yeh et al., 2003; Chang et al., 2006; Tseng et al., 2006; Tseng et al., 2007; MHIDAS, 2006).

Table 6.1 displays three accidents for MEKPO in Australia and UK from the Major Hazard Incident Data Service (MHIDAS)

6.2 CASE SELECTION

The case study in this chapter deals with a thermal explosion and runaway reaction of MEKPO occurred at Taoyuan County (the so-called Yung-Hsin explosion) that killed 10 and injured 47 people in Taiwan in 1996. Figures 6.1 (a) and (b) show the accident damage situation from the Institute of Occupational Safety and Health in Taiwan.

Accident development was investigated by a report from the High Court in Taiwan. Unsafe actions (wrong dosing, dosing too rapidly, errors in operation, cooling failure) caused an exothermic reaction and the first thermal explosion of MEKPO. Simultaneously, a great deal of explosion pressure led to the top of the tank bursting and the hot concrete broken and shot to the 10-ton hydrogen peroxide (H_2O_2) storage tank (d = 1, h = 3). Under this circumstance, the 10-ton H_2O_2 storage tank incurred the second explosion and conflagration that caused ten people to perish (including employers, fire-fighters) and 47 injuries. Many plants near Yung-Hsin Co. were also affected by the conflagration caused by the H_2O_2 tank. H_2O_2, dimethyl phthalate (DMP), and methyl ethyl ketone (MEK) were applied to manufacture the MEKPO product. To prevent any casualties from runaway reaction and thermal explosion events from occurring, the aim of this study was to simulate an emergency response

TABLE 6.1
Thermal Explosion Accidents Caused by MEKPO Globally

Year	Nation	Frequency	Injuries	Fatalities	Worst Case
1953–1978	Japan	14	115	23	114 (Injuries) 19 (Fatalities) in Tokyo
1980–2004	China	14	13	14	8 (Injuries) 5 (Fatalities) in Honan
1984–2001	Taiwan	5	156	55	49 (Injuries) 33 (Fatalities) in Taipei
2000	Korea	1	11	3	11 (Injuries) 3 (Fatalities) in Yosu
1973–1986	Australia	2	0	0	Not applicable
1962	UK	1	0	0	Not applicable

(Reproduced from Wu et al. 2008, *International Journal of Chemical Science*, Sadguru Publications)

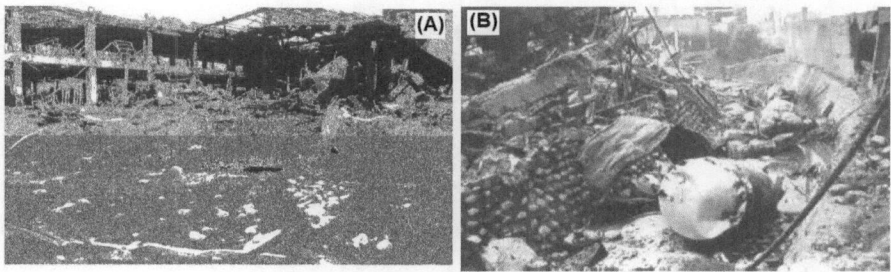

FIGURE 6.1 (a) The aftermath of the Yung-Hsin explosion which devastated the entire plant, including all buildings within 100 m (b) Reactor bursts occurred by thermal runaway (Reproduced from Wu et al. 2008, *International Journal of Chemical Science*, Sadguru Publications).

process. Differential scanning calorimetry (DSC) was employed to integrate thermal hazard development. The processing of experimental data and kinetics evaluation was implemented by applying the TDPro and For K software developed by CISP Ltd. The production method is described in detail by Kossoy and Akhmetshin (Yuan, Shu & Kossoy, 2005) for the creation of a kinetic model and the algorithms that are utilized. Due to MEKPO decomposing at low temperature (30–40°C) (Yuan, Shu & Kossoy, 2005; Fu et al., 2003) and exploding with exponential development, developing or creating an adequate emergency response procedure is very important for preventing it. The safety parameters, such as temperature of no return (TNR), time to maximum rate (TMR), self-accelerating decomposition temperature (SADT), maximum temperature (T_{max}), etc., were necessary and useful for studying emergency response procedure in terms of industrial applications. In view of loss prevention, the emergency response plan is mandatory and necessary for corporations to cope with reactive chemicals under upset scenarios.

6.3 THERMAL HAZARD ANALYSIS OF MEKP BY DSC

DSC has been employed widely for evaluating thermal hazards (ASTM E537-76, 1976; Ando, Fujimoto & Morisaki, 1991) in various industries. It is easy to operate, gives quantitative results, and provides information on sensitivity (exothermic onset temperature, T_0) and severity (heat of decomposition, ΔH_d) at the same time. DSC was applied to detect the fundamental exothermic behavior of 31 mass % MEKPO in DMP that was purchased directly from the Fluka Co. Density was measured and provided directly from the Fluka Co. ca. 1.025 g cm^{-3}. It was, in turn, stored in a refrigerator at 4 °C. DSC, as shown in Figures 6.2(a) and (b), involved two thermocouples, gold of test crucible (100 bar), STAR^e software, and so on. According to Figure 6.2(b), the S side put on sample crucible and the R side detects the blank crucible.

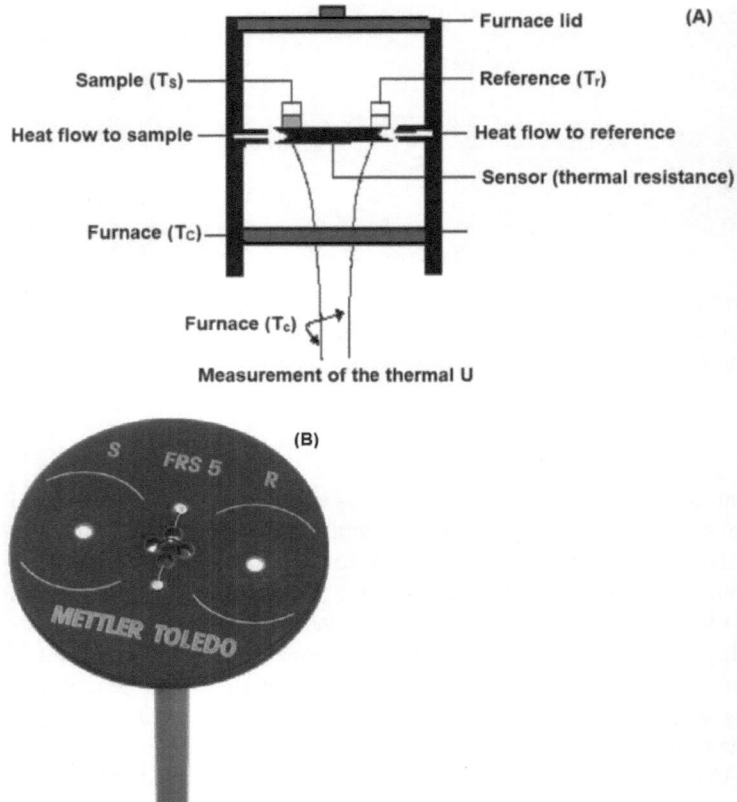

FIGURE 6.2 (a) Measurement principle of a heat flux DSC sensor with a single thermocouple (b) Measurement principle of the FRS5 interchangeable heat flux DSC sensor with 56 thermocouples(Reproduced from Wu et al. 2008, *International Journal of Chemical Science*, Sadguru Publications).

6.4 THERMAL ANALYSIS OF MEKP AND H_2O_2 THROUGH DSC ANALYSIS

6.4.1 THERMAL DECOMPOSITION ANALYSIS OF 31 MASS% MEKPO FOR DSC

Figure 6.3 demonstrates a comparison of thermal curves of decomposition of 31 mass % MEKPO with four types of H (H = 1, 2, 4, and 10°C min⁻¹) by DSC. Table 6.2 summarizes the thermodynamic data by the DSC STAR^e program for the runaway assessment. MEKPO could decompose slowly at 30–32°C, as disclosed by previous researches (Chen et al., 2006). We surveyed MEKPO decomposing at 30°C, shown in Figure 6.3. Various scanning rates by DSC were used to survey the initial decomposition circumstances. Under the scanning rate of 10°C min⁻¹ situation, the T_0 was measured at about 47°C and ΔH of the first peak was evaluated at about 96.87 J g⁻¹. As a result, a rapid rise of temperature may cause violent initial decomposition

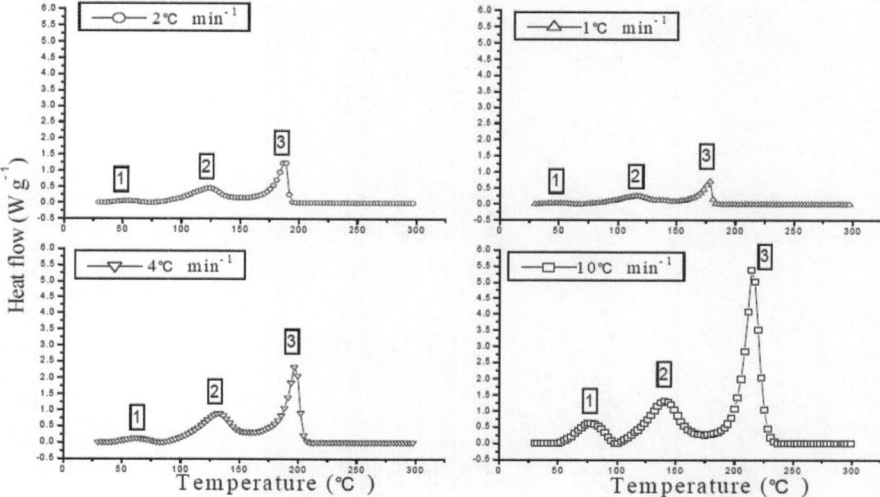

FIGURE 6.3 Heat flow vs. temperature of MEKPO 31 mass% under dynamic various scanning rates by DSC (Reproduced from Wu et al. 2008, *International Journal of Chemical Science*, Sadguru Publications).

(the first peak) of MEKPO under external fire conditions. Table 6.2 shows thermokinetic parameters, such as T_0, ΔH, T_{max}, of 31 mass% MEKPO by DSC under various scanning rates. The initial decomposition peak usually releases little thermal energy, so it is often disregarded. The T_2 of mainly decomposition was about 80°C. The total heat of reaction (ΔH_{total}) was about 1,200 J g^{-1}. DSC was applied to evaluate the Ea and frequency factor (A). The Ea under DSC dynamic test was about 168 kJ mol^{-1} and A was about 3.5.1019 (s^{-1}).

6.4.2 THERMAL DECOMPOSITION ANALYSIS OF 20 MASS% H₂O₂ BY DSC

Figure 6.4 displays the exothermic reaction of 20 mass % H_2O_2 under 4°C min^{-1} of β by the DSC. Due to H_2O_2 being a highly reactive chemical, operators must carefully control flow and temperature. H_2O_2 was controlled at 10°C, when it precipitated the reaction. H_2O_2 exothermic decomposition hazards are shown in Table 6.3.

6.5 KINETIC ANALYSIS OF THERMAL DEGRADATION

There are several well-known methods for evaluating simple autocatalytic models, i.e., for estimating the model parameters. One can derive complex multi-stage kinetic models that depict autocatalytic phenomena in more detail, but special numerical optimization methods are required to estimate parameters of such models as discussed by Kossoy and Koludarova (Kossoy & Koludarova, 1995) for a complex

TABLE 6.2
Thermokinetics and Safety Parameters of 31 Mass% MEKPO by DSC Under Various Scanning Rates

| β | Mass | Initial Decomposition | | Main Thermal Decomposition | | | | | | |
| | | 1st peak | | 2nd peak | | | 3rd peak | | | |
($^{\circ}$C min^{-1})	(mg)	T_1 ($^{\circ}$C)	ΔH_d (J g^{-1})	T_2 ($^{\circ}$C)	T_{max} ($^{\circ}$C)	ΔHd (J g-1)	T2 (oC)	T_{max} ($^{\circ}$C)	ΔH_d (J g^{-1})	ΔH_{total} (J g$_{-1}$)
1	4	30	29.16	70	115	382.76	142	180	825.29	1237.51
2	3.72	35	35.85	75	125	324.11	152	187	768.95	1128.91
4	4	42	41.35	83	135	304.05	160	200	768.13	1113.53
10	4.9	47	96.87	100	140	250.82	175	220	584	931.69

(Reproduced from Wu et al. 2008, *International Journal of Chemical Science*, Sadguru Publications)

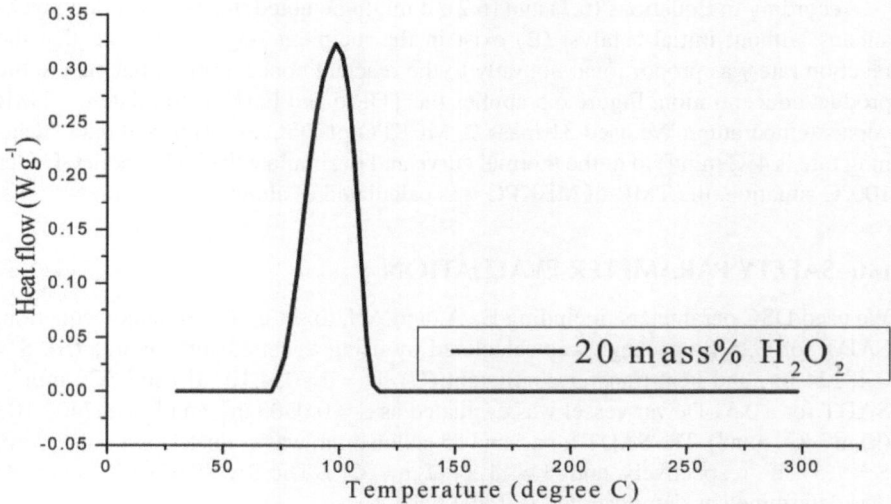

FIGURE 6.4 Heat flow vs. temperature of 20 mass% H_2O_2 under 4°C min^{-1} scanning rate by DSC (Reproduced from Wu et al. 2008, *International Journal of Chemical Science*, Sadguru Publications).

TABLE 6.3

Thermokinetics and Safety Parameters of 20 Mass% H_2O_2 by DSC Under Various Scanning Rates

Chemical	Mass (mg)	T_0 (°C)	T_{max} (°C)	ΔHd (J g^{-1})
H_2O_2	2.47	67	105	395
Reference[13]	2.2	69	99.6	409.7

(Reproduced from Wu et al. 2008, *International Journal of Chemical Science*, Sadguru Publications)

---

model of two consecutive reactions where the second stage is autocatalytic. The reaction mechanism of MEKPO could be represented by the following kinetic model:

$$\rightarrow \text{Initiation reaction}: A\ B+C+\ldots\ldots;\ \text{stage} \qquad (6.1)$$

$$\rightarrow \text{Autocatalytic reaction}: A\ 2B+C+\ldots\ldots;\ \text{stage} \qquad (6.2)$$

$$\frac{d\alpha}{dt} = r_1 + r_2 \qquad (6.3)$$

$$r_1 = \left(1-\alpha\right)^{n1} k_{01} \exp\left(-\frac{E_1}{RT}\right) \qquad (6.4)$$

$$r_2 = \alpha^{n21}\left(1-\alpha\right)^{n2} k_{02}\exp\left(-\frac{E_2}{RT}\right); \frac{dQ}{dt} = Q_1^\infty r_1 + Q_2^\infty r_2 \qquad (6.5)$$

According to Equations (6.1) and (6.2), it might be noted that two reaction mechanisms without initial catalyst (B) exist in the incipient stage, indicating that the reaction rate was proportional not only to the reactant concentration, but also to the product concentration. Figure 6.5 applies the TDPro and ForK to simulate the TMR versus temperature. We used 31 mass % MEKPO of DSC experimental data (scanning rate is 4°C min⁻¹) to fit the thermal curve and to simulate the TMR model. Under 100°C situation, the TMR of MEKPO was calculated as about three minutes.

6.6 SAFETY PARAMETER EVALUATION

We used DSC parameters, including E_a, A, and ΔH, to set up the Semenov equation. SADT for a 25 kg package was calculated by using a wetted surface area (S), S = 0.48124 m², and heat transfer coefficient (U), U = $1.7034.10^{-1}$ (kJ m⁻² °C⁻¹ min⁻¹). SADT for a 0.51 Dewar vessel was evaluated as S = 0.0303 m², and U = $8.7402.10^{-2}$ (kJ m⁻² °C⁻¹ min⁻¹). The SADT for a 5 and 55 gallon drum was evaluated as S = 0.137 m², S = 1.51 m², respectively, and U = 11.34 (J m⁻² °C⁻¹). The SADT of various vessels was determined as demonstrated in Table 6.4.

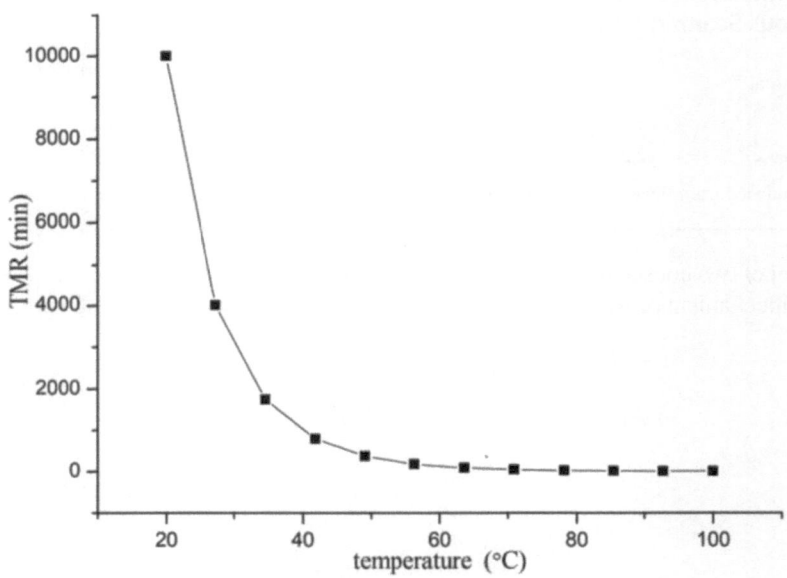

FIGURE 6.5 TMR vs. temperature (kinetic–based curve–fitting) for 31 mass% MEKPO by DSC for 4°C min⁻¹ of heating rate (Reproduced from Wu et al. 2008, *International Journal of Chemical Science*, Sadguru Publications).

TABLE 6.4

T_{NR} and SADT for Safety Storage and Transportation by Various Vessel Situations

Vessel Type	TNR (°C)	SADT (°C)
5 gallon drum	78	72
55 gallon drum	97.9	84.5
UN 25 kg package	62.7	54.2
UN 0.51 L Dessel vessel	55	50
Adiabatic test by VSP 2	100	80

(Reproduced from Wu et al. 2008, *International Journal of Chemical Science*, Sadguru Publications)

6.7 SUMMARY

According to the DSC experimental data, MEKPO decomposes at 30–40°C. If the H is high, the initial exothermic temperature could be delayed and ΔH would cause its temperature to rise quickly. Under external fire circumstances, MEKPO can decompose quickly and cause a runaway reaction and thermal explosion. During storage and transportation, a low concentration (< 40 mass %) and a small amount of MEKPO should be controlled. Under differential storage and transportation vessels, for the SADT there was a disparity. This chapter, with a view of predicting the SADT of a 55 gallon drum in Taiwan, came up with a value of about 85°C. H_2O_2 was controlled 10°C, when it joined a MEKPO manufacturing reaction. This is very dangerous for the MEKPO manufacturing process, so the reaction was a concern and controlled at less than 20°C in the reactor. Thermokinetics determined by an autocatalytic thermal curve could be used to assess the thermal explosion hazard for organic peroxides and to determine useful parameters such as T_0, SADT, temperature of no return (T_{NR}), and adiabatic time to maximum rate (TMR_{ad}). In practice, these data are necessary for the proper choice of safe conditions for application, storage, and transportation in terms of chemical products.

REFERENCES

Akaike, T., Sato, K., Ijiri, S., Miyamoto, Y., Kohno, M., & Ando, M. (1992). Bactericidal activity of alkyl peroxyl radicals generated by heme-iron-catalyzed decomposition of organic peroxides. *Arch. Biochem. Biophy.* 294 (1) 55–63.

Ando, T., Fujimoto, Y., & Morisaki, S. (1991). Analysis of differential scanning calorimetric data for reactive chemicals. *J. Hazard. Mat.* 28 (3), 251–280.

ASTME537-76. (1976). Thermal Stability of Chemicals by Methods of Differential Thermal Analysis.

Barbalace, K. (2009). Chemical Database: Methyl ethyl ketone peroxide. *Environ. Chem. Com.*

Bates, N., Driver, C.P., & Bianchi, A. (2001). Methyl ethyl ketone peroxide ingestion: toxicity and outcome in a 6-year-old child. *Ped.* 108 (2) 473–476.

Brigham, C.R., & Landrigan, P.J. (1985). Safety and health in boatbuilding and repair. *Am. J. Ind. Med.* 8 (3):169–182.

Chang, R. H., Tseng, J. M., Jehng, J. M., Shu, C. M., & Hou, H. Y. (2006). Thermokinetic model simulations for methyl ethyl ketone peroxide contaminated with 4 OR NaOH by DSC and VSP$_2$ *J. Therm. Anal. Calorim.* 83, 57–62.

Chen, K. Y., Lin, C. M., Shu C. M., & Kao, C. S. (2006). An evaluation on thermokinetic parameters for hydrogen peroxide at various concentrations by DSC. *J. Therm. Anal. Calorim.* 85, 87–89.

Eller, P.M., & Cassinelli, M.E. (1994). *NIOSH manual of analytical methods.* DIANE Publishing, Pennsylvania, United States.

Fu, Z. M., Li, X. R., Koseki, H. K., & Mok. Y. S. 2003. Evaluation on thermal hazard of methyl ethyl ketone peroxide by using adiabatic method. *J. Loss Prev. Process Ind.* 16 (5), 389–393.

Gooch, J.W. (2011). Methyl Ethyl Ketone Peroxide. Encyclopedic Dictionary of Polymers. 458.

Kossoy, A. A., & Koludarova, E. (1995). Specific features of kinetics evaluation in calorimetric studies of runaway reactions. *J. Loss Prev. Process Ind.* 8, 229–235.

Maria, G., & Heinzle, E. (1998). Kinetic system identification by using short-cut techniques in early safety assessment of chemical processes. *J. Loss Prev. Process Ind.* 11(3), 187–206.

MHIDAS, Mayor Hazard Incident Data Service. (2006) OHS_ROM, Reference Manual.

Mittleman, R.E., Romig, L.A., & Gressmann, E. (1986). Suicide by ingestion of methyl ethyl ketone peroxide. *J. For. Sci.* 31 (1) 312–320.

Prez-Martinez, A., Gutirrez-Junquera, C., Gonzlvez-Piera, J., Marco-Macin, A., Rubio-Guijarro, J., & Moya-Marchante, M. (1997). Oesophageal stenosis in a child caused by ingestion of methyl ethyl ketone peroxide. *Eur. J. of Ped.* 156 (12) 976.

STAR^e Software with Solaris Operating System, Operating Instructions. (2004). Mettler Toledo, Switzerland.

Tseng, J. M., Chang, R. H., Horng, J. J., Chang. M. K., & Shu, C. M. (2006). Thermal hazard evaluation for methyl ethyl ketone peroxide mixed with inorganic acids. *J. Therm. Anal. Calorim.* 85, 189–194.

Tseng, J. M., Chang, Y. Y., Su, T. S., & Shu, C. M. (2007). Study of thermal decomposition of methyl ethyl ketone peroxide using DSC and simulation. *J. Hazard. Mater.*, 142 (3), 765–770.

Wu, S.H., Su, C. H., & Shu, C. M. (2008). Thermal accident investigation of methyl ethyl ketone peroxide by calorimetric technique. *Int. J. Chem. Sci.* 6(2), 487–496.

Yeh, P. Y., Shu, C. M., & Duh, Y. S. (2003). Thermal hazard analysis of methyl ethyl ketone peroxide. *Ind. Eng. Chem. Res.*, 43 (1), 1–5.

Yuan, M. H., Shu, C. H., & Kossoy, A. A. 2005. Kinetics and hazards of thermal decomposition of methyl ethyl ketone peroxide by DSC. *Thermochimica Acta*, 430, 67–71.

7 Case Study on the Integrated Self-Assessment Module for Fire Rescue Safety in a Chemical Plant

7.1 INTRODUCTION

A wide variety of materials required in our daily lives are mainly provided by the chemical industry. These materials might have properties of toxicity, corrosiveness, flammability, and explosiveness and pose serious threats to the consumers. Hence, various safety laws and features have been set up by various government authorities specific to each country's management approach. Even after such rules and regulations have been laid down, it can be seen that Taiwan and the whole world from 2005–2015 had reported numerous accidents as shown in Table 7.1 and Table 7.2. Studies suggest that due to lack of awareness or failure in assessing safe distance for rescue operations during fire accidents and explosions, numerous fire-fighters have been injured, many of them severely (Liu et al., 2017). Hence, it became of the utmost necessity to investigate further in the safety operation distance to prevent such accidents and casualties. A few reliable disaster incident response systems, such as that produced by the California Specialized Training Institute (http://www.caloes.ca.gov) and a toxic chemical disaster emergency procedure card, were adhered to for preliminary safety. In addition, the Globally Harmonized System of Classification and Labelling of Chemicals, and the 2016 edition of the Emergency Response Guidebook (ERG), helped in determining the safe operating distance for fire-fighters to prevent any hazard to rescue workers (Mishra et al., 2014).

Risk analysis and the safety assessment of the domino effect need acute attention to prevent any accident as a result of explosion or fire accidents (Abdolhamidzadeh et al., 2010; Cozzani and Salzano, 2004a; Cozzani et al., 2006). Numerous analysis software is available to simulate the domino effect and thereby assess the safety parameter of a situation. Software packages such as ChemPlus, ALOHA, SLAB, SAFEGI, FDS, and FLACS are commonly used to simulate disasters, including pipeline leaks, fires, and chemical storage tank explosions (Cheng et al., 2009; Zhou and Liu, 2012). The simulations help identify the measures for preventing casualties

TABLE 7.1

Major Accidents with Casualties in Taiwanese Chemical Plants, 2005–2015

Items	Plants Location	Date	Chemicals	Type	Injured/Death	Remark
1	Tuku, Yunlin, Taiwan	04.02.2005	Trinitrotoluene	Explosion	3/2	Fire cracker plant
2	Nantun, Taichung, Taiwan	07.03.2005	Sodium nitrite	Fire	22/0	Three fire Fighters injured
3	Hsitun, Taichung, Taiwan	02.20.2006	Hydrofluoric acid	Splash	2/0	
4	Mailiao, Yunlin, Taiwan	03.06.2006		Fire	5/1	Improper operation
5	Shulin, New Taipei City	05.21.2007	Nickel sulfate	Leakage	4/2	
6	Luchu, Taoyuan, Taiwan	06.25.2007	Propylene glycol methyl ether	Explosion	5/1	
7	Xinwu, Taoyuan, Taiwan	08.16.2007	Nitrite	Explosion	6/0	Improper operation
8	Guanyin, Taoyuan, Taiwan	09.22.2009	Calcium fluoride	Explosion	3/1	
9	Mailiao, Yunlin, Taiwan	11.18.2009	Phosgene	Leakage	12/0	
10	Nantun, Taichung, Taiwan	06.11.2010	Xylene	Explosion	5/2	
11	Nantun, Taichung, Taiwan	06.18.2011	Foam	Fire	1/1	Grinder sparks
12	Mailiao, Yunlin, Taiwan	09.14.2011	Styrene	Explosion	2/0	
13	Lukang, Changhua, Taiwan	05.17.2012	Toluene	Explosion	13/1	
14	Nantun, Taichung, Taiwan	08.01.2012	Methyl isobutyl ketone	Explosion	3/0	Static spark
15	Dalin, Chiayi, Taiwan	08.10.2012	Hydrogen	Fire	3/0	
16	Minsyong, Chiayi, Taiwan	01.10.2014	Dodecyl benzene sulfonic acid	Explosion	1/1	Welding spark
17	Mailiao, Yunlin, Taiwan	03.05.2014	Hydrogen	Fire	2/0	
18	Guishan, Taoyuan, Taiwan	03.27.2014	Hydrofluoric acid	Leakage	1/0	One fire fighter injured
19	Cianjhen, Kaohsiung, Taiwan	03.31.2014	Propylene	Explosion	308/32	Twenty-four fire fighters injured, six fire fighters dead

(Data obtained with permission from Tsai et al. 2018, Copyright © Elsevier)

TABLE 7.2
Major Accidents with Casualties in Chemical Plants Globally, 2005–2015

Items	Plants Location	Date	Chemicals	Type	Injured/ Death	Remark
1	Texas, USA	03.23.2005	Light oil	Explosion	180/15	
2	Jilin, China	11.13.2005	Nitrobenzene	Explosion	60/8	
3	Buncefield, UK	12.11.2005	Gasoline	Explosion	43/0	
4	Osaka, Japan	12.19.2006	Aluminium	Explosion	2/0	
5	Hebei, China	05.11.2007		Fire	80/4	Improper operation
6	Florida, USA	12.19.2007	Trinitrotoluene	Explosion	14/4	
7	Istanbul, Turkey	01.31.2008	Trinitrotoluene	Explosion	68/17	
8	Markazi Province, Iran	05.25.2008		Explosion	38/30	Welding spark
9	Georgia, USA	07.02.2008	Dust	Explosion	42/13	
10	Liaoning, China	09.14.2008		Fire	2/3	
11	Penang, Malaysia	04.24.2009	Gas	Explosion	5/1	
12	Ahmedabad, India	07.06.2009	Trinitrotoluene	Explosion	100/30	
13	Kharg Island, Iran	07.24.2010		Explosion	0/4	Boiler pressure was too high
14	Liaoning, China	01.19.2011	Heavy oil	Explosion	30/0	
15	Guangxi, China	11.23.2011	Trinitrotoluene	Explosion	11/4	Fire cracker plant
16	Amuay, Venezuela	08.25.2012	Propane	Explosion	130/50	
17	Hyogo, Japan	09.29.2012	Acrylic acid	Explosion	30/1	One fire fighter dead
18	Jeollanam, Korea	03.14.2013	High density polyethylene	Explosion	11/6	Welding spark
19	Jiangsu, China	04.16.2014	Dust	Explosion	9/8	
20	Moerdijk, Netherlands	06.03.2014	Ethylbenzene	Explosion	2/0	

(Data obtained with permission from Tsai et al. 2018, Copyright © Elsevier)

among rescue workers with ChemPlus, ALOHA, and SLAB software being unable to calculate barrier and wind effects, thereby contributing to large discrepancies between simulated results and the actual accident conditions (Zhou and Liu, 2012). Another software package, FDS, lacked map integration and could not estimate the strain placed on fire rescue operators (I et al., 2009). Thus, it was established that all the software assessment models were incomplete, and that they were able to simulate only single accidents without being able to take in changes such as disaster type, disaster pattern and did not consider the safe working distance for the fire-fighters in such incidents.

As is already evident from the discussion, disastrous domino effects may overpower rescuers, even those attending only a very minor accident in a chemical plant such as leakages or explosions. It has also been observed after studying several cases that the firefighters are exposed to the tremendous threat to life during an accident such as a fire explosion because the safe working distance is often not brought into consideration. Hence, this chapter proposes a self-assessment model to predict a safer operating distance using SAFETI and simulate the effects of a possible domino effect, thereby visualizing the maps for fire and explosions (Cozzani et al., 2007, 2009).

7.2 RESEARCH METHODS

7.2.1 ENVIRONMENTAL DATA ANALYSIS

Average wind speed, relative humidity, and temperature data were collected from the Taiwan Central Weather Bureau for monitoring the situations (http://www.cwb.gov.tw/). The stability of the atmosphere was determined as per the sunlight intensity and several other parameters that were related to the speed of the wind. The geographic location of the chemical plant, including the neighboring areas, was considered to determine the surface roughness parameter (Patra, 2006).

7.2.2 IMPACT ANALYSIS PROGRAM

Simulation of the various types of accidents was done by using an impact analysis software package, SAFETI. Waves generated due to explosions (effect and physical) were utilized in determining the relationship between thermal radiation pressure and distance (Reniers et al., 2015).

7.2.2.1 Physical Mode

The identification of casualties was usually done after the occurrence of an accident, including the hazardous properties and areas. The present case study identified the mode as that of a fire explosion wherein the pressure waves and thermal radiations were determined after the discharge of the flammable substance and the exposure to a source of ignition (Cozzani et al., 2006).

7.2.2.2 Effect Mode

The effect mode investigates the extent to which the casualties were caused, including building damages by fires or explosions or any other hazardous material. The thermal radiation and the magnitude of damage were said to be inter-related (Cozzani et al., 2007, 2009).

$$Q = H_c MS \tag{7.1}$$

$$R = \sqrt{\frac{Q}{12\pi q_f}} \tag{7.2}$$

By contrast, pressure and the hazard due to the explosion were related (Cozzani and Salzano, 2004b).

$$\Delta P = 1.1 \frac{1}{\dfrac{R_1}{\sqrt[3]{\varpi}}} + 4.3 \frac{1}{\left(\dfrac{R_1}{\sqrt[3]{\varpi}}\right)^2} + 14 \frac{1}{\left(\dfrac{R_1}{\sqrt[3]{\varpi}}\right)^3} \tag{7.3}$$

7.2.3 INTEGRATED ASSESSMENT

7.2.3.1 Integrated Risk Frequency Analysis

The Major Accident Reporting System, the Major Hazard Incident Data Service, the Major Accident Database, the Failure and Accident Technical Information System, and the Major Accident Hazards Bureau Database were all used to determine the last two decades of data for secondary fire accidents, explosions due to thermal radiations, pressure waves, or explosion fragmentation (Darbra et al., 2010; Gómez-Mares et al., 2008; Zhang and Zheng, 2012).

7.2.3.2 Integrated Impact Analysis

To measure the increments of thermal radiations and pressure shock waves which may result in a secondary accident, a SAFETI simulation was done on a primary fire or explosion (Gómez-Mares et al., 2008; Spoelstra et al., 2015), the generalized flow chart is illustrated in Figure 7.1.

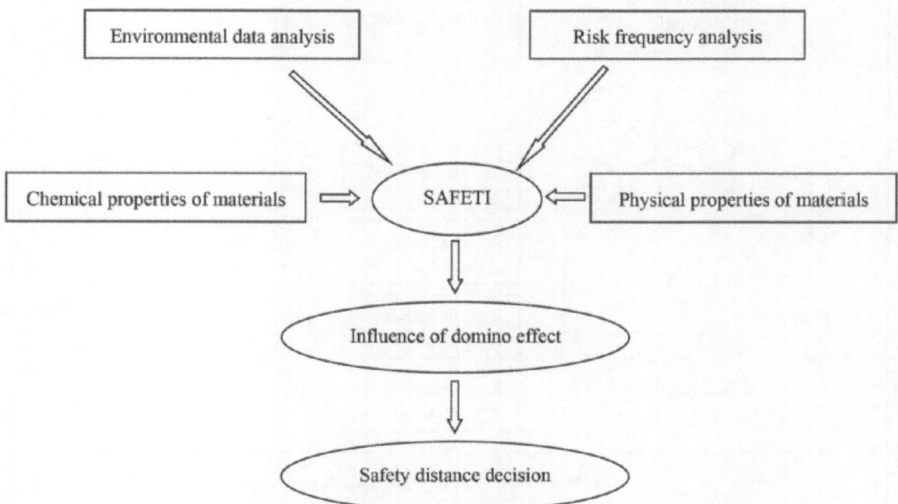

FIGURE 7.1 The general flow chart for this study (Reproduced with permission from Tsai et al. 2018, Copyright © Elsevier).

7.3 CASE STUDY

7.3.1 SITE LOCATION

The case study involved firms based on the Taichung Industrial Park, Taichung, Taiwan, a site which accommodates more than 800 companies and which is one of the largest industrial parks in Taiwan (Figure 7.2). The first two plants were three-storied buildings and the third plant was a five-storied building. Raw materials storage warehouses (A and B) were metal buildings with floor areas of 1300 m² and placed beside the old plant, having reinforced concrete with tin roofs above the third floor. The total area of the plant is 4300 m². Both the warehouses were 2.4 m apart and approximately 4.8 m from the new plants.

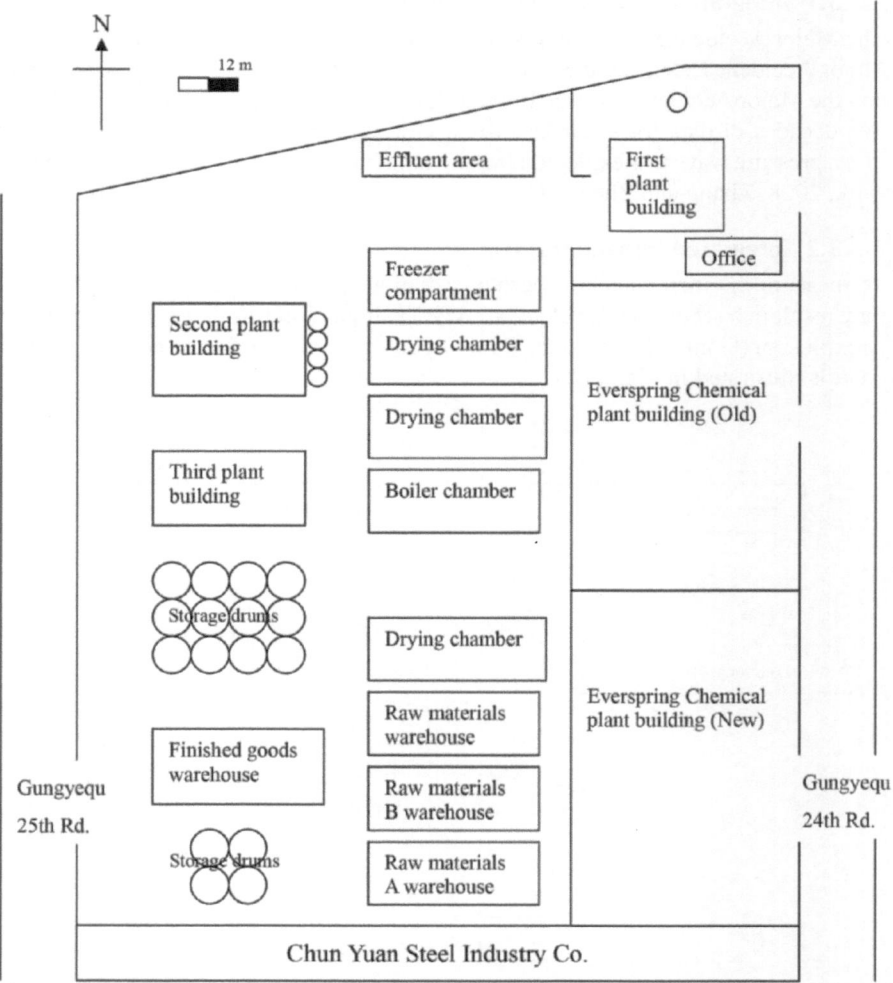

FIGURE 7.2   Schematic plot of the plant for this study (Reproduced with permission from Tsai et al. 2018, Copyright © Elsevier).

7.3.2 STORAGE OF HAZARDOUS MATERIALS

Hazardous goods were stored in the three main locations of the chemical company, viz. warehouse, sump area, and outdoor storage. The list of hazardous materials stored is as per Table 7.3. The warehouse in addition to the storage of hazardous materials was also utilized for the storage of raw materials (Table 7.4) which were burned out during the fire. Flammable liquids were mainly stored in the reservoir which was located next to plants 1 and 2. Toxic substances were stored in the outdoor storage area located next to the drying chamber on the west side of the plant, which remained unaffected by the fire.

7.3.3 CASE STUDY ON THE FIRE RESCUE UNIT

An incident of fire was reported at 12:10 p.m., July 3, 2005, with an actual occurrence time of 22.33 the day before. In between the occurrence and the reporting of the event, a series of explosions at every two to three minutes on average. The relative humidity of 68% and a wind speed of 1.5 m/s accelerated the spread of the fire, which was caused mainly by the explosion of a 53-gallon drum in the plant. Three floors of warehouse A were completely gutted by the fire, which spread to nearly 300 m^2 of the plant. The rescue operation took nearly 10 hours. The accident resulted in the injury of 22 people, of which two were professional fire-fighters and one was a volunteer firefighter (Figure 7.3) (Chen et al., 2010).

TABLE 7.3
Types and Amounts of Hazardous Material in Storage Tanks

Number	Category	Chemical	Location	Capacity(kg)
1	Flammable liquid	Benzene	1st plant	19000
2	Flammable liquid	Methanol	2nd plant	12000
3	Flammable liquid	Isopropanol	2nd plant	3800
4	Flammable liquid	Acetonitrile	2nd plant	1700
5	Flammable liquid	N-heptane	2nd plant	3400

(Data obtained with permission from Tsai et al. 2018, Copyright © Elsevier)

TABLE 7.4
Hazardous Material in Storage

Number	Category	Chemical Stored	Capacity (kg)
1	Pyrophoric substances and substances which in contact with water emit Flammable gas	Ammonia	4380
2	Flammable liquids	Hydrazine	5800

(Data obtained with permission from Tsai et al. 2018, Copyright © Elsevier)

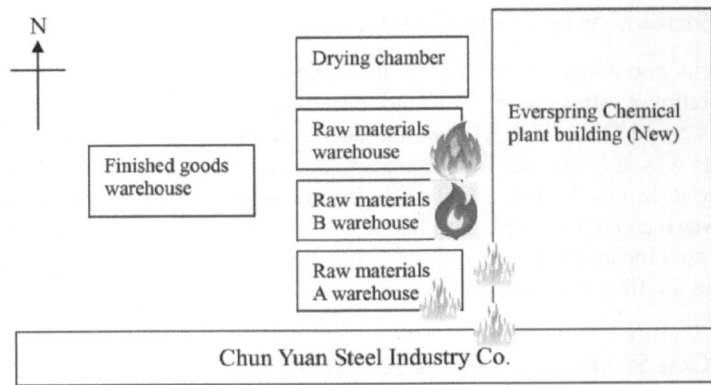

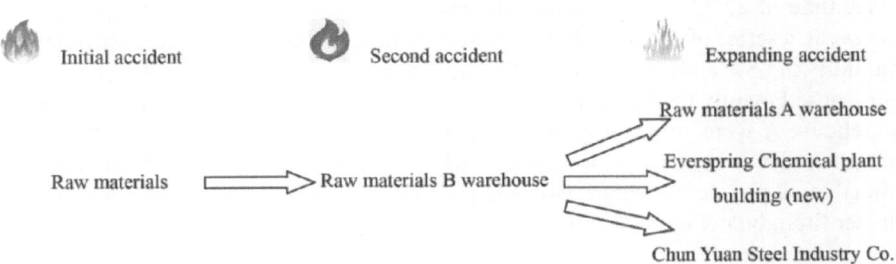

FIGURE 7.3   Accident scenario at case study (Reproduced with permission from Tsai et al. 2018, Copyright © Elsevier).

7.3.4 ENVIRONMENTAL DATA ANALYSIS

Taiwan's Central Weather Bureau provided the data for maximum temperature, wind speed, monthly average temperature, and relative humidity for July 2005. These are listed in Table 7.5.

7.3.5 INTEGRATED IMPACT ANALYSIS

Fires or explosions as the initial event for an accident had been occulting for 20 years, according to the records in the databases. A plant similar to the present case was considered to use the failure rate and event data within risk assessments. A value of 1×10^{-3} (times per year) was considered as the risk frequency.

The improper storage of sodium nitrite resulted in the ignition of flammable liquids of 5800 kg of hydrazine and 4380 kg of ammonia water. The inherent risks of the hazardous chemicals, including the physical and chemical properties, were fed into the SAFETI software which simulated the disasters involving a domino effect.

TABLE 7.5
Weather in the Disaster Zone in July 2005 (http://www.cwb.gov.tw/)

Item	Parameter	Numerical Value
Temperature(°C)	Maximum temperature	35.4
	Lowest temperature	22.4
	Average temperature	28.8
Wind velocity(m/s)	Maximum wind velocity in ten minutes	10.4
	Maximum instantaneous wind velocity	34
	Average wind velocity	1.5
Wind direction(360°)	Maximum wind in ten minutes	360
	Maximum instantaneous wind	350
	Average wind	220
Relative humidity(%)	Minimum humidity	49
	Average humidity	76
Rainfall(mm)		378.1
Precipitation days(day)	≥ 0.1mm	11

(Data obtained with permission from Tsai et al. 2018, Copyright © Elsevier)

7.4 RESULTS AND DISCUSSIONS OF THE INITIAL ACCIDENT AND THE SECOND ACCIDENT

7.4.1 INITIAL ACCIDENT

The warehouse had already caught fire by the time the firefighters had arrived at the scene. It was later identified that the cause of the explosion was the inappropriate storage of the chemical sodium nitrite. This also resulted in the ignition of the hydrazine stored nearby, releasing around 28.5 kW of thermal radiation within a radius of 13 m. At about 14 m from the source of ignition, the thermal radiation strength had reduced to 18 kW/m². For a normal human being, exposure to the thermal radiation of strength 4 kW/m² for 2 s at a distance of 32 m would result in severe body pain. Thus, the rescue personnel was directed to wear appropriate protective gear to perform their duties. Prolonged exposure to 9.5 kW/m² of thermal radiation would end up in a domino effect, causing the ignition of all the combustible materials.

Hydrazine ignition resulted in a blast overpressure of 0.1379 bar within 23 m radius and 0.2068 bar within 16 m radius. The pressure wave radius of the hydrazine tank is shown in Figure 7.4. The nominal 5% probability of lethal effect was found to be at a radius of 87 m, which had a pressure of 0.0207 bar. Hence, it was recommended that all rescue teams should wear the required personal protective equipment (PPE) within the aforesaid work radius of 87 m.

Hydrazine has been classified as a highly corrosive and inflammable liquid as per 2012 ERG and instructions have been given that it should be kept at an isolation radius of at least 50 m in case of any leakage or damage. As per the thermal radiation formula, thermal radiation was found to be 28.5, 18, 4, and 9.5 kW/m² for the corresponding radii of 24.1, 30.3, 32, and 41.76 m, respectively. The explosion pressure

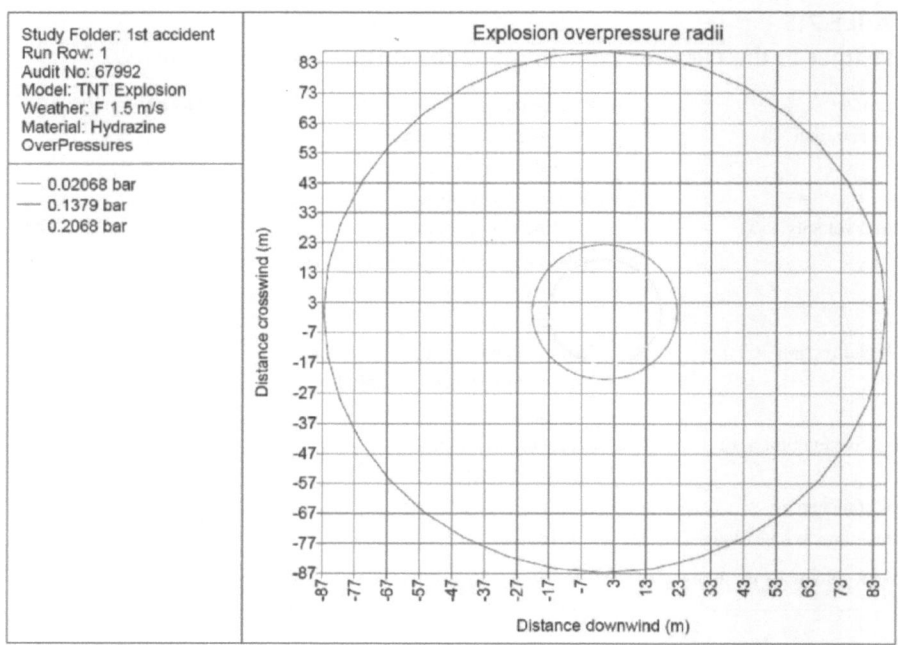

FIGURE 7.4 Pressure wave radii of hydrazine tank release (Reproduced with permission from Tsai et al. 2018, Copyright © Elsevier).

wave formulae indicated possible pressures of 0.2068, 0.1379, and 0.0207 bar at radii of influence 15.0, 17.5, and 31.0 m, respectively.

7.4.1.1 Domino Effect

1. Thermal radiations at a distance of 22.5 m from the ignition source were found to be 9.5 kW/m^2.
2. The blast overpressure was found to decrease to 0.2068 bar at a distance of 16 m from the source.

7.4.1.2 Safety Distance for Rescue Personnel

1. The thermal radiation value decreased to 4 kW/m^2 at a distance of 32 m from the ignition source.
2. The overpressure also decreased to 0.0207 bar at a distance of 87 m from the ignition source.

7.4.2 Second Accident

Figure 7.5 and Figure 7.6 suggested that ammonia ignited due to hydrazine combustion in the warehouse, causing a thermal radiation release of 8.3 kW/m^2 within a 10m radius. With an increasing radius of 23 m, the radiation level decreased to 6.3 kW/m^2. As per the studies, the human body can resist radiations of up to 4 kW/m^2 for

○ : Thermal radiation 28.5 kW/m² of hydrazine   ○ : Thermal radiation 8.3 kW/m² of ammonia

FIGURE 7.5 Thermal radiation of hydrazine and ammonia tanks release (Reproduced with permission from Tsai et al. 2018, Copyright © Elsevier).

duration of 1 min which would persist within a 37 m radius. Thus, it was assumed that the rescue team could have carried out the operation provided they were well equipped with personal protective equipment; however, thermal radiation beyond 8.3 kW/m² would have resulted in damage, as well as the domino effect without proper protection. The ammonia combustion of 0.2068 bar within 29 m radius, blast overpressure of 0.1379 bar within 35 m radius as indicated in Figure 7.7 which resulted in the combustion of adjacent materials. The lethal effect probability decreased to 5% with an overpressure of 0.2017 bar at a distance of 145 m. As per the previous suggestions, with appropriate PPE the rescue team could perform their duties within the above-mentioned range.

As per ERG classification, ammonia being a corrosive gas must be isolated at a minimum radius of 150 m as a preventive measure. The thermal radiation levels as per the thermal radiation formula were calculated to be 8.3, 6.3, and 4 kW/m² with the corresponding radius of influence as 21.82, 25.05, and 31.43 m, respectively. The pressures as per the blast overpressure formula were 0.2068, 0.1379, and 0.0207 bar at the corresponding radius of influence as 16.8, 20, and 58 m, respectively.

To carry out the simulations in SAFETI caused by the fire accident as a result of ammonia water ignition, causing an explosion with a chain reaction as a result of thermal, radiation, and overpressure effects must be brought into consideration.

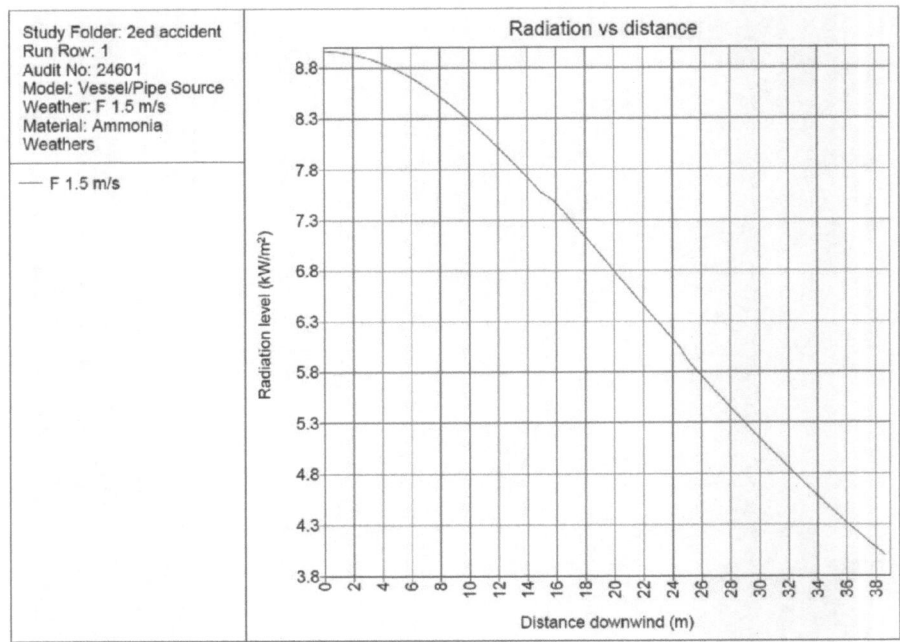

FIGURE 7.6 Thermal radiation and distance of ammonia tank release (Reproduced with permission from Tsai et al. 2018, Copyright © Elsevier).

7.4.2.1 Domino Effect

A pressure wave was caused by an explosion of 29.0 m away from the ignition sources, with an overpressure of 0.2068 bar.

7.4.2.2 Safety Distance for Rescue Personnel

(1) The thermal radiation decreased to less than 4 kW/m² at a distance of 37 m from the source of ignition.
(2) The overpressure decreased to 0.0207 bar at a distance of 145 m from the source of ignition.

7.5 SUMMARY

For fire-fighters to perform their duties safely, ERG has to be followed as the prevailing reference. The present chapter established a self-assessment model to determine the safe distance for fire-fighters to perform their duties in case of an explosion in a chemical plant. As per the simulations, a safety distance of 87 m from the source of ignition was calculated, and with the presence of protective equipment, the acceptable risk might be anywhere between 32 and 87 m at the time of the initial accident. It was also suggested that a second accident would have been caused as a result of the

○ : Blast overpressure 0.2068 bar of hydrazine   ○ : Blast overpressure 0.1379 bar of hydrazine

○ : Blast overpressure 0.2068 bar of ammonia   : Blast overpressure 0.1379 bar of ammonia

FIGURE 7.7   Pressure wave radii of hydrazine and ammonia tanks release (Reproduced with permission from Tsai et al. 2018, Copyright © Elsevier).

domino effect if the combustibles at a range of 22.5 m could not be removed or the thermal radiations could not be reduced below 22.5 kW/m^2. On the occurrence of the second accident, the fire-fighters with protective gear could work at an acceptable risk of 37 to 145 m. It was also suggested that if combustions were in the vicinity of 29 m of the ignition source, a third accident could occur.

SAFETI was used to simulate the domino effects as a result of the fire and explosions. The study provides a visual impact of the domino effect on the people in charge of the plant so that a pre-accident safety measure could be laid out to prevent any major devastation. The simulation also helped in determining the safe working distance for the fire-fighters and rescuers to reduce casualties and fatalities in case of an emergency crisis. In addition, the following should be implemented:

Flammable material should be removed from the incident area to minimize the occurrence of the domino effect. To overcome ignition hazards, measures such as foam covering, fire-fighting, and water jet cooling must be used. Furthermore, vehicles and working staff must be restricted to a safe distance to minimize the occurrence of casualties and fatalities.

REFERENCES

Abdolhamidzadeh, B., Abbasi, T., Rashtchian, D., & Abbasi, S.A. (2010). A new method for assessing domino effect in the chemical process industry. *J. Hazard Mater.* 182 (1–3), 416–426.

Chen, C.C., Wang, T.C., Chen, L.Y., Dai, J.H., & Shu, C.M. (2010). Loss prevention in the petrochemical and chemical-process high-tech industries in Taiwan. *J. Loss Prev. Process. Ind.* 23 (4), 531–538.

Cheng, S., Chen, G., Chen, Q., & Xiao, X. (2009). Research on 3D dynamic visualization simulation system of toxic gas diffusion based on virtual reality technology. *Process Saf. Environ. Protect.* 87 (3), 175–183.

Cozzani, V., Gubinelli, G., & Salzano, E. (2006). Escalation thresholds in the assessment of domino accidental events. *J. Hazard Mater.* 129 (1-3), 1–21.

Cozzani, V., & Salzano, E. (2004a). The quantitative assessment of domino effect caused by overpressure. *Part II. Case studies. J. Hazard Mater.* 107 (3), 81–94.

Cozzani, V., & Salzano, E. (2004b). The quantitative assessment of domino effects caused by overpressure. Part I. Probit models. *J. Hazard Mater.* 107 (3), 67–80.

Cozzani, V., Tugnoli, A., & Salzano, E. (2007). Prevention of domino effect: from active and passive strategies to inherently safer design. *J. Hazard Mater.* 139 (2), 209–219.

Cozzani, V., Tugnoli, A., & Salzano, E. (2009). The development of an inherent safety approach to the prevention of domino accidents. *Accid. Anal. Prev.* 41 (6), 1216–1227.

Darbra, R.M., Palacios, A., & Casal, J. (2010). Domino effect in chemical accidents: main features and accident sequences. *J. Hazard Mater.* 183 (1–3), 565–573.

Gómez-Mares, M., Zárate, L., & Casal, J. (2008). Jet fires and the domino effect. *Fire Saf. J.* 43 (8), 583–588.

I, Y.P., Shu, C.M., & Chong, C.H. (2009). Applications of 3D QRA technique to the fire/explosion simulation and hazard mitigation within a naphtha-cracking plant. *J. Loss Prev. Process. Ind.* 22 (4), 506–515.

Kyunghyun, R., George, E., Zacharakis, J., & Kong, S.K., (2014). Performance enhancement of ammonia-fueled engine by using dissociation catalyst for hydrogen generation. *Int. J. Hydrogen Energy* 39, 2390–2398. *J. Loss Prev. Process Ind.*, 8, 229 (1995).

Liu, X., Zhang, L., Guo, S., & Fu, M. (2017). A simplified method to evaluate the fire risk of liquid dangerous chemical transport vehicles passing a highway bridge. *J. Loss Prev. Process. Ind.* 48, 111–117.

Mishra, K.B., Wehrstedt, K.D., & Krebs, H. (2014). Amuay refinery disaster: the aftermaths and challenges ahead. *Fuel Process. Technol.* 119, 198–203.

Patra, A.K. (2006). Influence of wind speed profile and roughness parameters on the downwind extension of vulnerable zones during dispersion of toxic dense gases. *J. Loss Prev. Process. Ind.* 19 (5), 478–480.

Reniers, G., Van Lerberghe, P., & Van Gulijk, C. (2015). Security risk assessment and protection in the chemical and process industry. *Process Saf. Prog.* 34 (1), 72–83.

Spoelstra, M., Mahesh, S., Kooi, E., & Heezen, P., (2015). Domino effects at LPG and propane storage sites in The Netherlands. *Reliab. Eng. Syst. Saf.* 143, 85–90.

Tsai, S.-F., Huang, A.-C., & Shu, C.-M. (2018). Integrated self-assessment module for fire rescue safety in a chemical plant – A case study. *J. Loss Prev. Process. Ind.* 51, 137–149.

Zhang, H.D., & Zheng, X.P. (2012). Characteristics of hazardous chemical accidents in China: a statistical investigation. *J. Loss Prev. Process. Ind.* 25 (4), 686–693.

Zhou, Y. & Liu, M. (2012). Risk assessment of major hazards and its application in urban planning: a case study. *Risk Anal.* 32 (3), 566–577.

8 Chemical Releases in a Semiconductor Plant
Emergency Response Study

8.1 INTRODUCTION

A large number of chemicals and gases are used in the semiconductor and photoelectric panel plants. Although sealed rooms are usually used as the maturating areas, nonetheless a leak or emission would spread into all such rooms via the internal air-conditioning system. Hence, a large number of health hazards are said to be associated with the operation of such plants. In order to minimize the probability of any hazardous incident occurrence, few allowable measures at par with environmental protection are required (Tseng et al., 2008; Wang et al., 2009; Yun et al., 2007) as well as the poisonous gases commonly used in such processes have been listed in Table 8.1.

Large tanks and steel cylinders are used to supply and store the toxic gases used in the semiconductor and photoelectric panel industries. It was suggested by a few researchers that each plant required its own individual ERP system because under severe emergency conditions, even experienced industry personnel might be unable to tackle serious problems. Hence, it is necessary for an organization to have proper communication in order to deal with such emergency situations (Ramabrahmam et al., 1996). It was further suggested by Ramabrahmam et al. (2000) that a disaster management plan would always be required, even though a well-planned plant might make the correct use of personal protective equipment (PPE). Safety operating procedures for its staff and working equipments were explained by citing the scenario of chlorine leakage wherein a Hazard and operation stability for Emergency response process (ERP) was required. Each of these analyses studied the rescue and evaluation, the usage of PPE, the reporting of accidents and the provision of training and study for handling emergency situations (Ramabrahmam and Swaminathan, 2000; Tseng et al., 2008). In order to expand productivity or for routine replacement, the process industry would always demand the assembly and disassembly of machinery and equipment. Thereby, giving rise to the probability of specialty gas leakage which

TABLE 8.1

List of Gases Commonly Used in Semiconductor Facilities With Potential Health Hazards Commonly Used in Semiconductor Facilities

Material Chemical Formula	Application	TLV (Threshold Limit Value)	IDLH (Immediately Dangerous to Life or Health)	Potential Health Hazard
HCl	Deposition	5 ppm	100 ppm	Strongly corrosive and irritable to skin and eyes.
AsH$_3$	Diffusion, ion implantation	0.05 ppm	6 ppm	Damage to livers, kidneys and cells; carcinogen.
B$_2$H$_6$	Diffusion, ion implantation	0.1 ppm	40 ppm	Strong stimulus and damage to respiratory and central Nervous systems, livers and kidneys.
SiCl$_2$H$_2$	Diffusion, ion implantation	0.5 ppm	–	Irritable to mucous membrane of the respiratory tract.
PH$_3$	Diffusion, deposition, ion implantation	0.3 ppm	200 ppm	Dizziness, coma and liver damage.
SiH$_4$	Diffusion, ion implantation	5 ppm	–	Nausea and irritable to the wind pipe.
POCl$_3$	Deposition, ion implantation	0.1 ppm	–	Irritable, chest pain sand liver and kidney damage.

(Data obtained with permission from Lin et al. Copyright © Elsevier, 2009)

may be due to the expansion of new machines for increased productivity, the expansion or replacement of pipelines, the inappropriate handling and transfer of gas-filled cylinders, the deterioration of supply equipment and a lack of proper maintenance. The occurrence of such an incident would cause fires and explosions, thereby suspending production and risking many lives in an attempt to control the incident.

Several researchers have already suggested that it is absolutely essential to train rescue personnel to handle disaster situations, including the provision of medical services as required (Ford and Schmidt, 2000). The efficient way to do so is to systematically approach the extent and depth of training, emphasizing that it is always necessary for rescue personnel to revise their skills in order to enhance their ability to actually apply things in real-life situations as per the induction training, and to work together as a team in order to properly understand the disaster scenario and implement their ERP training successfully (Ford and Schmidt, 2000). It has also been suggested that without proper cooperation and amicability between the teams responsible for rescue operations, the mission in itself would be a big failure. Hence, human errors, as a result of pressure to carry out a mission in a complex plant such as a semiconductor plant, must also be considered as a liability for unsuccessful missions (Carol et al., 2002; Gangopadhyay and Das, 2008).

8.2 SEMICONDUCTOR PROCESS OVERVIEW

Semiconductors may have integrated circuits with diodes and transistors on wafers. These are implemented onto the wafers by processes such as etching, diffusion, and ion implantation. Wafers are further made up of smaller chips to achieve a certain function. The basic procedure for the fabrication of a wafer is shown in Figure 3.1.

The thin film type of semiconductor is mainly composed of a variety of thicknesses and materials. These thin films are embedded onto the surface of the chips through various techniques such as thin film deposition and thin film growth. Thin film deposition is mainly of two types: physical vapor deposition and chemical vapor deposition. Different patterns can also be cast on the wafer surface by a process

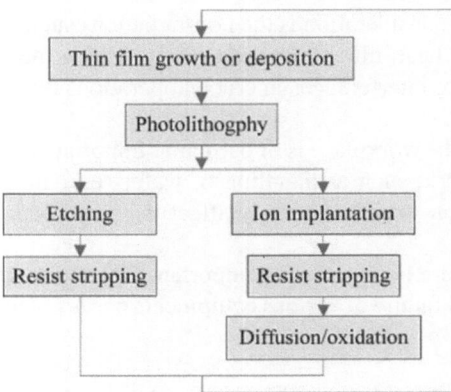

FIGURE 8.1 Wafer fabrication procedure for semiconductor circuit (Reproduced with permission from Lin et al. Copyright © Elsevier, 2009).

known as photolithography. Etching is another important process required in a semiconductor industry and is basically of two types: viz. dry etching and wet etching.

The specialty of a semiconductor is its ability to conduct electricity between a conductor and a non-conductor. This process is mainly impacted by the energy gap and the presence of doping agents dopants. It has been observed that the dopants are able to generate an energy gap equivalent to the level of silicon. There are several methods approached for the doping, which mainly include diffusion, ion implantation wherein an ion beam is used to ionize the dopant that would be embedded in the semiconductor. The concentration of the ion to be embedded can be controlled by controlling the ion beam current; there is also the option of adjusting the distribution of the dopants. The wafers are cleansed during the wet process to remove various dirt and contaminants that may deposit during the processes of diffusion, oxidation, etching, chemical vapor deposition or physical vapor deposition.

8.3 HAZARDS OF THE SEMICONDUCTOR INDUSTRY

It has been observed that the semiconductor industry posed serious threats of fatality or possible occupational injury due to the use of the various chemicals, machinery and the operating environment leading to various forms of hazards such as physical, chemical or those as a result of human engineered flaws (Chang, 2008)

(1)   The extensive usage of a wide variety of chemicals might result in explosion or fire as a result of exhaust gas venting, the maintenance of the chemical heat sink or the failure of the waste gas treatment system.

(2)   Any machinery which deals with the use of toxic chemicals such as hydrofluoric acid (HF) or toxic by-products may expose serious threats for explosion or fire during maintenance, operation, disassembly and design errors.

(3)   Radioactive radiations also pose some serious threats in the semiconductor industries. This includes, in particular, the ion-implanter, which yields ion beams using highly radioactive elements.

(4)   Another crucial consideration is the non-radiations such as radio frequency, infrared radiant heat, ultraviolet radiant heat which may hamper the eyes and skin or cause interferences in crucial operations by creating sparks and static electricity.

(5)   Ergonomics in the workplace is of particular importance. The repetitive performance of work, such as handling of wafers etc., may result in muscle discomfort which would ultimately affect the performance of the workers in the long run.

(6)   Mechanical failure is also another important aspect that needs to be considered. Functional failure of various equipments poses potential hazards to the operators.

Different hazards existing in a semiconductor plant has been classified in Table 8.2 (Chang, 2008; Shih and Hwang, 1997) along with the different process areas and the respective characteristics.

TABLE 8.2

Potential Process Hazards Exposed in Semiconductor Industry

Hazard Process	Chemical Hazard				Physical Hazard					Human Factors Engineering Hazard
	Irritable/Toxic Gas	Metal	Acid/Alkaline Solution	Organic Solvent	Radiation	RF	UV	High Temp	Noise	
Photoresist/photolithography	–	–	–	–	–	–	–	–	⊕	⊕
Exposure	–	–	–	–	–	–	⊕	–	⊕	⊕
Dry etching	–	–	⊕	–	–	⊕	⊕	–	⊕	⊕
Wet etching	–	–	⊕	–	–	–	–	–	⊕	⊕
Furnace	⊕	–	⊕	–	–	–	–	–	⊕	⊕
Vapor deposition	⊕	⊕	⊕	–	–	⊕	⊕	–	⊕	⊕
Ion implantation	⊕	⊕	⊕	–	⊕	–	–	–	⊕	⊕

Remarks: ⊕ refers to be effective; RF: radio frequency; UV: ultraviolet radiant; –: not applicable. (Data obtained with permission from Lin et al. Copyright © Elsevier, 2009)

8.4 CHEMICAL HAZARDS IN A SEMICONDUCTOR PLANT

Numerous specialty gases are utilized in a semiconductor industry, which mainly find application in the film deposition, ion implantation, dry etching, etc. It has already been specified that such specialty gases are highly corrosive and toxic in nature with a great affinity towards inflammability and explosions. Some gases may have combination behaviors like being toxic and corrosive or toxic and flammable at the same time. The material safety data sheet (MSDS) of such specialty gases plays an important role in understanding the safety precautions needed to prevent major incidents. Table 8.3 can be used as a reference for the same.

A few of the specialty gases are as mentioned below:

(1) Flammable gases such as B_2H_6, SiH_2Cl_2, PH_3, NH_3, CO, etc.
(2) Pyrophoric gases such as SiH_4, PH_3, Si_2H_6, etc.
(3) Oxidizing gases such as N_2O, O_2, NF_3, Cl_2, etc.
(4) Corrosive gases such as Cl_2, HBr, BCl_3, WF_6, PH_3, NH_3, etc.
(5) Toxic gases such as SiH_4, SiH_2Cl_2, PH_3, Si_2H_6, ClF_3, WF_6, etc.
(6) Inert gases such as N_2, CF_4, C_2F_6, C_4F_8, SF_6, CO_2, Ne, Kr, He, etc.

8.5 EMERGENCY RESPONSE PROCEDURES

In determining the safe working distances to be observed in a semiconductor industry, the Emergency Response Guidebook (ERG) is one of the key texts. It basically describes a module wherein the safety working distances could be determined in a chemical plant disaster under different scenarios. As per the reference in ERG, the safety working distance without PPE becomes 87 m and if the rescue personnel are equipped well with PPE, the safe distance comes between 32 m and 87 m at the initial stage of the explosion or fire. In addition, any combustible material near the accident area must be removed at the earliest opportunity. It has been suggested that a distance of 22.5 m is good enough to prevent a catastrophe; however, the inability to do so results in the severe domino effect and thereby a second accident. Under the conditions of a second incident, the safety working distance without PPE is 145 m and with PPE ranges between 37 m and 145 m. Also, any combustible material within a 29 m vicinity of the second accident may result in a third accident if proper precautions are not taken into consideration.

SAFETI simulations for a pre-assessment model of an incident appear to be of great help for those in charge of the plant. The simulations not only helps to get a hold of all the preventive measures required in order to handle a emergency incident or a domino effect but also helps in providing the safety working distance for the fire rescue personnel, thereby improving their efficiency and reducing the number of casualties. The following need to be implemented to achieve the same.

Removal of any combustible or flammable materials from the accident area, subduing ignition using foam, fire-fighting, water jet cooling etc. In addition, the staff and any vehicles must be directed to a safe assembly point which would eliminate the likelihood of severe casualties.

TABLE 8.3
Comparisons of Gas Characteristics

Chemical Formula	Appellation	Flammable	Non-Flammable	Inert	Toxicity	Oxidation	Corrosiveness
H_2	Hydrogen	□					
N_2	Nitrogen		□	□			
O_2	Oxygen		□			□	
He	Helium gas		□	□			
CH_4	Methane	□					
Ar	Argon			□			
NO	Nitric oxide				□	□	
NF_3	Nitrogen trifluoride				□	□	□
SiF_4	Silicon hydride				□		□
WF_6	Tungsten hexafluoride				□		□
PH_3	Phosphine	□			□		
B_2H_6	Diborane	□			□		
SiH_4	Silane	□			□		
GeH_4	Germanium tetrahydride	□			□		
NH_3	Ammonia	□			□		
SF_6	Sulfur hexafluoride		□	□			
HCl	Hydrogen chloride				□		□
Cl_2	Chlorine				□	□	□
HBr	Hydrogen bromide				□		□
N_2O	Nitrous oxide						
BCl_3	Boron trichloride				□		□
SiH_2Cl_2	Dichlorosilane	□			□		□
CHF_3	Trifluoromethane		□	□			
C_2F_6	Hexafluoroethane		□	□			

(Data obtained with permission from Lin et al. Copyright © Elsevier, 2009)

8.6 PROBLEMS FACED IN AN EMERGENCY RESPONSE

The emergency response required to tackle an incident varies from industry to industry. Hence, different ERP have to be adopted to tackle an actual circumstance that may arise as a result of incidents. Tagging a definite ERP to a specific type of incident would help the rescue units to respond to emergencies quickly and in an effective manner. Basic knowledge and judgments required to tackle such a situation must be taught to the people in charge of handling the emergency situation.

The different types of emergency response procedures are as follows

(1)   In order to report an emergency a brief report should be prepared which would include the unit name, the name and extension number of the reporting person, the time and place of accident and type of assistance required, if any.

(2)   Various monitoring systems are available in a plant to detect an emergency, such as the fire monitoring system, very early smoke detection apparatus and control systems. The information related to each hazards, are collected and an evaluation is done on the same to identify different rescue approaches.

(3)   Evacuation in case of an incident plays a very important role in safeguarding the staff. The first stage of evacuation can be carried out under the supervision of the top executive whereas the second phase of evacuation is the responsibility of the commander or the deputy of ERT. Every personnel engaged during the evacuation process must be reported to the commander so that a detailed list of personnel can be accounted for.

(4)   Based on the type of accident or incident, a suitable action plan needs to be laid out to tackle the scenario. Any item which may create an increased hazard must be removed from the accident site to a safe distance. The rescue personnel must also be well equipped with PPE and proper evacuation plans in order to avoid any further mishap.

(5)   In order to handle an emergency situation an ERT is formed consisting of staff and rescue team, safety and first aid team, etc. Each of these teams have their own duties and responsibilities and are supposed to report to a commanding officer. Under each commander there would be a safety supervisor, an information officer and a few selected staff members to aid the commander to plan out the rescue operation. The tasks of the ERT include searching and saving trapped, rescue team members, safeguarding the accident site, the removal of items which pose a threat, and so on. The first aid team is mainly responsible for taking care of the people injured during an accident by providing the basic medical needs before sending them for further treatment. The safety control team is responsible for sealing and isolating the accident area and monitors the disaster area with the progress of the rescue mission. They are also responsible for setting up a decontamination area wherein rescue personnel and rescues can decontaminate themselves. A general ERT is as shown in Figure 8.2. Different zones such as the hot, warm and cold zones are shown in Figure 8.3.

(6)   Before embarking on an evacuation procedure rescue personnel must be well informed about the various hazards existing in the accident area, such

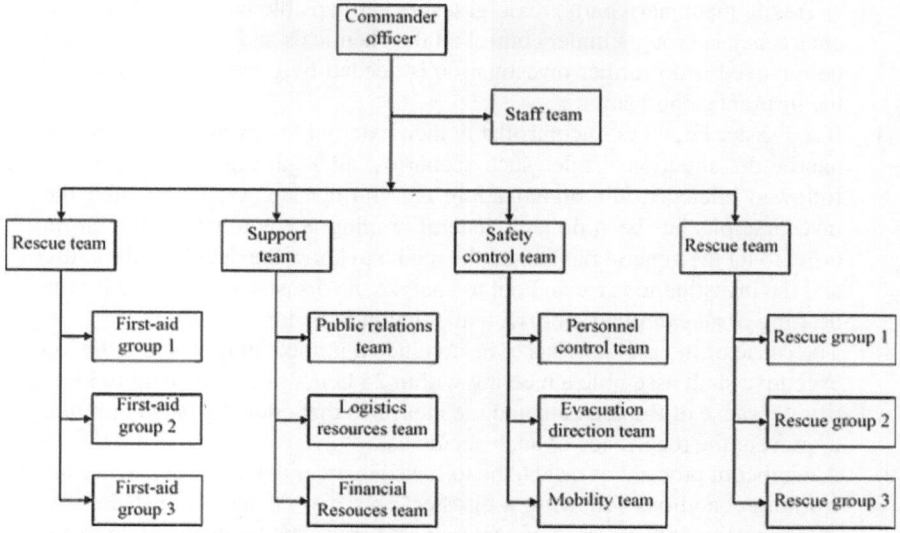

FIGURE 8.2   A hierarchic organizational chart for emergency response (Reproduced with permission from Lin et al. Copyright © Elsevier, 2009).

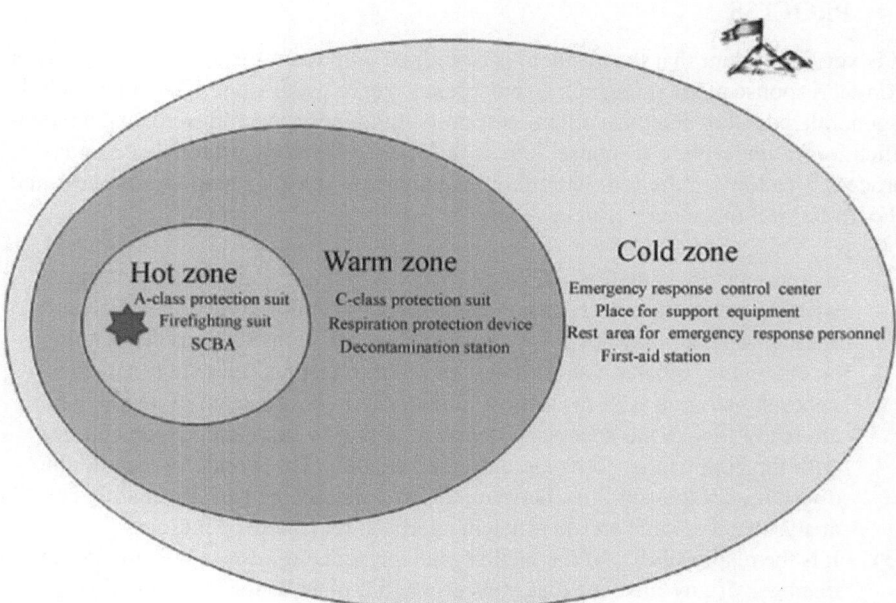

FIGURE 8.3   Schematic of zoning partition (Reproduced with permission from Lin et al. Copyright © Elsevier, 2009).

as fragile machinery parts, toxic gases and collapsible buildings. Once the emergency is brought under control all the chemicals and equipment should be removed if no further investigation is needed by government agency or the insurance company.

(7)   If a disaster becomes uncontrollable then external teams may be invited to handle the situation. Under such scenarios, all legal regulations must be followed related to the organization. It is further suggested that after the investigations has been done, a general briefing needs to be carried out in order to let the general public and the media to know the details of the cause and the investigations carried out to analyze the disaster which includes the drafting of an external report with all legal jurisdiction.

(8)   The cause of the accident has to be conducted immediately wherein the top executive shall assemble a meeting within 24 hours of the accident to identify the cause of the accident and the measures that could be taken in future to prevent the recurrence of such accidents.

(9)   A number of procedures need to be followed in order to return back to normal working conditions following a disaster. This may include re-installations, reconstruction, the cleaning of contaminated areas, the provision of compensation for injured staff or deceased staff and their families, legal proceedings that needs to be followed for the start up of the industry after an incident, and so on. All these procedures have been summarized as in Figure 8.4.

8.7 COMMON PROBLEMS DURING THE EMERGENCY RESPONSE PROCESS

It is very important that the accident is examined with the utmost diligence so that a correct response method for such an emergency can be established. They may include a generalized list of response orders, response guidelines to be followed, and an identification of underlying response defects in order to correctly establish a responsive process. In addition, the following failures may recur during a response process and should be well taken care of in advance.

(1)   Evaluating the situation of an accident is very important to determine whether or not the condition can be brought under control by an ERT team.

(2)   The commander may not always be available to give instructions on handling an emergency situation at all times. In order to respond quickly to an emergency situation, it is of the utmost importance to have a team of people who are ready to take the necessary steps and action to share the responsibilities with the commander to tackle a severe situation. The work can include the division of responsibilities between different people gather information and analyse on a reliable and an efficient approach for rescue operations.

(3)   It is the responsibility of the facility management engineer to inform the site manager of any incident that may occur in a plant at the very initial stage such that a regional ERT team may be gathered to handle the situation. If the situation is much more severe and beyond the control of the regional ERT, an immediate decision must be made on whether an external team would be required to eliminate the chances of any catastrophe.

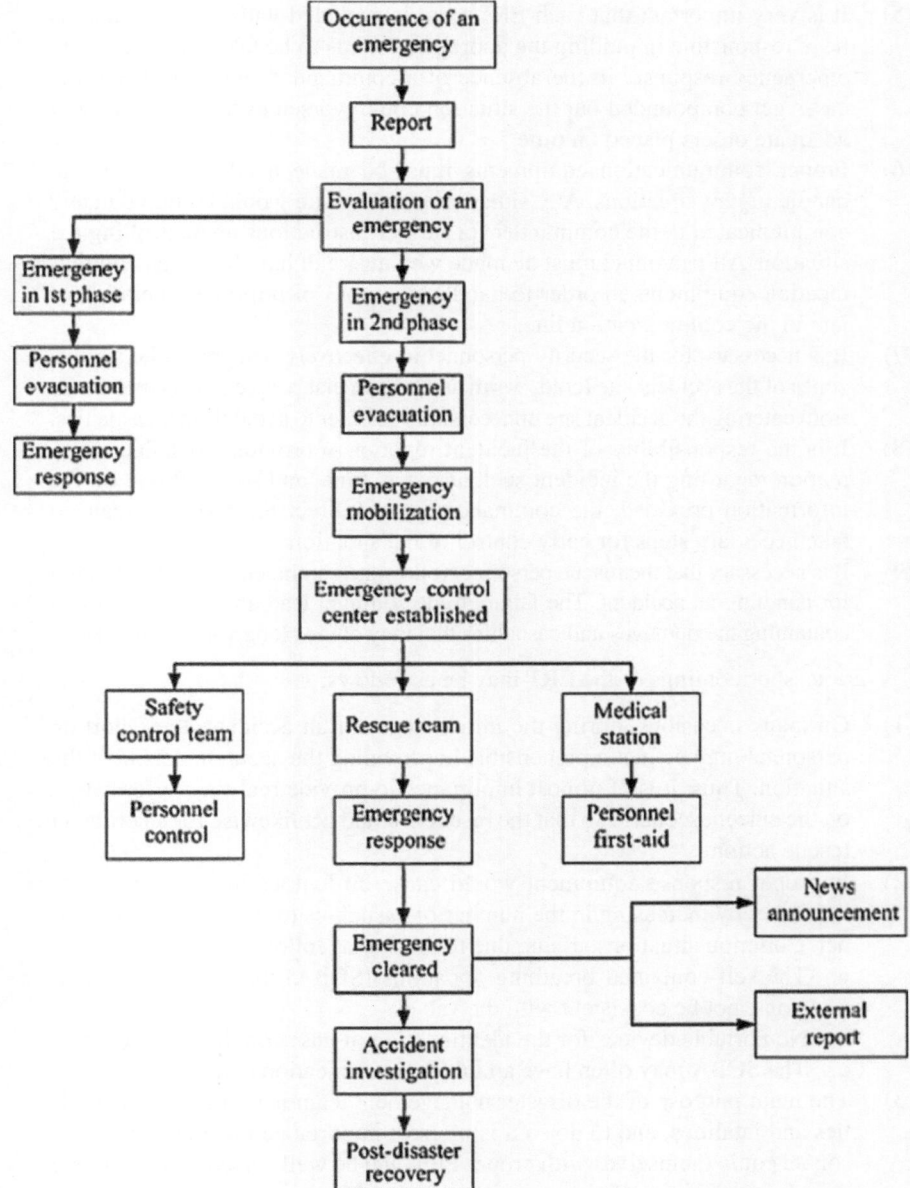

FIGURE 8.4 Emergency response procedures for high-tech plant (Reproduced with permission from Lin et al. Copyright © Elsevier, 2009).

(4) Many times due to the extreme pressure and chaotic situation in an emergency site, the rescue personnel responsible for the evacuation process may become confused with their roles, thereby delaying the overall response to the emergency scenario.

(5)   It is very important that each ERT team is provided with a commander as he is responsible in guiding the appropriate steps to be taken in tackling an emergency response. In the absence of a commander, not only would the cases get compounded but the situation could worsen even further without adequate orders placed on time.

(6)   Proper communication equipments must be made available in order to handle urgent situations. Any situation faced on site would be immediately communicated to the commander for further instructions on controlling the situation. All personnel must be made well aware of handling such communication equipments in order to handle situations of improper operation or jam in the communication line.

(7)   It is necessary for the security personnel to effectively demarcate the various zones of the accident site (cold, warm and hot) so that people may be prevented from entering the accident site unnecessarily in order to avoid further casualties.

(8)   It is the responsibility of the incident reporter to provide a detailed information regarding the incident such as place, time, and so on. Based on the information provided, the commander would direct the response team to take necessary steps for early control of the situation.

(9)   It is necessary that the rescue personnel follow the contingency actions required for handling an accident. The failure to do so might lead up to the inability in containing the damages and casualties that may come along with the incident.

Various shortcomings of the ERT may be as follows:

(1)   On many occasions, during the initial stages of an accident, the reporting personnel may be incomprehensible in providing the accurate details of the situation. Thus, it is of utmost importance to provide real-time information on the current scenario so that the rescuers could act likewise for an efficient rescue action.

(2)   Improper response equipment would cause difficulties in a rescue operation, thereby increasing in the number of casualties to the response personnel. Common situations arising due to this are as follows:
        a.   The self-contained breathing apparatus (SCBA) may, on many occasions, not be consistent with the valves.
        b.   No portable devices for the identification of gases on the leakage site.
        c.   The SCBA may often have a faulty communication connector.

(3)   The main purpose of the disaster management team is to minimize casualties and fatalities, and to do so it is of great importance that the rescue personnel equip themselves with proper PPE, and be well versed with the scene in order to elicit an efficient response.

(4)   With the expansion of high-end technology plants, the staff turnover also increases. With such a rapid increase, proper training on response procedures must be provided to the newly recruited staff so that all of them can be well prepared to handle emergencies during times of necessity.

(5)   The training of ERT staff and rescue personnel helps to carry out rescue operations more efficiently. In the absence of training on ERP, the people in charge would not be aware of their responsibilities in handling the accident scenario.

8.8 SUMMARY

As discussed in this chapter, the semiconductor industry utilizes a wide variety of gases. Based on the type of gases and chemicals used, the ERP for each type of reaction would be different. The emergency response team should always be taken as the last line of protection when a particular incident goes beyond control. Failure of machinery and equipment, and human errors, would unquestionably be always present, hence in order to limit the extent of damage the accident must be controlled at the very earliest stages. The ERP in semiconductor plants are highly complex, mainly including an overall understanding of the problem, the simulation of various scenarios to identify the possible difficulties. Once a proper evaluation is carried out, regular emergency response drills would be done without any fixed schedule to review problems that may occur in a real-life scenario and improve the procedures or steps for an improved emergency response for different simulated conditions.

REFERENCES

Carol, S., Vilchez, J.A., & Casal, J. (2002). Study of the severity of industrial accidents with hazardous substances by historical analysis. *J. Loss Prev. Proc. Ind.* 15(6), 517–524.

Chang, H.K. (2008). Emergency response procedures for specialty gas leak in high-tech plants, Master Thesis, National Yunlin University of Science and Technology, Yunlin, Taiwan, ROC.

Ford, J.K., & Schmidt, A.M. (2000). Emergency response training: strategy for enhancing real-world performance. *J. Hazard. Mat.* 75, 195–215.

Gangopadhyay, R.K., & Das, S.K. (2008). Ammonia leakage from refrigeration plant and the management practice. *Proc. Safety Prog.* 27, 15–20.

Lin, C.P., Chang, H.K., Chang, Y.M., Chen, S.W., & Shu, C.M. (2009). Emergency response study for chemical releases in the high-tech industry in Taiwan—A semiconductor plant example. *Proc. Safety Environ. Protec.* 87, 353-360.

Ramabrahmam, B.V., Sreenivasulu, B., & Mallikarjunan, M.M., (1996). Model on-site emergency plan. Case study: toxic gas release from an ammonia storage terminal. *J. Loss Prev. Proc. Ind.* 9, 259–265.

Ramabrahmam, B.V., & Swaminathan, G., (2000). Disaster management plan for chemical process industries. Case study: investigation of release of chlorine to atmosphere. *J. Loss Prev. Proc. Ind.* 13, 57–62.

Shih, T.S., & Hwang, W.W., (1997). Preliminary study of potential hazards for the semiconductor manufacturing process in Taiwan. *J. Occ. Saf. Heal.*, 24(1): 1–7. (Institute of Occupational Safety and Health (IOSH), Council of Labor Affairs, Executive Yuan, Taipei, Taiwan, ROC)

Tseng, J.M., Kuo, C.Y., Liu, M.Y., & Shu, C.M. (2008). Emergency response plan for boiler explosion with toxic chemical releases at Nan-Kung industrial park in central Taiwan. *Proc. Safety Environ. Prot.* 86, 415–420.

Wang, H., Chen, B., He, X., Tong, Q., & Zhao, J. (2009). SDG-based HAZOP analysis of operating mistakes for PVC process. *Proc. Safety Environ. Prot.* 87, 40–46.

Yun, R.L., Wan, T.J., Lin, C.H., Chang, Y.M., & Shu, C.M. (2007). Fire and explosion characteristics of 3-methyl pyridine at 270 °C with high oxygen concentrations. *Proc. Safety Prog.* 85, 251–255.

9 Thermal Hazard and Safety during Combustion of 1-Butylimidazolium Nitrate

9.1 INTRODUCTION

With the continuous expansion of industries, environmentally friendly norms for the functioning of various industries have been suggested. The use of green chemicals is one such drive to make the environment more sustainable. The ionic liquids are attracting the industries due to their special properties of low vapour pressure, and low melting and boiling points to replace other volatile solvents frequently used in various industries. Several researchers have suggested different ILs which has considerable thermal stability (Götz et al., 2015; Smiglak et al., 2006). However, despite the available advantages it has been observed that the decomposition of the ionic liquids into a gaseous phase poses threats of spontaneous combustion and ignition (Chen et al., 2014a, b; Dai et al., 2016; Wang et al., 2015). A generalized understanding on the development of ionic liquids from the above research papers can be summarized as follows (Götz et al., 2015; Smiglak et al., 2006):

(a) Thermodynamics and chemistry of the process needs more understanding.
(b) The operating conditions may be insufficient.
(c) Unavailability of safety device.
(d) Improper training of the operators.

Hence, to know more about the thermal and toxic properties of the ILs, rigorous research has been done in recent years. This in turn would help in determining the safety of a process and the safety measure required to be undertaken for the use of ILs in various industrial processes.

9.2 UNDERSTANDING IONIC LIQUIDS

ILs are very interesting due to the fact that thousands of varieties can be produced by altering the combination of cations and anions (Diallo et al., 2012). It is of utmost necessity to determine the thermal properties of the ILs prior to their use. Generally, TGA is used to determine the thermal stability of ILs as it is one of the simple measurement options easily available, which is not sufficient enough to determine the detailed thermal properties and the associated hazards of ILs. In addition to the TGA, a quantitative structure-property relationship (QSPR) can also be used to determine the thermal properties (Rybinska et al., 2016; Yu et al., 2013). QSPR cannot be taken as the ultimate analysis medium because the data obtained through the QSPR analysis are limited. Thus arises the need of a superior analysis technique which could well determine the thermal properties and give details about the hazards associated with the different ILs. The present chapter describes a methodology to determine the thermal behavior of 1-butylimidazolium nitrate ([BIM][NO$_3$]; C$_7$H$_{13}$N$_3$O$_3$). The main features of the method include the following:

(a) A novel method for evaluating the thermal stability of [BIM][NO$_3$] during the production process along with the associated hazards.
(b) Thermal hazards were determined by TGA, DSX, FTIR, GC/MS (Sharifi et al., 2013; Madria et al., 2013; Rewar et al., 2016; Pieczy'nska et al., 2015; Ying et al., 2015).
(c) Thermokinetic parameters of a thermal runaway for [BIM][NO$_3$] and the model-free prediction for an adiabatic runaway excursion was done by the Vent sizing package 2 (VSP2).
(d) Thermokinetic methods were also used to determine the safe working conditions for [BIM][NO$_3$].

9.3 EXPERIMENTAL STUDIES ON 1-BUTYLIMIDAZOLIUM NITRATE

9.3.1 APPARATUS AND MATERIALS

The thermal properties of [BIM][NO$_3$], mainly the combustion and exothermic properties, were meticulously determined using methods such as TGA, FTIR, VSP2, GC/MS, DSC and FPA. The adiabatic runaway reaction was determined based on the DSC data obtained, whereas a runaway was confirmed by the VSP2 test. Combustion properties were determined by TGA, FPA and combustion experiments. The flammability of the gases evolved during decomposition was determined by GC-MS and FTIR. A nitrate-based ionic liquid, [BIM][NO$_3$] has been used in performing all the studies (Ionic Liquids Technologies, GmbH, Salzstrasse 184. D-74076 Heilbronn, Germany).

9.3.2 Preliminary Combustion Experiment

An unsealed ampule was loaded with 0.5 g of [BIM][NO$_3$] and heated with the help of an alcohol burner. The opening of the ampoule was mounted with an ignition apparatus to determine the decomposed gases arising out of the IL and access their flammability, thereby determining the boiling point and the ignition temperature.

9.3.3 Thermogravimetry and Differential Scanning Calorimetry

The thermal properties of the ILs were determined using TGA (Perkin Elmer) with a temperature range of 30-700 °C and a heating arte of 1 and 5.6 K/min. exothermic behavior of the IL was determined by DSC analysis (Mettler Toledo) with a temperature range of 35-350°C. Each of these analyses used a 3.0 ± 0.1 mg of [BIM][NO$_3$] (Moukhina, 2015; Roduit et al., 2015; Nascimento et al., 2015). The activation energy and the pre-exponential factor was analyzed using a differential iso-conversional analysis at the heating rates of 0.5, 1, 2, 4 and 8 K/min thermokinetic parameters, such as the pre-exponential factor ($A(a)$) and apparent activation energy ($E(a)$) could be expressed by means of a differential iso conversional analysis based on five constant heating rates (0.5, 1.0, 2.0, 4.0, and 8.0 K/min) (Tseng and Lin, 2011).

9.3.4 Adiabatic Runaway Reaction – Experiment and Prediction

Thermal investigation of [BIM][NO$_3$] such as maximum rate at adiabatic conditions were determined using the VSP2 (Fauske & Associates) which helps in creating adiabatic conditions for chemical reactions. Model-free prediction was used to estimate the adiabatic runaway progress.

9.3.5 Flash Point Analyzer

The ignition temperature of gases which were generated by the decomposition of the ILs were analyzed using the flash point analyzer (FPA) (Petroleum Analyzer Company). A preliminary guess on the ignition temperature from the TGA must be provided in order to use as a reference for FPA (Liaw et al., 2012). The flash point was determined by ASTM D93A method (Liaw et al., 2014).

9.3.6 Qualitative Investigation

Chemical composition was analyzed by performing FTIR and GC/MS. Thermal composition was carried out to understand the ignition of ILs and identify the gas products. The environment of the approximate flash point was imitated by heating a small amount of [BIM][NO$_3$] (3.0 ± 0.01 g), sealed in a proof-pressure glass vessel, in an isothermal furnace (Li and Kobayashi, 2016; Balogh et al., 2015; He et al., 2015). After heating the ILs for about 90 min, 5 ml and 25 ml of the gas products produced were used for analysis in GC-MS and FTIR with operating parameters as mentioned in Table 9.1.

TABLE 9.1

Operating Parameters of Apparatus

Apparatus	Brand Name	Heating Rate ($K\ min^{-1}$)	Temperature Range (°C)	Sample Mass (mg)	Run Time (h)
TG	PerkinElmer Pyris 1 TG	1.0–5.6	30.0–350.0	3±0.1	5.5–1.0
DSC	Mettler-Toledo 821e	1	30.0–350.0	3±0.1	5.5
			Temperature range (°C)	Sample mass (g)	Run time (h)
FPA	Pensky-Martens HFP 360	5.6	30.0–197.0	3±0.1	0.5
Isothermal furnace	Carbolite CWF 1100	Isothermal	150.0–200.0	3±0.1	1.5
VSP2	Fauske & Associates, LLC	Adiabatic	120.0–350.0	3±0.1	120
			Wavenumber (cm^{-1})	Sample volume (mL)	Run time (h)
			Mass range (amu)		
FTIR	PerkinElmer Spectrum 100 FT-IR Spectrometer	–	450–4000	5000±10	0.5
GC/MS	Agilent Technologies 6890N Network GC system 5973N Mass selective detector	–	35.0–250.0	25±11	1.2

(Data obtained with permission from Liu et al. Copyright © 2018, Elsevier)

9.4 RESULTS AND DISCUSSION

9.4.1 COMBUSTION EXPERIMENT

A preliminary thermal stability test similar to the FPA was done with 0.5 g of [BIM] [NO₃] heated in an unsealed ampoule. The test helped in determining the flammability of the decomposed gases. At a temperature of 105 ºC, white smoke seems to appear which later turns into yellow on reaching a temperature of 150 ºC, with a further increase in temperature to 160-180 ºC a more intense yellow-coloured flame was produced. During each of these decomposition stages, the gas generated were regularly analyzed for their flammability with the help of an ignitor.

Tests suggested that the gases produced beyond 176 ºC produced a flash which then progressed to a flame at 200 ºC. It was, however, seen that although the same sequence followed on replicating each test yet a slight variation in the temperature was observed which was mainly due to the unstable heating. Thus, the main target of the experiment was to confirm the gases which could be ignited.

9.4.2 INHERENT THERMAL HAZARDS FOR TGA AND DSC

The graphs obtained by TGA and DSC analysis of [BIM][NO₃] were combined (Figure 9.1). The TGA graph clearly showed the mass loss in weight % with time, thereby providing details on the thermal stability of the ionic liquids. The onset of decomposition was found to be at around 125 ºC, 150–200 ºC showed a very unstable

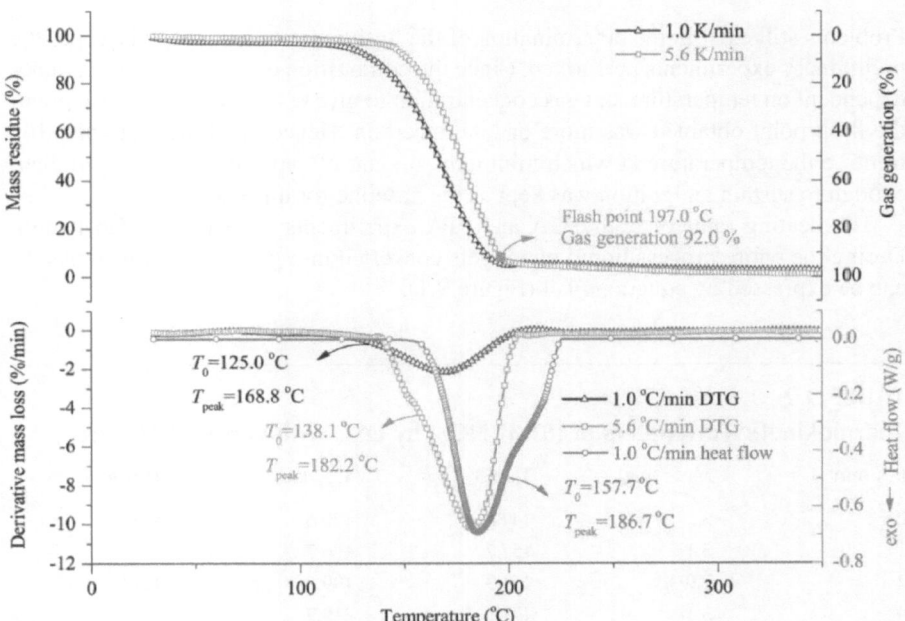

FIGURE 9.1 Heat and mass loss combined with flash point by TG, DSC, and FPA (Reproduced with permission from Liu et al. Copyright © 2018, Elsevier).

state of the [BIM][NO$_3$] and at a temperature beyond 200 °C, the liquid disappeared completely. The exothermic peak by DSC analysis was observed at 160–220 °C with the same heating rate of TGA. The dis-uniformity was mainly due to the pressure accumulation inside the sealed sample crucibles which delayed the onset of the exo-thermic reaction. Thermokinetic parameters at different heating rates are shown in Table 9.2. Tests were done at a heating rate of 0.5 K/min to examine if there was any leak in the system. It was found that negligible mass changes were observed, proving the efficiency of the machine and the analysis method.

9.4.3 FPA Test

Once the initial experiments were performed, a more precise test to determine the flash point of the generated gases needs to be done. This could be done using FPA analysis. The ASTM D93A test method was performed to determine the flash point to be at 197 °C. This temperature was not similar to the one obtained by previous experiments mainly because of difference in vapour concentration and different heat-ing rates. The gas which is ignited is in the form of a unsaturated air-vapor mixture (Albahri, 2015). The concentration of the gases changed with temperature due to the fact that reactions occurred both during the evaporation and the decomposition of the ionic liquid. It serves as one of the primary safety features to know the flash point of the ILs to know the thermal hazards associated with the handling of ILs.

9.4.4 Concentration of Ignition

Problems still exist in the determination of the flash point of ionic liquids as per the preliminary experiments performed. Since the combustion experiments were mainly dependent on temperature and gas concentration to give results about the flash point, the flash point obtained was more or less uncertain. Hence, the lower flammability limit i.e the temperature at which minimum amount of vapor produced is sufficient enough to sustain an ignition was kept as the baseline for this test.

The heating rates of both TGA and FPA experiments were kept at 5.6 K/min. During the entire process liquid phase gets converted into gas phase. This transition can be expressed by equations 1-4 (Figure 9.1).

TABLE 9.2
Thermokinetic Parameters of [BIM][NO$_3$] by DSC with Various β Values

β(K min^{-1})	Mass (mg)	T$_o$ (°C)	T$_{peak}$ (°C)	ΔH (J g^{-1})
0.5	3	147.6	170.6	1464.1
1	3.1	157.7	186.7	1452
2	2.9	164.3	199.7	1467.9
4	3.1	171.8	210.9	1480.2
8	3.1	182.6	223.8	1463.1

(Data obtained with permission from Liu et al. Copyright © 2018, Elsevier)

$$m_{ILs,im} - m_{ILs,rm} = m_{ILs,gp} \tag{9.1}$$

$$\frac{dm_{ILs,im}}{dT} - \frac{dm_{ILs,rm}}{dT} = \frac{dm_{ILs,gp}}{dT} \tag{9.2}$$

$$0 - \frac{dm_{ILs,rm}}{dT} = \frac{dm_{ILs,gp}}{dT} \quad 6 \tag{9.3}$$

$$\frac{dm_{ILs,gp}}{dT} = -\frac{dm_{ILs,rm}}{dT} \tag{9.4}$$

Where, $m_{ILs,im}$= Initial mass of ionic liquid
$m_{ILs,rm}$= Residual mass of ionic liquid
$m_{ILs,gp}$=Mass of gas produced

Equation (9.1) is basically a general hypothesis on the generation of gases which, in combination with the results obtained from FPA analysis, estimated with 92% confidence of gas production by the decomposition of [BIM][NO₃]. A mass of 2.76 g of gas is said to be generated at the initial flash occurrence. Assuming a completely sealed method and that no gas escaped for the FPA analysis, it was found that 27.6 g/L of gas was produced as a result of decomposition of [BIM][NO₃]. The LFL for the same experiment were also calculated. The gas production was calculated as $\int_{T_2}^{T_1} (dm_{ILs,g}/dT)dT$, where T_2 was the flash point temperature and T_1 was the self-heating onset temperature since the experiments were performed in an open vessel. The simulated data for the two methods are shown in Table 9.3 and the heating process is shown in Figure 9.2. The flash point was said to occur with maximum heating rate, whereas the real LFL value was said to be much lower than the obtained value of 27.6 g/L.

9.4.5 Prediction of Adiabatic Runaway Reaction

A runaway reaction is an important detail that needs to be determined to underline the precautions associated with an ionic liquid. It is basically a condition where an initial increase in temperature results in the further increase of the same. For [BIM][NO₃] the runaway reaction was determined by using 5 varying heating rates, ranging from 0.5-8 K/min. The corresponding results were used in the determination of $E(\alpha)$ and $A'(\alpha)$

$$A'(\alpha) = A(\alpha) f(\alpha) \tag{9.5}$$

TABLE 9.3
Simulation Results of LFL Data

Test Method	Heating Rate	Flash Point (°C)		LFL (g L⁻¹)
		T_{flash}	T_{onset}	
1 st combustion test	728.7	177.2	155.8	22.2
2nd combustion test	254.5	179.6	169.3	19.3
3rd combustion test	401.8	190.4	172.8	19.7

(Data obtained with permission from Liu et al. Copyright © 2018, Elsevier)

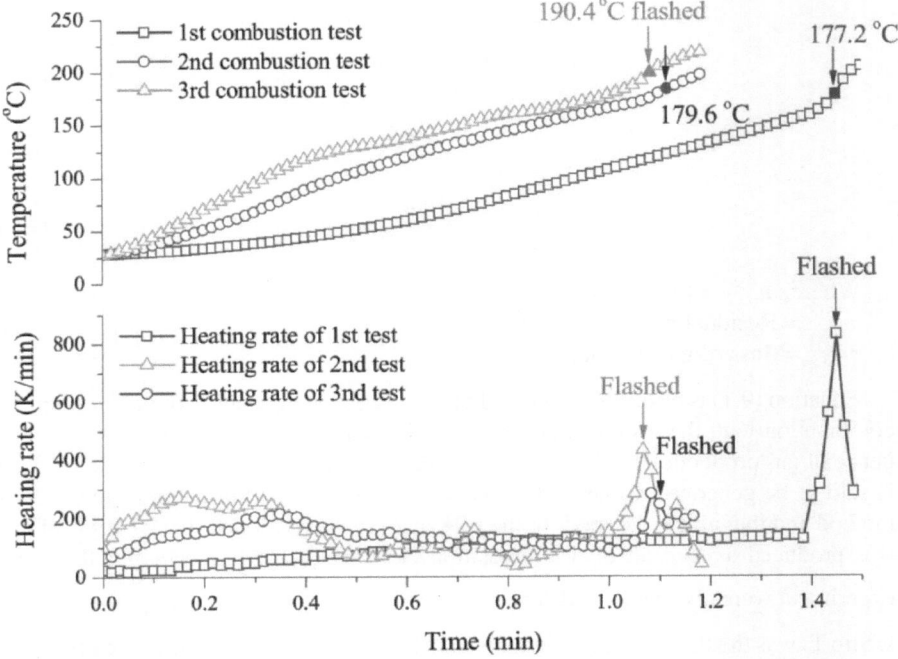

FIGURE 9.2 Preliminary combustion experiments with three times on heating process (Reproduced with permission from Liu et al. Copyright © 2018, Elsevier).

$$\frac{d\alpha}{dt} = A'(\alpha)\exp\left(\frac{-E(\alpha)}{RT(t)}\right) \tag{9.6}$$

$$\ln\left(\frac{d\alpha}{dt}\right)_{\alpha i} = \ln\left(A'(\alpha)\right)_{\alpha i} - \left(\frac{E(\alpha)}{RT(t)}\right)_{\alpha i} \tag{9.7}$$

where, αi = Conversion from 0 to 1

The method assumes both the parameters to be variables without a definite $f(\alpha)$ and the obtained values are shown in Figure 9.3. The model-free prediction is expressed in Equation (9.8):

$$t_{\alpha i} = \int_{\alpha i}^{\alpha 0} \frac{d\alpha}{A'(\alpha)\exp\left(\dfrac{-E(\alpha)}{RT(t)}\right)} \tag{9.8}$$

$$\Phi = \frac{M_c C_{p,c} + M_s C_{p,s}}{M_s C_{p,s}} \tag{9.9}$$

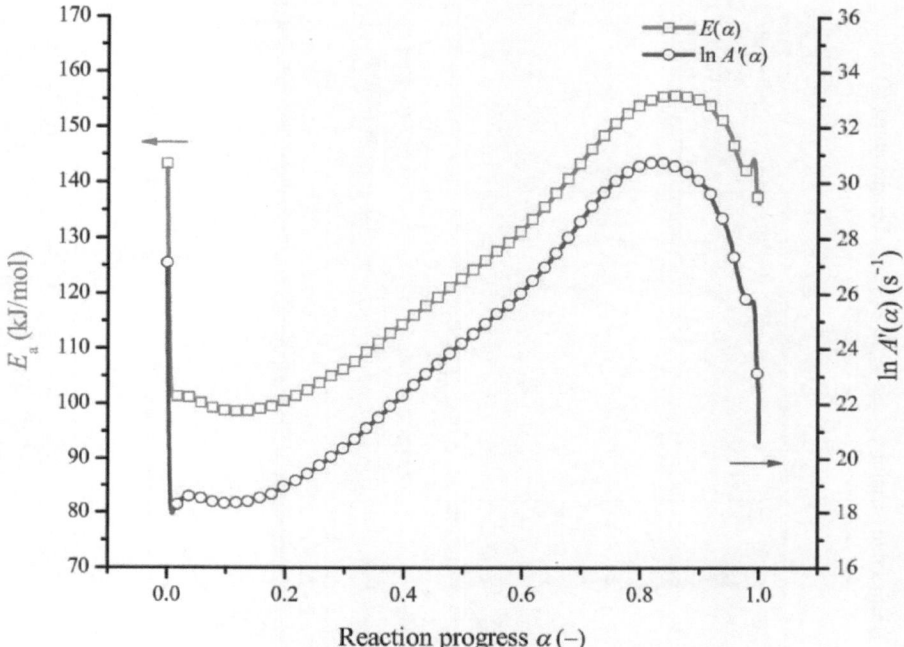

FIGURE 9.3 Thermokinetic parameters for $E(\alpha)$ and $\ln A'(\alpha)$ by iso-conversional method (Reproduced with permission from Liu et al. Copyright © 2018, Elsevier).

$$\Delta T_{ad} = \Phi \frac{-\Delta H}{C_{p,s}} \qquad (9.10)$$

$$\frac{dT}{dt} = \Delta T_{ad} \frac{d\alpha}{dt} \qquad (9.11)$$

The initial temperature was first fixed under adiabatic conditions and then the temperature was increased gradually. The values of ΔT_{ad} and dT/dt were expressed as per equations (9.10, 9.11). For the determination of runaway reactions under adiabatic conditions the VSP2 was used wherein around 3 g of [BIM][NO$_3$] was taken in a stainless steel container with a stirrer. The gas produced as a result of decomposition of [BIM][NO$_3$] and by the VSP2 test had a specific heat capacity values of 2 and 0.5 J/(gK). Calculating the Φ value using Equation (9.9) gives a value of 5.57. To get close to the process condition an extremely low value of Φ which means a corresponding low mass of IL taken in the test cell or the test cell itself must be of low weight. However, experiments have suggested having a thin cell was not viable as such a test cell cannot withstand the high pressure during the runaway reaction. Thus reducing the sample mass was considered and the runaway reaction was carried out as per the normal heat–wait–search (H–W–S) procedure (Figure 9.4) with an initial temperature of 120.0°C. The exothermic temperature (T_{onset}) was found to be 180 °C

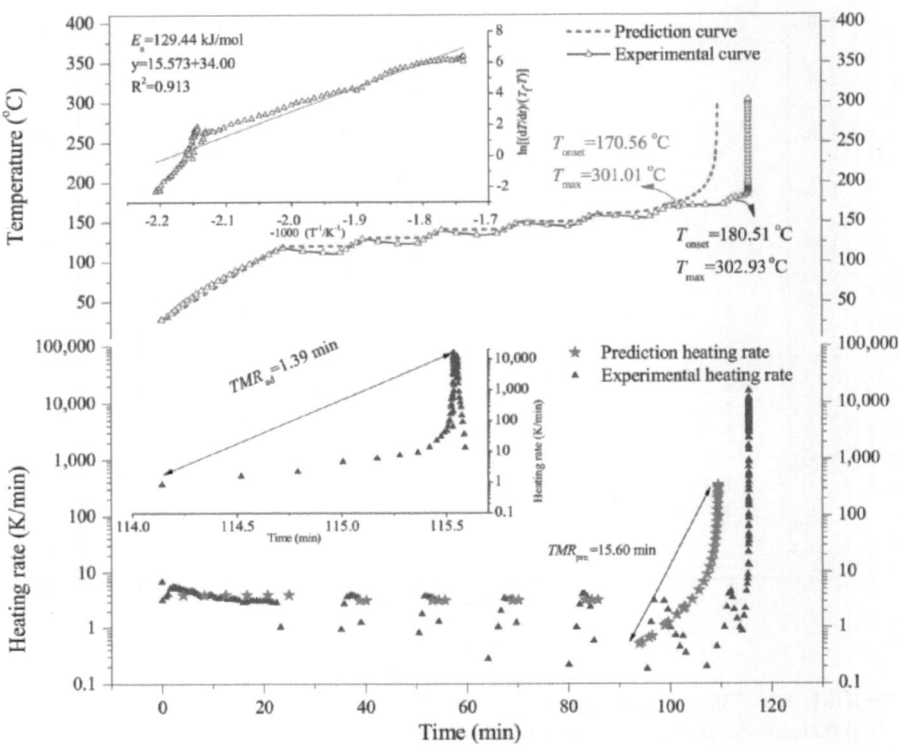

FIGURE 9.4 Prediction and experiment for adiabatic runaway curves followed by H–W–S procedure (Reproduced with permission from Liu et al. Copyright © 2018, Elsevier).

TABLE 9.4
Simulation Results of LFL Data

			Thermokinetic Parameters		
Methods	T_{onset}(°C)	T_{max}(°C)	**TMR (min)**	E_a(kJ mol⁻¹)	**A (s⁻¹)**
Prediction	170.56	301.01	15.6	125.06	3.3×10^{12}
Experiment	180.51	302.93	1.36	129.44	5.83×10^{14}

(Data obtained with permission from Liu et al. Copyright © 2018, Elsevier)

as per the VSP2 equipment, whereas the predicted value was 170.65 °C. Similarly the predicted and experimental T_{max} were 301.01°C and 302.93°C, respectively. Heating rate vs. time graphs for predicted and experimental values are shown in Figure 9.4. TMR_{ad} value was found to be significantly different for experimental and predictive values at 1.39 min and 15.60 min, respectively. Such a difference was mainly accounted for by the inconsistent exothermic temperature. Thus, as per the current experiments, the exothermic temperature changed with time unless the rate of change of temperature was greater than 0.2 K/min.

TABLE 9.5
Estimation of SADT with the Different Masses of the Package

Package Size	SADT (°C)
50	91.4
100	90.8
200	90.1
300	89.6
500	88.8

(Data obtained with permission from Liu et al. Copyright © 2018, Elsevier)

The Townsend and Tou approach was used for the determination of apparent activation energy.

$$\ln k = \ln \frac{dT/dt}{C_0^{n-1}\left(\frac{T_{max}-T}{T_f-T_0}\right)^n (T_{max}-T_0)} = lnA - \frac{E_a}{RT} \quad (9.12)$$

$$\ln \frac{dT/dt}{(T_{max}-T)} = lnA - \frac{E_a}{RT} \quad (9.13)$$

Equation (9.12) can be shown as in Figure 9.4 considering n=1. Table 9.4 shows all the thermokinetic parameters determined during the present experiment.

9.4.6 Estimating safety limits

Advanced Thermal Analysis Software was used to determine the self-accelerating decomposition temperature (SADT) and TMR_{ad}. SADT is the lowest temperature at which a self-accelerating decomposition occurs in a container. It is thus a very important parameter as per the UN "Recommendations on the Transport of Dangerous Goods". For the purpose of calculation, different parameters were considered and the results obtained are enlisted in Table 9.5. A cylinder package of masses 50-500 kg was used with a height to diameter ratio of 1.6, a density of 1000 kg/m³, specific heat capacity of 2 J/g K and a heat transfer coefficient of 5 W/ m² K.

The mass of the sample did not have a profound effect in altering the results. SADT studies suggested values greater than 90 °C; hence [BIM][NO₃] is not tagged as self-reacting. However, since the SADT is greater than 75 °C for a 50 kg pack, a safe use of the IL must be carried out keeping in mind the different health hazards associated with it. Another important parameter, TMR_{ad}, was also calculated as described in section 3.5, but the initial temperature was set with a definite value. Hence the initial temperature was a specifically designated value instead of being the variable T_{onset} (Figure 9.5). As per the experiments, it was found that, when the initial temperature is kept at 108-109 °C, the maximum reaction temperature under adiabatic conditions could well be reached within a day's time.

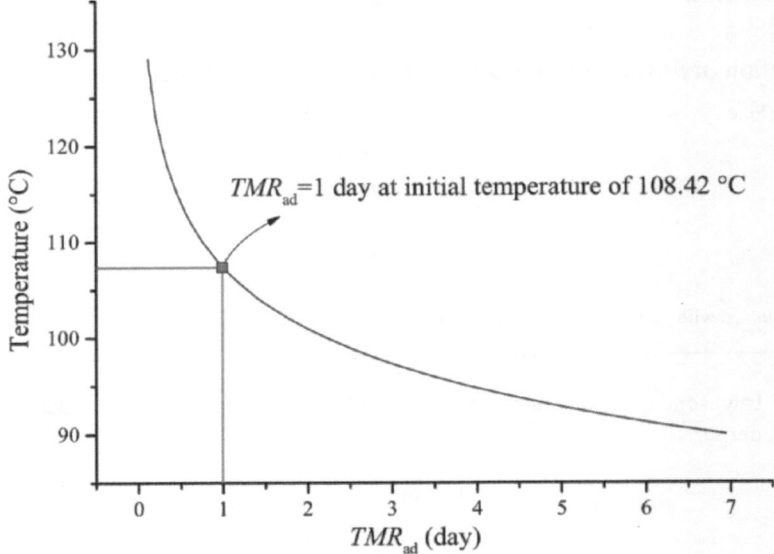

FIGURE 9.5   The estimated result of **TMR_{ad}** with different initial temperature (Reproduced with permission from Liu et al. Copyright © 2018, Elsevier).

9.5 SUMMARY

The chapter mainly focused on determining the thermal properties of [BIM][NO₃], which is considered to be a very promising ionic liquid. The GC-MS and FTIR studies suggested that a large quantity of C, H and O were the main elemental composition of [BIM][NO₃], which rightly explained the combustible nature of the gases produced. Unstable nature was reported at 160 °C, which is said to promote the production of toxic gases. An enthalpy of 1480 J/g was reported during the DSC analysis. A drastic runaway reaction was detected at a temperature of 180 °C and up to 302 °C. This was mainly detected by VSP2 and a conclusion that a detail on the runaway reaction could be well explained by both VSP2 and DSC analysis. The SADT and TMR$_{ad}$ were calculated as 91.5 °C and 108. 4 °C, respectively. The properties may vary with changes in the type of ionic liquids, hence ILs in general must not be considered as a safe solvent. To ensure their thermal stability and safety a combination of tests must be performed which will include TGA, DSC, FPA< VSP2, FTIR etc.

REFERENCES

Albahri, T.A., 2015. MNLR and ANN structural group contribution methods for predicting the flash point temperature of pure compounds in the transportation fuels range. *Process Saf. Environ. Prot.* 93, 182–191.

Balogh, R.K., et al., 2015. Determination and quantification of 2'-O-fucosyllactose and 3-O-fucosyllactose in human milk by GC–MS as O-trimethylsilyl-oxime derivatives. *J. Pharm. Biomed. Anal.* 115, 450–456.

Chen, Y.T., et al., 2014a. Auto-ignition characteristics of selected ionic liquids. *Procedia Eng.* 84, 285–292.

Chen, Y., et al., 2014b. Fabrication of clean nanostructured metal materials on ionic liquid/water interface. *Mater. Lett.* 132, 153–156.

Dai, Z., et al., 2016.Combination of ionic liquids with membrane technology: a new approach for CO_2 separation. *J. Membr. Sci.* 497, 1–20.

Diallo, A.O., et al., 2012. Revisiting physico-chemical hazards of ionic liquids. *Sep. Purif. Technol.* 93, 228–234.

Götz, M., et al., 2015. Long-term thermal stability of selected ionic liquids in nitrogen and hydrogen atmosphere. *Thermochim. Acta* 600, 82–88.

He, Q., et al., 2015. TG-GC–MS study of volatile products from Shengli lignite pyrolysis. *Fuel* 156, 121–128.

Li, K., Kobayashi, T.A., 2016. A FT-IR spectroscopic study of ultrasound effect on aqueous imidazole based ionic liquids having different counter ions. *Ultrason. Sonochem.* 28, 39–46.

Liaw, H.J., et al., 2012. Relationship between flash point of ionic liquids and their thermal decomposition. *Green Chem.* 14, 2001–2008.

Liaw, H.J., et al., 2014. Effect of heating temperature on the flash point of ionic liquids. *Procedia Eng.* 84, 293–296.

Liu, S.-H., Lin, W.-C., Xia, H., Hou, H.-Y., Shu, C.-M., 2018. Combustion of 1-butylimidazolium nitrate via DSC, TG, VSP2, FTIR, and GC/MS: An approach for thermal hazard, property and prediction assessment. *Proc. Safety Environ. Protec.* 116, 603–614.

Madria, N., et al., 2013. Ionic liquid electrolytes for lithium batteries: synthesis, electrochemical, and cytotoxicity studies. *J. Power Sources* 234, 277–284.

Moukhina, E., 2015. Thermal decomposition of AIBN part C: SADT calculation of AIBN based on DSC experiments. *Thermochim. Acta* 621, 25–35.

Nascimento, L.C.S., et al., 2015. Thermal study and characterization of nicotinates of some alkaline earth metals using TG–DSC–FTIR and DSC-system. *Thermochim. Acta* 604, 7–15.

Pieczy´nska, A., et al., 2015. A comparative study of electrochemical degradation of imidazolium and pyridinium ionic liquids: a reaction pathway and ecotoxicity evaluation. *Sep. Purif. Technol.* 156, 522–534.

Rewar, S., et al., 2016. Polybenzimidazole based polymeric ionic liquids possessing partial ionic character: effects of anion exchange on their gas permeation properties. *J. Membr. Sci.* 497, 282–288.

Roduit, B., et al., 2015. Thermal decomposition of AIBN, Part B: simulation of SADT value based on DSC results and large scale tests according to conventional and new kinetic merging approach. *Thermochim. Acta* 621, 6–24.

Rybinska, A., et al., 2016. Filling environmental data gaps with QSPR for ionic liquids: modeling n-octanol/water coefficient. *J. Hazard. Mater.* 303, 137–144.

Sharifi, A., et al., 2013. Ionic liquid [b_{mim}][NO_3], an efficient medium for green and one-pot synthesis of benzo thiazinones at room-temperature. *Sci. Iran.* 20, 555–560.

Smiglak, M., et al., 2006. Combustible ionic liquids by design: is laboratory safety another ionic liquid myth? *Chem. Commun.* 24, 2554–2556.

Tseng, J.M., Lin, Y.F., 2011. Evaluation of a tert-butyl peroxy benzoate runaway reaction by five kinetic models. *Ind. Eng. Chem. Res.* 50, 4783–4787.

Wang, Y., et al., 2015. Adsorption of imidazolium-based ionic liquids from aqueous solution onto cellulose-derived activated carbon materials. *J. Environ. Chem. Eng.* 3, 2426–2434.

Ying, A.G., et al., 2015. Novel multiple-acidic ionic liquids: green and efficient catalysts for the synthesis of bis-indolyl methanes under solvent-free conditions. *J. Ind. Eng. Chem.* 24, 127–131.

Yu, G., et al., 2013. QSPR study on the viscosity of bis (trifluoromethylsulfonyl) imide-based ionic liquids. *J. Mol. Liquid* 184, 51–59.

10 Safety and Flammability Analysis for Fuel–Air– Diluent Mixtures Plant

Safety and Flammability Analysis

10.1 INTRODUCTION

A series of explosions blamed on propylene leaks from underground pipelines killed 32 people and injured more than 300in the southern Taiwanese city of Kaohsiung in 2014. The explosions resulted in a series of major fires and significantly damaged property and roads. Flammability limits are key characteristics applied to determine the fire and explosion (F&E) dangers of gases and vapors. The lowest and highest concentrations capable of sustaining the propagating flame are the lower (LFL) and upper flammability limits (UFL), respectively (Mannan, 2005; Crowl and Louvar, 2011). The flammability limits of a given fuel depend on several factors, including initial temperature, initial pressure, system scale and configuration, ignition energy, flame stretch, radiation reabsorption, the presence of inert gases, and others (Mannan, 2005; Ju et al., 2001). In industrial processes, an inert gas is often added to combustible mixtures to prevent the F&E hazard, especially in oxidation processes. Nonetheless, much more flammability limit data is published for pure fuel substances than for fuel/inert gas mixtures. Deriving such data in Taiwan is costly. Therefore, a process to estimate the flammability limits of fuel/inert gas mixtures would provide practical benefits.

10.2 UNDERSTANDING THE FLAMMABILITY OF INERT GAS MIXTURES

Kondo et al. (2006a, 2006b, 2007) modified Le Chatelier's formula to provide an empirical equation for evaluating the flammability boundaries of several fuels mixed with inert gases. The parameters of the empirical equation depend on the tested flammable and inert gases and must be experimentally determined. Theories relating flammability limits to heat loss predict a minimum flame temperature, below which the flame cannot propagate (Spalding, 1957; Buckmaster, 1976; Joulin and Clavin,

121

1976). When estimating the flammability limits of fuels mixed with inert gases, investigators often assume the adiabatic flame temperature is constant for a particular fuel (Shebeko et al., 2002; Vidal et al., 2006; Melhem, 1997; Hansen and Crowl, 2010). When using Melhem's (1997) method or Hansen and Crowl's (2010) equations to estimate the flammability boundaries, the adiabatic flame temperatures at the LFL and UFL are required; unfortunately, these temperatures are mostly unknown. A theoretical linear equation was derived describing the flammability boundaries of fuels mixed with inert gases (Chen et al., 2009a, 2009b). The slope of this linear equation must be determined by regressing experimental data (Chen et al., 2009a, 2009b). Recently, a model describing the flammability limits of fuel and diluent mixtures by considering thermal radiation loss was validated (Liaw et al., 2012). The estimated flammability envelopes depend upon the assumed combustion products of the carbon in the fuels, especially at the UFL (Liaw et al., 2012). Chen et al. (2008) observed carbon monoxide (CO) and carbon dioxide (CO_2) in the burned gas at the UFL of pure propane. Ju et al. (2001) and Britton (2002) indicated that the measured flammability limits are different when using different experimental apparatuses. The apparatus used to measure the flammability limits includes vertical glass tubes, spherical glass flasks, and spherical explosion vessels. The vertical glass tube and spherical glass flask are open to the atmosphere after the explosion and are regarded as constant pressure systems. However, the spherical explosion vessel is an intrinsic constant volume system. The methods for the estimation of flammability limits are based on the theory of enthalpy change equaling zero. Thus, they were developed for constant pressure systems. In this work, a model to calculate the flammability envelopes of mixtures containing inert gases for a constant volume system is derived. The combustion products at the LFL and UFL were analyzed to verify the assumption of the model. Methyl formate can be used in a variety of reactions to produce industrial products, such as acetic acid, methyl acetate, and other chemicals (Gérard et al., 1998). In semiconductor manufacturing, acetone, methanol, and isopropyl alcohol (IPA) are often used in clean room procedures (Liaw and Chiu, 2003). Thus, acetone, methanol, IPA, and methyl formate diluted with either steam or nitrogen (N_2) were selected as samples for model validation.

10.3 EXPERIMENTAL PROCEDURES

10.3.1 APPARATUS AND MATERIALS

American Chemical Society (ACS) standards (Pharmco Products Inc., USA) were used to verify the purity of IPA (99.8 %).Methanol (99.9 %) and acetone (99.8 %) (Mallinckrodt, USA) met ACS specifications. Methyl formate (97.0 %) was purchased from Alfa Aesar (UK). Water was purified using the Milli-Q Plus filtration purification method from the Millipore Corporation (USA).

10.3.2 SPHERICAL EXPLOSION VESSEL

The flammability measurements were conducted in a 20-lspherical explosion vessel (Adolf Kühner, Switzerland) standardized to American Society for Testing and

Materials (ASTM) E1226-05 (2005). The mixtures tested included acetone + steam, methanol + steam, methyl formate + steam, IPA + steam, IPA + nitrogen, and acetone + nitrogen. Figure 10.1 shows the system configuration. The hollow sphere is made of stainless steel. A permanent spark with ignition energy of about 10 J is located in the center of the sphere. The KSEP 332, a unit delivered with the sphere, controls the ignition system and measures the pressure with piezoelectric sensors. Thermal insulation covers the test chamber to reduce energy loss. The measurements were conducted based on American Society for Testing and Materials (ASTM) E681 (1994) and American Society for Testing and Materials (ASTM) E918-83 (2005) standards. A pressure of 1 atm and a temperature of 150°C, which is greater than the normal boiling points of the studied samples, were selected as the initial pressure and temperature settings. Flammability was defined as a pressure increase of 7% or more than the initial pressure in the vessel (ASTM E918-83). The vapor phase composition of the fuel and steam was determined from its mass using a digital scale (EL-410D: sensitivity 0.001 g, Setra Systems, USA), and the sample was stirred for 10 min before being introduced to the vessel. The liquid sample was injected using a syringe. The composition of the nitrogen was determined using a partial pressure procedure (pressure sensor sensitivity = 0.01 torr). The test chamber was heated to the initial temperature (150°C), evacuated, and flushed with air three times before the fuel, or fuel + water, were added. Nitrogen was then loaded into the chamber followed by the air.

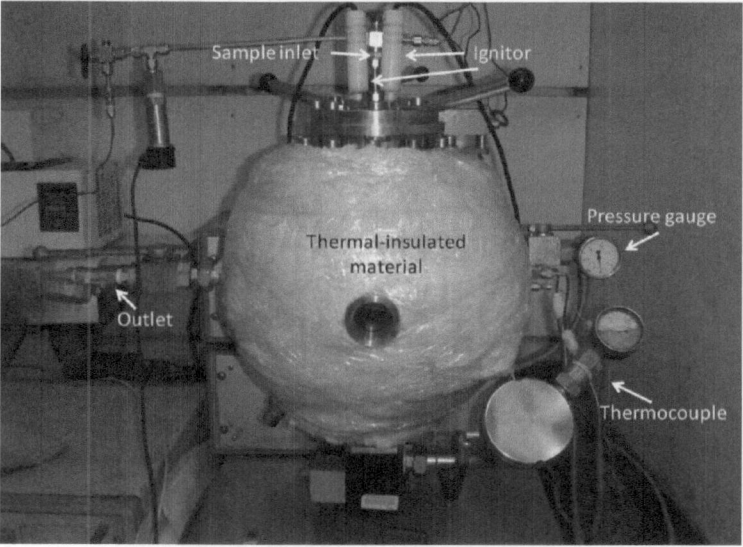

FIGURE 10.1 The basic system configuration of the 20-l, spherical explosion vessel (Reproduced with permission from Liaw et al. 2016, Copyright © Elsevier).

10.3.3 Fourier Transform Infrared Spectroscopy

The explosion vessel containing the burned gas was cooled to room temperature after each test. Then the combustion products were collected by Flex Foil grab bags, which are strong and flexible with good stability for hydrogen (H_2), CO, CO_2, and methane (CH_4) (SKC, 2015). The products were analyzed within 24 h using Fourier Transform Infrared Spectroscopy (FTIR; FTIR Spectrum 100, Perkin Elmer, USA) at a wavenumber resolution of 0.5 cm^{-1} on the transmission setting. A background spectrum was used to eliminate the environmental signals of the sample spectrum.

10.3.4 Theory

10.3.4.1 Mathematical Model

The model for estimating the flammability envelope of mixtures containing inert gas in a constant volume system was derived based on the energy balance equation. The estimations of the flammability boundaries are based on the flame temperatures at the LFL and UFL. These two temperatures were assumed to be constant, except at the upper flammability boundary around the LOC. It was assumed that the fuel was completely converted into CO_2 and steam at the LFL while, at the UFL, it was converted to CO_2, CO, steam, and H_2. When the mole ratio of the fuel, x, is less than x_U, where the fuel to oxygen (O_2) ratio at the UFL equals the stoichiometric ratio, more fuel converts to CO_2 at the UFL. Additionally, it was assumed that the flame temperature at the UFL was linear with the quantity of CO. Where

$$x = \frac{fuel}{fuel + added\ inert} \tag{10.1}$$

The nitrogen in the air is excluded from the inert expression in Equation (10.1). The effect of nitrogen in the air is expressed in other parameters, including P_L in Equation (10.2) and P_U in Equation (10.6). The complete derivation of the model is presented in Appendix A. The model is based on the equations for estimating the flame temperatures, values of x_L (the value of x at LOC) and x_U, and the lower and upper flammable boundaries.

At the LFL, the constant flame temperature is calculated by Equation (10.2) while, at the UFL, it is calculated by Equation (10.3):

$$L_0\left(\Delta H_c^o\right) + \int_{298}^{T}\left[L_0\left(Q_L - P_L\right) + P_L\right]dT +$$
$$\int_{T_0}^{298}\left[L_0\left(C_{pf} - P_L\right) + P_L\right]dT + RT_0 - \left(1 + L_0 P_{PL}\right)RT = \tag{10.2}$$
$$-\alpha e \underline{A_s}\sigma\left(T^4 - T_0^4\right)\Delta t$$

and
$$0.21f_U\left(1 - U_0\right)\left(\Delta h_c\right)_U + \int_{298}^{T}\left[U_0\left(C_{pf} - P_{U0}\right) + P_{U0}\right]dT$$
$$+ \int_{T_0}^{298}\left[U_0\left(C_{pf} - P_L\right) + P_L\right]dT + RT_0$$
$$-\left[0.79 + 0.21U_0 + 0.21\left(1 - U_0\right)P_{PU0}\right]RT = -\alpha e \underline{A_s}\sigma\left(T^4 - T_0^4\right)\Delta t \tag{10.3}$$

The value of x_L, the value of x at the LOC, is obtained:

$$x_L = -\frac{0.21\left[\int_{T_0}^{T} C_{pI}dT + R(T_0-T)+\alpha e\underline{A_s}\sigma\left(T^4-T_0^4\right)\Delta t\right]}{f_L\left[\int_{T_0}^{T} P_L dT + R(T_0-T)+\alpha e\underline{A_s}\sigma\left(T^4-T_0^4\right)\Delta t\right]}$$
$$+0.21\int_{298}^{T}\left(Q_L-C_{pI}\right)dT + \int_{T_0}^{T298}\left(C_{pf}-C_{pI}\right)dT+\Delta h_c^o - P_{PL}RT]$$

(10.4)

The lower flammability boundary, L, is calculated:

$$L = -\frac{\int_{T_0}^{T} P_L dT + R(T_0-T)+\alpha e\underline{A_s}\sigma\left(T^4-T_0^4\right)\Delta t}{\int_{298}^{T} Q_L dT + \int_{T_0}^{298} C_{pf}dT - \int_{T_0}^{T}\left[C_{pI}-\left(C_{pI}-P_L\right)\frac{1}{x}\right]dT+\Delta h_c^o - P_{PL}RT}$$

(10.5)

The x_U value is calculated as:

$$x_u = -\frac{0.21fu\left[\int_{T_0}^{T} C_{pI}dT + R(T_0-T)+\alpha e\underline{A_s}\sigma\left(T^4-T_0^4\right)\Delta t\right]}{\int_{298}^{T} P_u dT + \int_{T_0}^{298} P_L dT + RT_0 -\left(0.79+0.21P_{PU}\right)RT}$$
$$+\alpha e\underline{A_s}\sigma\left(T^4-T_0^4\right)\Delta t + 0.21fu[\int_{T_0}^{T}\left(C_{pf}-C_{pI}\right)dT+\left(\Delta h_c\right)_u$$

(10.6)

For $1-x \leq 1-x_U$, the upper flammability boundary, U, is obtained as:

$$U = -\frac{\int_{298}^{T} P_u dT + \int_{T_0}^{298} P_L dT + 0.21f_u\left(\Delta h_c\right)_u RT_0 -\left(0.79+0.21P_{Pu}\right)RT}{\int_{T_0}^{T}\left(C_{Pf}-C_{PI}\right)dT + \int_{298}^{T}\left(C_{PI}-P_u\right)\frac{1}{x}dT + \int_{T_0}^{298}\left(C_{PI}-P_L\right)\frac{1}{x}dT}$$
$$\phantom{U = -\frac{}{}}+\alpha e\underline{A_s}\sigma\left(T^4-T_0^4\right)\Delta t$$
$$\phantom{U = -\frac{}{}}-0.21f_u\frac{1}{x}\left(\Delta h_c\right)_u - \frac{0.21}{x}\left(1-P_{Pu}RT\right)$$

(10.7)

For $1-x > 1-x_U$, the upper flammability boundary is calculated as:

$$U = -\frac{\int_{298}^{T}\left[P_L-0.41\left(Cp_{CO}-Cp_{CO_2}\right)-0.21Cp_{O_2}\right]dT}{\int_{298}^{T}\left[Q_u+\frac{1}{x}\left(C_{PI}-P_L\right)+\left(Cp_{CO}-Cp_{CO_2}\right)\left(p+\frac{0.42}{x}\right)+\frac{0.21}{x}Cp_{CO_2}\right]dT}$$
$$+\int_{T_0}^{298} P_L dT + 0.42\left(\Delta h_{CO/CO_2}^o\right)+R(T_0-0.79T)+\alpha e\underline{A_s}\sigma\left(T^4-T_0^4\right)\Delta t$$
$$+\int_{T_0}^{298}\left[\left(Cp_f-Cp_I\right)+\frac{1}{x}\left(Cp_I-P_L\right)\right]dT+\left(\Delta h_p\right)+\left(P_{P2u}-\frac{0.21}{x}\right)RT$$

(10.8)

Similar equations for a constant pressure system were derived previously (Liaw et al., 2012). The differences in the equations between the constant pressure and constant volume systems originate from the different forms of the energy balance equation used. The former is presented in enthalpy change and the latter, in internal energy change. Thus, the terms relevant to the gas constant, R, as presented in Equation (10.2–10.8), are not given for the constant pressure system (Liaw et al., 2012) but the constant volume system.

Estimation Procedure

Equations (10.2, 10.3) were used to calculate the flame temperatures at the LFL and UFL, respectively, and x_L and x_U were obtained using Equations (10.4, 10.6). Then, the LFL was estimated from $1\text{-}x = 0$ to $1\text{-}x = 1\text{-}x_L$ by Equation (10.5), and Equations (10.7, 10.8) were used to estimate the UFL from $1\text{-}x = 0$ to $1\text{-}x = 1\text{-}x_U$ and from $1\text{-}x = 1\text{-}x_U$ to $1\text{-}x = 1\text{-}x_L$, respectively. Figure 10.2 shows the estimation procedure.

10.4 RESULTS AND DISCUSSIONS

10.4.1 COMBUSTION PRODUCTS OF ACETONE AND METHYL FORMATE

The assumptions of the model were verified by combustion product analyses of methyl formate or acetone diluted with steam by FTIR (Figures 10.3–10.6). Figures 10.3 and 10.5 reveal that in the lower flammability envelope, only CO_2 (2364 cm^{-1}, 2338 cm^{-1}) and acetone or methyl formate with steam. For larger values of $1\text{-}x$, 0.895 for acetone and 0.875 for methyl formate, a miniscule quantity of CO was detected. This quantity

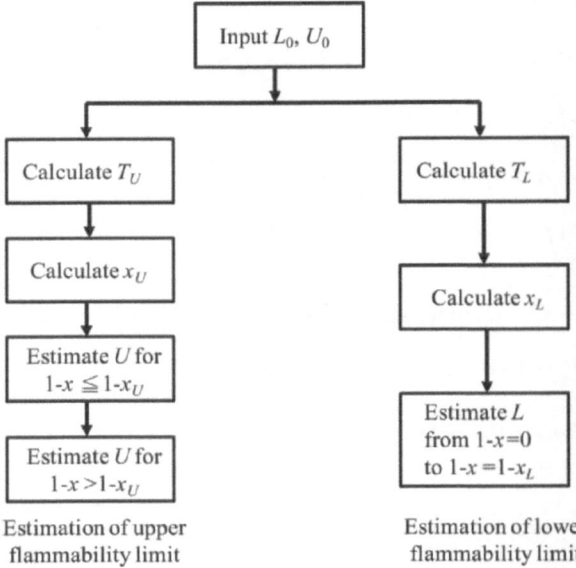

FIGURE 10.2   The procedure to estimate the flammability envelope (Reproduced with permission from Liaw et al. 2016, Copyright © Elsevier).

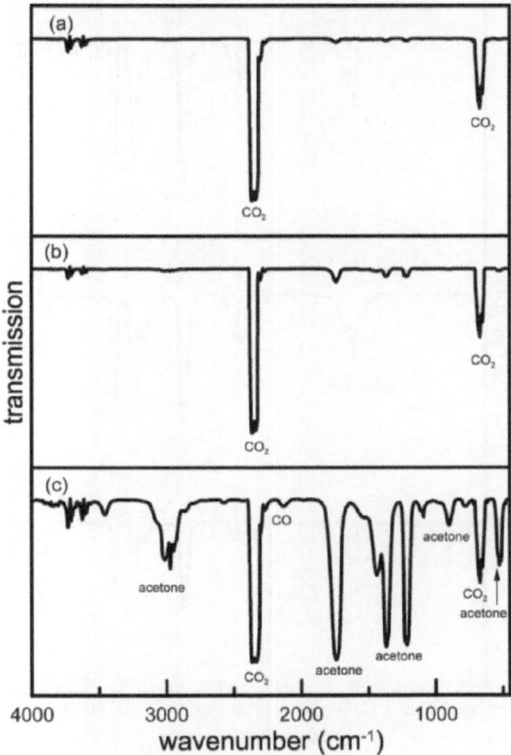

FIGURE 10.3 FTIR spectra for combustion product of acetone diluted with steam at the LFL, (a) 1 - x = 0, (b) 1 - x = 0.5, and (c) 1 - x = 0.895 (Reproduced with permission from Liaw et al. 2016, Copyright © Elsevier).

was negligible compared to that of CO_2. This observation confirmed the assumption that the carbon in the fuel was almost entirely converted into CO_2 during the combustion of a lean limit mixture. For acetone diluted with steam, only CO was observed in the burned gas at the UFL for small values of $1-x = 0$ and $1-x = 0.5$ (2120 cm^{-1} and 2172 cm^{-1} [Figure 10.4(a) and (b)]). No CO_2 was found. For methyl formate, in addition to CO, a negligible quantity of CO_2 was present in the combustion products (Figure 10.6(a) and (b)). However, remarkable amounts of CO_2 were observed for acetone and methyl formate at greater values of $1-x$, 0.9 and 0.875, respectively (Figure 10.4(c) and 10.6(c)). The estimated values of $1 - x_U$ for acetone and methyl formate diluted with steam were 0.805 and 0.759, respectively, based on the adiabatic conditions, and were 0.776 and 0.738, based on greatest heat loss, with the four values being greater than 0.5 and less than 0.9 and 0.875. The carbon in the fuel was almost exclusively converted to CO for $1 - x < 1 - x_U$ for acetone and methyl formate. Thus, the value of $a1$, moles of CO produced for 1 mol of fuel ($C_aH_bO_c$) burned, in Equation (A17, A denotes appendix) can be approximated to a. In addition, the assumption that the CO_2/CO ratio would increase for values of $1 - x > 1 - xU$ was validated. Unburned fuel (acetone or methyl formate), in addition to CO and a negligible quantity of CO_2, was also observed in the

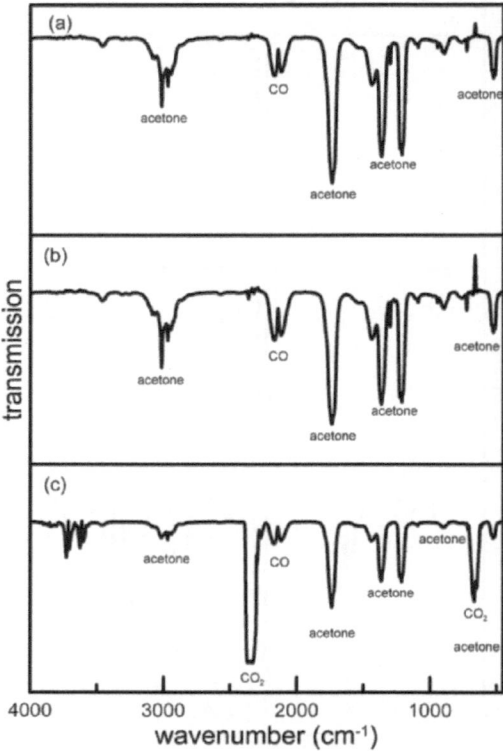

FIGURE 10.4 FTIR spectra for combustion product of acetone diluted with steam at the UFL, (a) 1 - x = 0, (b) 1 - x = 0.5, and (c) 1 - x = 0.9 (Reproduced with permission from Liaw et al. 2016, Copyright © Elsevier).

combustion products at the UFL by FTIR. Absorbance bands at 3016 cm^{-1} (C–H stretching in alkyl groups), 1740 cm^{-1} (C–O stretching in ketones), 1438 cm^{-1} (C–H bending vibration in alkyl groups), 1368 cm^{-1} (C–H bending vibration in alkyl groups), and wavenumbers 1228 cm^{-1} and 516 cm^{-1} both in the fingerprint region were observed for acetone (Figure 10.4). Absorbance bands at 3016 cm^{-1} (C–H stretching in alkyl groups), 1766 cm^{-1} (C–O stretching in esters), 1744 cm^{-1} (C–O stretching in esters), 1450 cm^{-1} (C–H bending vibration in alkyl groups), 1220 cm^{-1}, 1194 cm^{-1}and 1172 cm^{-1} (C–O stretching vibrations), and 934 cm^{-1} were observed for methyl formate (Figure 10.6). No other hydrocarbons resulting from incomplete combustion were detected. Thus, the fuel was present in excess, which is consistent with the assumption of the model. The combustion products of acetone and methyl formate followed the stoichiometry of the proposed combustion reaction at the UFL.

10.4.2 Experimental and Estimated Flammability Limits

10.4.2.1 Condition for the Simulation

The flammability limit measurements validated the predictive model for acetone, methyl formate, methanol, and IPA diluted with either steam or nitrogen. Table 10.1

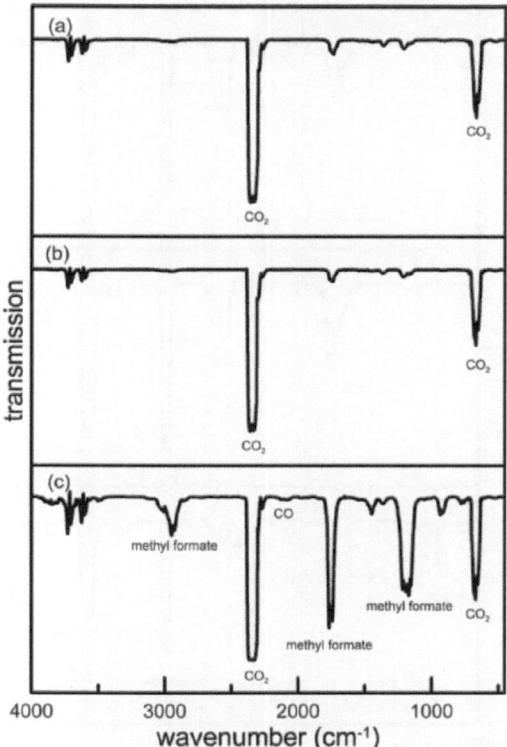

FIGURE 10.5 FTIR spectra for combustion product of methyl formate diluted with steam at the LFL, (a) $1 - x = 0$, (b) $1 - x = 0.5$ and (c) $1 - x = 0.875$ (Reproduced with permission from Liaw et al. 2016, Copyright © Elsevier).

compares the measured flammability limits for the pure substances with the corresponding published data. Although the test conditions of the published data were not entirely the same as ours, Table 10.1 provides a rough overview of the similarity of the results.

The experimental LFL and UFL for acetone, 2.60% and 13.87%, respectively, are close to those published by the Design Institute for Physical Properties (DIPPR) (2012), Kuchta (1985), and Zabetakis (1965), 2.6% and 13.0%, respectively. However, they differ from those of Coward and Jones (1952), 3.0% and 11.8%, respectively. The differences between the measurements and published data (Design Institute for Physical Properties (DIPPR), 2012; Kuchta, 1985; Zabetakis, 1965; Coward and Jones, 1952) are small and acceptable for IPA at the LFL and UFL and methyl formate at the LFL, although there are greater deviations for methyl formate at the UFL. There are greater deviations between the experimental results and literature values (Design Institute for Physical Properties (DIPPR), 2012; Kuchta, 1985; Zabetakis, 1965; Coward and Jones, 1952; Brooks and Crowl, 2007; Rowley et al., 2010) for methanol. The differences in flammability limits between the measured data and those in the literature are attributed to differences in the initial temperature,

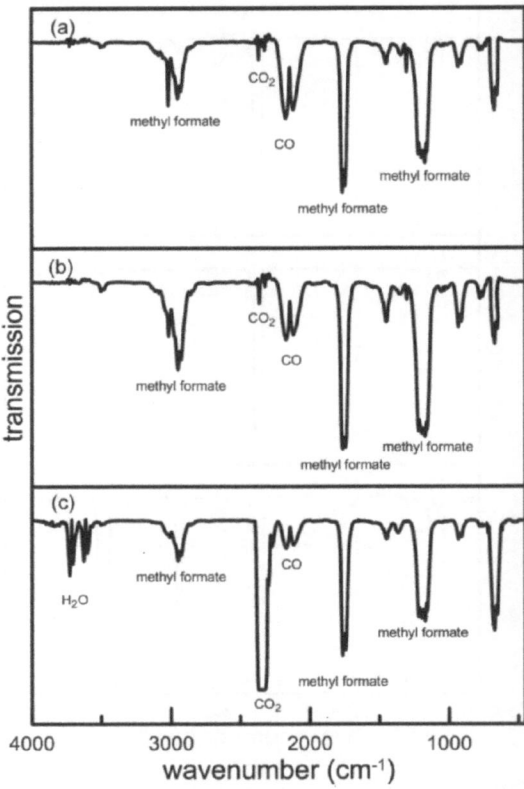

FIGURE 10.6   FTIR spectra for combustion product of methyl formate diluted with steam at the UFL, (a) 1 - x = 0, (b) 1 - x = 0.5, and (c) 1 - x = 0.875 (Reproduced with permission from Liaw et al. 2016, Copyright © Elsevier).

ignition energy, and test apparatuses. The LFL and UFL decreases and increases with the initial temperature, respectively (Coward and Jones, 1952). The initial temperature used in this study, 150°C, is greater than that of Kuchta (1985), Zabetakis (1965), Coward and Jones (1952), and Brooks and Crowl (2007), which was 25°C. The initial temperatures used to generate the Design Institute for Physical Properties (DIPPR) (2012) are not described. Only the initial temperature used by Rowley et al. (2010), 420 K, is close to 150°C. Since the spherical explosion vessel is not the typical apparatus for the determination of flammability limits of gases, such data are scarce. However, 20-l explosion vessels are commonly used for testing the LOC of gases and vapors (ASTME 2079-07 (2007)). Thus, the data in the literature used for comparison are derived from the different apparatuses than ours. The standard heat of formation and the temperature dependency heat capacity constants for the components used in this article were adopted from the Design Institute for Physical Properties (DIPPR) (2012).

TABLE 10.1

Comparison of Flammability Limits Adopted from the Literature with Measured Data

Substance	Experimental Data		Literature	
	L (vol.%)	U (vol.%)	L (vol.%)	U (vol.%)
Acetone	2.60	13.87	2.6[1,2,3]	13.0[1,2,3]
			3.04	11.8[4]
Methanol	5.38	41.10	7.18[1]	36.5[1]
			6.7[2,3,5]	36.0[2,3]
			7.35[4]	
			6.08[6]	
Isopropanol	2.26	13.19	2.0[1]	12.7[1]
			2.2[2,3]	12.0[2]
			2.5[4]	
Methyl formate	4.86	28.27	5.2[1]	23.0[1,2,3]
			5.0[2,3]	20.4[4]
			5.9[4]	

[1] Design Institute for Physical Properties (DIPPR) (2012).
[2] Kuchta (1985).
[3] Zabetakis (1965).
[4] Coward and Jones (1952).
[5] Brooks and Crowl (2007).
[6] Rowley et al. (2010).
(Data obtained with permission from Liaw et al. 2016, Copyright © Elsevier)

The longest flame propagation distance was equivalent to the radius (R_0) of the spherical, 20-l explosion vessel. If the flame propagated spherically, then the efficiency factor relative to the radiation heat transfer surface area, α, may be calculated as:

$$\alpha = \frac{\int_0^{R_f} \sigma e 4\pi r^2 \left(T^4 - T_0^4\right) dr / u}{\sigma e 4\pi R_0^2 \left(T^4 - T_0^4\right) \Delta t} \leq \frac{\int_0^{R_0} \frac{\sigma e 4\pi r^2 \left(T^4 - T_0^4\right) dr}{u}}{\sigma e 4\pi R_0^2 \left(T^4 - T_0^4\right) \Delta t} = \frac{1}{3}$$

where R_f is the exact distance the flame propagated, u is the burning velocity, and $\Delta t = R_0/u$. That is, the value of α is less than or equal to 1/3. The σA_s value is 2.468 × 10^{-8} W K^{-4} mol^{-1} as established in our spherical, 20-l explosion receptacle. The emissivity, e, is 0.074 for the stainless steels material of the vessel (Welty et al., 1976). The burning velocities of the tested fuels in the air near the flammability limits were around 10 cm/s, when extrapolating from the literature data (Chong and Hochgreb, 2011; Egolfopoulos et al., 1992; Dooley et al., 2010; Veloo and Egolfopoulos, 2011) to the flammability limits in a linear manner. Such values of burning velocity are close to the corresponding values for other fuels, such as paraffin hydrocarbons, acetylene, and hydrogen, with values of a few centimeters per second (Mannan, 2005; Zabetakis, 1965; Ronney and Wachman, 1985). A burning velocity of 10 cm/s was used to estimate the flame propagation duration, Δt, which was estimated to be in the order of 1 s. Thus, the values of $\alpha\sigma eA_s\Delta t$ were set at 0 and 10^{-9} J K^{-4} mol^{-1}, signifying the adiabatic condition, with the greatest heat loss through thermal radiation. Because the test apparatus was covered with insulation, the experiment approached the adiabatic condition. For simplicity, it was assumed that there is no formation of CO_2 at the UFL for $1 - x \leq 1 - x_U$. It is also expected that the amount of H_2 was negligible at the UFL, based on the assumption of the hydrogen atoms of the fuels were completely converted into H_2O. Thus, the value of b_1, as presented in Appendix A, was set at zero.

10.4.2.2 Radiation Heat Loss Effect on the Estimated Flammability Boundaries

The estimation differences for the upper flammability boundaries were unremarkable and negligible for acetone and methyl formate mixed with steam (Figures 10.7 and 10.8) when considering the different heat losses. Additionally, the estimated lower flammability boundaries closely overlapped for the two compounds. The negligible estimation difference of the upper flammability boundary for various heat losses was also seen for methanol–steam mixtures (Figure 10.9). Small analogous differences were found for IPA diluted with steam and acetone or IPA diluted with nitrogen (Figures 10.10–10.12). The estimated lower flammability boundaries also overlapped for different heat losses for the other flammable compounds mixed with inert gases (Figures 10.9–10.12). The results indicated that the heat loss by radiation is not a significant parameter for the estimation of flammability boundaries for the compounds studied under the study conditions.

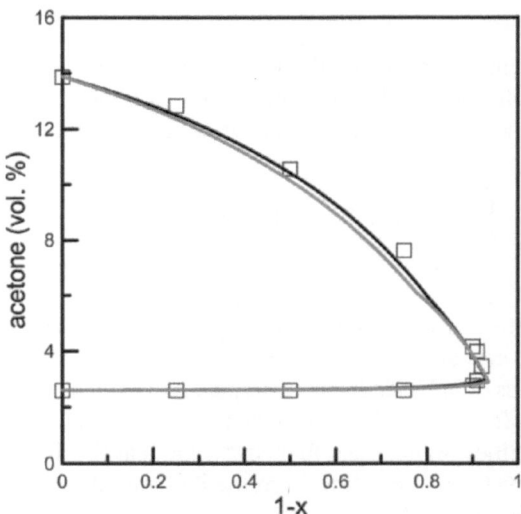

FIGURE 10.7 Effect of steam dilution on the flammability limits of acetone. At the point of $x = x_U$, where more fuel is converted to CO_2, the predicted UFLs showed a dent. This behavior is more significant for acetone considering heat loss (grey curve) □ denotes experimental data where black dash, considers $\alpha e\sigma As\Delta t = 0$ JK^{-4}mol^{-1} and grey dash considers $\alpha e\sigma As\Delta t = 10^{-9}$ J K^{-4} mol^{-1} (Reproduced with permission from Liaw et al. 2016, Copyright © Elsevier).

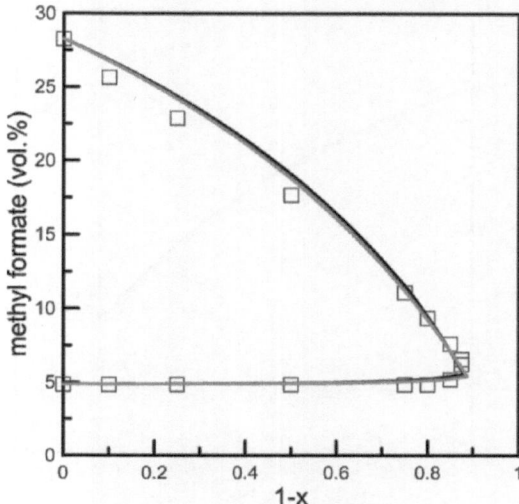

FIGURE 10.8 Effect of steam dilution on the flammability limits of methyl formate. At the point of $x = x_U$, where more fuel is converted to CO_2, the predicted UFLs showed a dent. $\square$ denotes experimental data where black dash, considers $\alpha e \sigma As \Delta t = 0 \ JK^{-4}mol^{-1}$ and grey dash considers $\alpha e \sigma As \Delta t = 10^{-9} \ J \ K^{-4} \ mol^{-1}$ (Reproduced with permission from Liaw et al. 2016, Copyright © Elsevier).

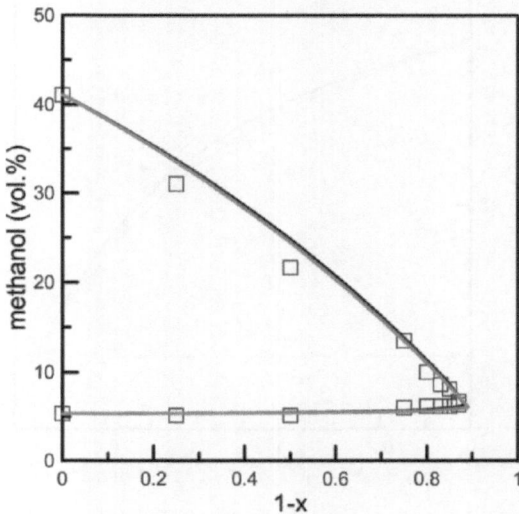

FIGURE 10.9 Effect of steam dilution on the flammability limits of methanol. $\square$ denotes experimental data where black dash, considers $\alpha e \sigma As \Delta t = 0 \ JK^{-4}mol^{-1}$ and grey dash considers $\alpha e \sigma As \Delta t = 10^{-9} \ J \ K^{-4} \ mol^{-1}$ (Reproduced with permission from Liaw et al. 2016, Copyright © Elsevier).

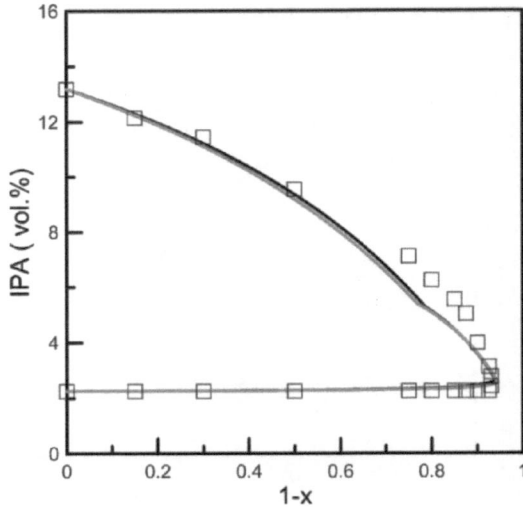

FIGURE 10.10 Effect of steam dilution on the flammability limits of isopropanol. At the point of $x = x_U$, where more fuel is converted to CO_2, the predicted UFLs showed a dent. $\square$ denotes experimental data where black dash, considers $\alpha e \sigma As \Delta t = 0$ $JK^{-4}mol^{-1}$ and grey dash considers $\alpha e \sigma As \Delta t = 10^{-9}$ J K^{-4} mol^{-1} (Reproduced with permission from Liaw et al. 2016, Copyright © Elsevier).

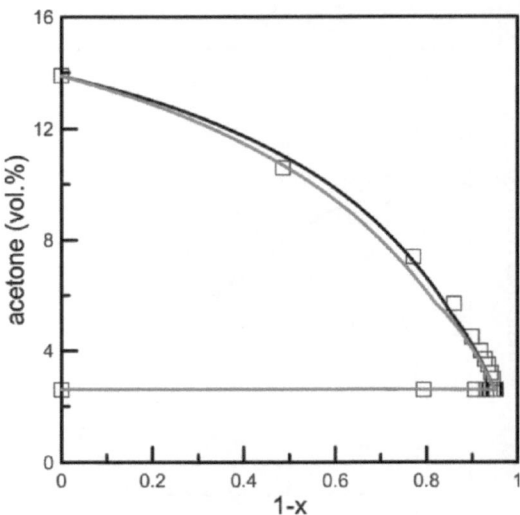

FIGURE 10.11 Effect of nitrogen dilution on the flammability limits of acetone. At the point of $x = x_U$, where more fuel is converted to CO_2, the predicted UFLs showed a dent. $\square$ denotes experimental data where black dash, considers $\alpha e \sigma As \Delta t = 0$ $JK^{-4}mol^{-1}$ and grey dash considers $\alpha e \sigma As \Delta t = 10^{-9}$ J K^{-4} mol^{-1} (Reproduced with permission from Liaw et al. 2016, Copyright © Elsevier).

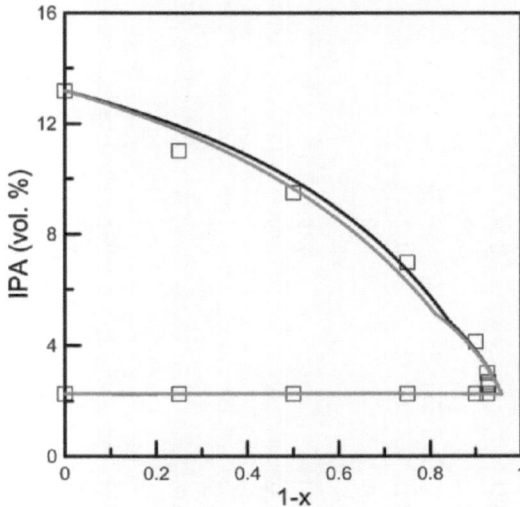

FIGURE 10.12 Effect of nitrogen dilution on the flammability limits of isopropanol. At the point of $x = x_U$, where more fuel is converted to CO_2, the predicted UFLs showed a dent. $\square$ denotes experimental data where black dash, considers $\alpha e\sigma As\Delta t = 0$ $JK^{-4}mol^{-1}$ and grey dash considers $\alpha e\sigma As\Delta t = 10^{-9}$ J K^{-4} mol^{-1} (Reproduced with permission from Liaw et al. 2016, Copyright © Elsevier).

10.4.2.3 Steam Dilution Effect on Flammability Envelope

The predicted LFL increased slightly with the value of $1 - x$, the mole ratio of steam based on free air, regardless of taking heat loss into account or not. The model describes the experimental data well for acetone and methyl formate (Figures 10.7 and 10.8 and Table 10.2). At the point of $x = x_U$, where more fuel began to convert to CO_2, the predicted UFLs showed a dent. This behavior is more significant for acetone considering the heat loss. Nonetheless, the UFLs, measurements, and the predicted curves for acetone and methyl formate show agreement (Figures 10.7 and 10.8, Table 10.2).

The agreement in the estimated upper and lower flammability boundaries for acetone and methyl formate diluted with steam are attributed to two causes. One is the consistency of the assumptions of the proposed model, that is, the combustion stoichiometry at the flammability limits, and CO_2/CO ratio increases at the UFL for $1 - x > 1 - x_U$, as shown by FTIR for the LFL and UFL. The other is that $a1$ was set to a constant, a, which assumes that the carbon in the fuel was entirely converted into CO. This is consistent with the FTIR spectra in which none or an immaterial quantity of CO_2 was found in the combustion products of a rich limit mixture for $1 - x < 1 - x_U$. The estimated values of the UFLs depend on the combustion product distribution at the UFL for the burned fuel (Liaw et al., 2012). The ratio of CO/CO_2 in the combustion products at the UFL does not remain constant as it does with acetone and methyl formate for $1 - x < 1 - x_U$. For example, the CO/CO_2 ratio of methane varied along the upper flammability boundary, 12.22 and 4.15 for $1 - x = 0$ and 0.372, respectively. The predictive flexibility of the model with methanol and IPA diluted with

TABLE 10.2

Estimated Flame Temperatures at the LFL and UF$_L$, T$_L$ and T$_U$, and the Deviations Between Estimated and Measured Flammability Limits[a], ΔL, and ΔU, for the Tested Mixtures

System	T$_L$ (K)	T$_U$ (K)	ΔL (vol.%)	ΔU (vol.%)
Acetone + steam				
$\alpha\sigma As\Delta t = 0$ J K^{-4} mol^{-1}	2063	2214	0.11	0.29
$\alpha\sigma As\Delta t = 10^{-9}$ J K^{-4} mol^{-1}	1759	1924	0.12	0.39
Methyl formate + steam				
$\alpha\sigma As\Delta t = 0$ J K^{-4} mol^{-1}	1946	1792	0.21	0.72
$\alpha\sigma As\Delta t = 10^{-9}$ J K^{-4} mol^{-1}	1694	1656	0.19	0.60
Methanol + steam				
$\alpha\sigma As\Delta t = 0$ J K^{-4} mol^{-1}	1831	1696	0.24	1.19
$\alpha\sigma As\Delta t = 10^{-9}$ J K^{-4} mol^{-1}	1612	1582	0.33	0.99
Isopropanol + steam				
$\alpha\sigma As\Delta t = 0$ J K^{-4} mol^{-1}	2021	2348	0.10	0.51
$\alpha\sigma As\Delta t = 10^{-9}$ J K^{-4} mol^{-1}	1734	1982	0.06	0.56
Acetone + N$_2$				
$\alpha\sigma As\Delta t = 0$ J K^{-4} mol^{-1}	2063	2214	0.05	0.31
$\alpha\sigma As\Delta t = 10^{-9}$ J K^{-4} mol^{-1}	1759	1924	0.05	0.57
Isopropanol + N$_2$				
$\alpha\sigma As\Delta t = 0$ J K^{-4} mol^{-1}	2021	2348	0.05	0.38
$\alpha\sigma As\Delta t = 10^{-9}$ J K^{-4} mol^{-1}	1734	1982	0.05	0.85

[a] Deviation of flammability limits: $\Delta L = \Sigma_N |L\text{exp.} - L\text{pred.}|/N$; $\Delta U = \Sigma_N |U\text{exp.} - U\text{pred.}|/N$. (Data obtained with permission from Liaw et al. 2016, Copyright © Elsevier)

TABLE 10.3

Comparison of $1 - x_L$ and Flammability Limit Values at the LOC Between the Estimations and Measured Data

System	Estimation			Experimental Data		
	$1 - x_L$	Fuel (vol.%)	Oxygen (vol.%)	$1 - x_L$	Fuel (vol.%)	Oxygen (vol.%)
Acetone + steam						
$\alpha\varepsilon\sigma\Delta s\Delta t = 0\ \mathrm{J\ K^{-4}\ mol^{-1}}$	0.930	3.00	12.00	0.92	3.47	11.89
$\alpha\varepsilon\sigma\Delta s\Delta t = 10^{-9}\ \mathrm{J\ K^{-4}\ mol^{-1}}$	0.935	2.91	11.60			
Methyl formate + steam						
$\alpha\varepsilon\sigma\Delta s\Delta t = 0\ \mathrm{J\ K^{-4}\ mol^{-1}}$	0.881	5.59	11.14	>0.875	6.24–6.59	<9.93–10.5
$\alpha\varepsilon\sigma\Delta s\Delta t = 10^{-9}\ \mathrm{J\ K^{-4}\ mol^{-1}}$	0.887	5.44	10.89			
Methanol + steam						
$\alpha\varepsilon\sigma\Delta s\Delta t = 0\ \mathrm{J\ K^{-4}\ mol^{-1}}$	0.885	6.31	9.48	>0.87	6.42–6.76	<10.1–10.6
$\alpha\varepsilon\sigma\Delta s\Delta t = 10^{-9}\ \mathrm{J\ K^{-4}\ mol^{-1}}$	0.891	6.13	9.19			
Isopropanol + steam						
$\alpha\varepsilon\sigma\Delta s\Delta t = 0\ \mathrm{J\ K^{-4}\ mol^{-1}}$	0.940	2.62	11.83	>0.93	2.63–2.78	<12.7–13.1
$\alpha\varepsilon\sigma\Delta s\Delta t = 10^{-9}\ \mathrm{J\ K^{-4}\ mol^{-1}}$	0.945	2.53	11.34			
Acetone + N_2						
$\alpha\varepsilon\sigma\Delta s\Delta t = 0\ \mathrm{J\ K^{-4}\ mol^{-1}}$	0.949	2.58	10.38	0.946	3.00	9.33
$\alpha\varepsilon\sigma\Delta s\Delta t = 10^{-9}\ \mathrm{J\ K^{-4}\ mol^{-1}}$	0.949	2.59	10.34			
Isopropanol + N_2						
$\alpha\varepsilon\sigma\Delta s\Delta t = 0\ \mathrm{J\ K^{-4}\ mol^{-1}}$	0.957	2.24	10.06	>0.928	2.58–2.67	<13.2–13.5
$\alpha\varepsilon\sigma\Delta s\Delta t = 10^{-9}\ \mathrm{J\ K^{-4}\ mol^{-1}}$	0.957	2.25	10.01			

(Data obtained with permission from Liaw et al. 2016, Copyright © Elsevier)

steam was tested for which the combustion products were unknown. Just as for acetone and methyl formate, the values of $a1$ and $b1$ were set at a and 0, respectively, for simplicity. The predicted LFLs for methanol diluted with steam agreed with the measured LFLs, and the predicted UFLs agreed with the measured UFLs (Table 10.2 and Figure 10.9). For IPA, the predicted UFLs were less than the measurements for $1 - x$ values between 0.7 and 0.9, although the predicted LFLs agreed with the measurements well. To extend the curves based on the trends of the estimations, Figure 10.10 indicates that the predicted UFLs are less than the measurements for $1 - x >$ $1 - x_U$, if neglecting Equation (10.8) for IPA. This deviation remarkably decreased when accounting for the effect of more fuel being converted into CO_2 (Equation (10.8)), which resulted in the dent in Figure 10.10. An increase in CO_2 elevates the estimated UFLs. It is suspected that the estimated UFLs falling below the measured UFLs for IPA is associated with the fact that the CO/CO_2 ratios were not constant for $1 - x < 1 - x_U$ but decreased with the value of $1 - x$.

10.4.2.4 Nitrogen Dilution Effect on the Flammability Envelope

The experimental LFL values for acetone or IPA diluted with nitrogen remained almost constant (Figures 10.11 and 10.12), which differed from the effect of steam. The model predicts this behavior. For other compounds diluted with nitrogen such as methane, propane, isobutane, ethylene, propylene, and methyl formate, the measured LFL values remained almost constant (Liaw et al., 2012), as did those of acetone and IPA. The constant LFL, when diluting with nitrogen, is because the heat capacity approaches that of air (Liaw et al., 2012). Alternatively, different mean values of heat capacity between steam and air caused the values of LFL to vary when using steam as the inert gas. Using acetone as an example, with the adiabatic flame temperature being 2063 K, the mean values of heat capacity for nitrogen and steam were 33.4 and 43.8 J mol^{-1}K^{-1}, respectively, with the former being close to that of air (33.8 J mol^{-1}K^{-1}). This result was similar for other fuels. To maintain a constant flame temperature, the greater heat capacity of steam increased the LFL. The predicted LFLs and UFLs for acetone and IPA diluted with nitrogen agreed remarkably well with their measurements, similar to diluting with steam (Table 10.2).

10.4.2.5 Location of the LOC and Flammability Limit at the LOC

Inerting design and the resolution of practical purging problems are based on the LOC data. This section compares the predicted flammability limits at the LOC with the corresponding experimental results. In this study, some of the tests terminated when the measured values of the LFL and UFL were extremely close. In such cases, the exact value of $1 - x_L$ was slightly greater than the measured value, and the exact value of the flammability limit at the LOC occurred within the range of the measured LFL and UFL, i.e., the exact value of oxygen concentration at the LOC was less than the measured range, as shown in Table 10.3.

The different heat losses had a small or negligible effect on the estimated values of $1 - x_L$ for the steam and nitrogen dilution. This effect of nitrogen dilution is also observed in a constant pressure system (Liaw et al., 2012). Table 10.3 shows that the estimated $1 - x_L$ values, regardless of heat loss, were close to the experimentally derived values for all gas–inert mixtures. The estimated values of $1 - x_L$ for acetone + steam were 0.930 and 0.935, respectively, for the adiabatic condition and accounting for heat loss. Both values are similar to the corresponding measurement of 0.92. In contrast to the LFL estimations closely matching the measured values over most of the composition region for all of the tested samples, the estimated values of the flammability limit were less than the measured values at the LOC (Table 10.3). This phenomenon was also observed for methane, propane, isobutane, ethylene, propylene, and methyl formate tested in a constant pressure system (Liaw et al., 2012). Maček (1979) conjectured that the carbon of the fuel was converted to CO around the LOC based on the observation of the CO stoichiometric curve crossing the LOC point. The FTIR spectra did not support this conjecture (Figures 10.3(c) and 10.5(c)) although a minuscule quantity of CO was observed around the LOC for acetone or methyl formate mixed with steam. Therefore, the explanation for elevated LFLs at the LOC due to the formation of CO at the LOC is incorrect. In the LFL estimation, it was assumed the fuel was totally consumed. The FTIR spectra showed that only a negligible amount of fuel existed in the burned gas at the LFL, except when close to the LOC. A remarkable quantity of fuel was observed at the LOC for acetone and methyl formate (Figures 10.3 and 10.5). Because the fuel was not totally expended, a greater quantity of fuel was needed to maintain a constant flame temperature, which caused an elevated LFL. Thus, the remaining unburnt fuel was the most significant cause of an increase in the LFL around the LOC (Figures 10.3(c) and 10.5(c)). The flammability limits of gases diluted with inert gases estimated using the proposed model are based on the values of the pure gases. The flammability limits of a given pure substance are not universal physical properties. They strongly depend on external factors (Mannan, 2005; Ju et al., 2001), such as initial temperature, ignition energy, and others. Therefore, the estimated values of the flammability limits for fuel/inert mixtures should be applied with careful attention to such external factors, when considering F&E safety. Many fuels that remain a long time in a vessel become partially oxidized at low temperatures and they produce cool flames that widen the flammability range (Pekalski et al., 2002). Since the proposed model is based on the assumption that the flame temperature remains constant, the flammability range resulting from cool flames is beyond the application of the model. That is, the proposed model is limited to, fuel-inert mixtures residing for a short time in a vessel.

10.5 SUMMARY

In this chapter a model was developed to describe the flammability boundaries of mixtures containing an inert gas for a constant volume system. The assumptions of the model were validated by combustion product analysis at the flammability boundaries. The differences in the estimated flammability envelope were small when considering the different heat losses. The predicted LFLs described the measured LFLs well, excluding the region around the LOC, where the assumption of total fuel depletion fails. All of the predicted UFLs agreed well with the experimental UFLs. Overall, the flammability boundary estimations are substantially in agreement with the measurements.

A. APPENDIX

A.1 Energy Balance Equation

For a constant volume, closed system, such as a spherical explosion vessel, the energy balance equation is:

$$\Delta U = Q_r \tag{A1}$$

where the heat is considered positive upon transfer from the surrounding to the system. In this study, only radiation heat loss was considered because of the high temperature during combustion. At the flammability limits, the pressure increase is not significant; the determination of flammability in this study was consistent with this assumption. The combustion mixture before and after burning is approximated as an ideal gas when the initial pressure is set at atmospheric pressure. If the temperature before and after combustion is T_0 and T, respectively, Equation (A1) is re-formulated as:

$$\Delta H^{\circ}_{298} + \sum_{reactants} \int_{T_0}^{298} n_i Cp_i dT + \sum_{products} \int_{298}^{T} n_i Cp_i dT + \sum_{reactants} n_i RT_0 - \sum_{products} n_i RT_0 = Q_r \tag{A2}$$

where ΔH°_{298} is the standard reaction heat at 298 K. Since the temperature dependence of specific heat at constant pressure, C_p, is easier to access than that of specific heat at constant volume, C_v, we approached this constant-volume problem with C_p rather than C_v.

A.2 Lower Flammability Boundary

Only fuels made up of C, H, and O atoms, $C_aH_bO_c$, are considered in this analysis. The combustion reaction at the LFL is assumed to be: (Liaw et al., 2012)

$$C_aH_bO_c + \left(a + \frac{b}{4} - \frac{c}{2}\right)O_2 \rightarrow aCO_2 + \frac{b}{2}H_2O$$

$$f_L = a + \frac{b}{4} - \frac{c}{2}$$

$$g_L = a \tag{A3}$$

$$h = \frac{b}{2}$$

A.2.1 Estimation of LFL

When consuming 1 mol of a combustible mixture at its LFL, where the mole of fuel is L, the standard heat of reaction at 298 K is (Liaw et al., 2012)

$$\Delta H^0_{298} = L(\Delta h^0_C) \tag{A4}$$

Where Δh_C^0 is the fuel's standard combustion heat. The heat capacities before and after burning, respectively, are: (Liaw et al., 2012)

$$\sum_{reactants} n_i C_{pi} = L\left(C_{pf} - C_{pi}\right) + \left(C_{pi} - P_L\right)\frac{L}{x} + P_l \tag{A5}$$

$$\sum_{products} n_i C_{pi} = L\left(C_{QL} - C_{pi}\right) + \left(C_{pi} - P_L\right)\frac{L}{x} + P_l \tag{A6}$$

Where,

$$x = \frac{fuel}{fuel + added\ inert} \tag{A7}$$

The nitrogen from air is excluded from the inert expression

$$P_L = 0.79 C_{pN_2} + 0.21 C_{pO_2} \tag{A8}$$

$$Q_L = -f_L C_{pO_2} + g_L C_{pCO_2} + h_L C_{pHO_2} \tag{A9}$$

And,

$$\sum_{reactants} n_i RT_O = RT_O \tag{A10}$$

$$\sum_{product} n_i RT_O = \left[1 + L\left(g_L + h_L - f_L - 1\right)\right]RT = \left(1 + LP_{PL}\right)RT \tag{A11}$$

Where

$$P_{P_L} = g_L + h_L - f_L - 1 \tag{A12}$$

Radiation heat loss of 1 mol of the mixture is:

$$Q_r = -\alpha e\, A_S \sigma\left(T^4 - T_o^4\right)\Delta t \tag{A13}$$

Inserting Equations A4–A6, A10, A11 and A13 into Equation A2:

$$\int_{T_0}^{298}\left[L\left(Cp_f - Cp_l\right) + \left(Cp_l - P_L\right)\frac{L}{X} + P_L\right]dT$$
$$+ \int_{298}^{T}\left[L\left(Q_L - Cp_l\right) + \left(Cp_l - P_L\right)\frac{L}{X} + P_L\right]dT$$
$$+ L\left(\Delta h_c^\circ\right) + RT_0 - \left(1 + LP_{PL}\right)RT = -\alpha e\underline{A_s}\sigma\left(T^4 - T_0^4\right)\Delta t \tag{A14}$$

Reformulated the above equation:

$$L = -\frac{\int_{T_0}^{T} P_L dT + R(T_0 - T) + \alpha e \underline{A}_s \sigma (T^4 - T_0^4) \Delta t}{\int_{298}^{T} Q_L dT + \int_{T_0}^{298} Cp_f dT - \int_{T_0}^{T} \left[(Cp_l - P_L)\frac{1}{x} \right] dT + \Delta h_c^\circ - P_{PL}RT} \tag{A15}$$

A.2.2 Estimation of x_L

At the point of the LOC, where $x = x_L$ (Liaw et al., 2012):

$$\frac{fuel}{oxygen} = \frac{L}{0.21\left(1 - \dfrac{L}{x_L}\right)} = \frac{1}{f_l}$$

Inserting Equation (A15) into the above equation results in:

$$X_L = -\frac{0.21\left[\int_{T_0}^{T} Cp_l dT + R(T_0 - T) + \alpha e \underline{A}_s \sigma (T^4 - T_0^4)\Delta t\right]}{\begin{aligned} f_L&\left[\int_{T_0}^{T} P_L dT + R(T_0 - T) + \alpha e \underline{A}_s \sigma (T^4 - T_0^4)\Delta t\right] \\ &+0.21\left[\int_{298}^{T}(Q_L - Cp_l)dT + \int_{T_0}^{298}(Cp_f - Cp_l)dT + \Delta h_c^0 - P_{PL}RT\right]\end{aligned}} \tag{A16}$$

A.3 Upper Flammability Boundary

The combustion reaction at the UFL is described as (Liaw et al., 2012):

$$C_a H_b O_c + \left(a - \frac{a_1}{2} + \frac{b - b_1}{4} - \frac{c}{2}\right)O_2 \rightarrow a_1 CO + (a - a_1)CO_2 + \frac{b_1}{2} H_2 + \frac{b - b_1}{2} H_2O$$

Thus, the method is limited to fuels with burned gases at the UFL that follow the above stoichiometry. Applicable definitions follow:

$$f_U = \frac{1}{a - \left(\dfrac{a_1}{2}\right) + \dfrac{(b - b_1)}{4} - \left(\dfrac{c}{2}\right)}$$

$$q_U = \frac{a_1}{a - \left(\dfrac{a_1}{2}\right) + \dfrac{(b - b_1)}{4} - \left(\dfrac{c}{2}\right)}$$

$$g_U = \frac{a - a_1}{a - \left(\dfrac{a_1}{2}\right) + \dfrac{(b - b_1)}{4} - \left(\dfrac{c}{2}\right)} \tag{A17}$$

$$t_U = \frac{\dfrac{b_1}{2}}{a - \left(\dfrac{a_1}{2}\right) + \dfrac{(b - b_1)}{4} - \left(\dfrac{c}{2}\right)}$$

$$h_U = \frac{\dfrac{b - b_1}{2}}{a - \left(\dfrac{a_1}{2}\right) + \dfrac{(b - b_1)}{4} - \left(\dfrac{c}{2}\right)}$$

A.3.1 UFL Estimate for $1 - x \leq 1 - x_U$

In the region of $1 - x \leq 1 - x_U$, the standard heat of reaction is (Liaw et al., 2012):

$$\Delta H^0_{298} = 0.21 f_U \left(1 - \frac{U}{x}\right) \left(\Delta h^0_c\right)_U \tag{A18}$$

$$\left(\Delta h^0_c\right)_U = \Delta h^0_c - a_1 \left(\frac{\Delta h^0_{CO}}{CO_2}\right) - \frac{b_1}{2}\left(\Delta h^0_{H_2O}\right) \tag{A19}$$

And

$$\frac{\Delta h^o_{CO}}{CO_2} = \Delta H^o_{f.CO_2,298} - \Delta H^o_{f.CO.298}$$

The heat capacities before and after burning, respectively, are (Liaw et al., 2012):

$$(A20) \sum_{reactants} n_i Cp_i = U\left(Cp_f - Cp_I\right) + \frac{U}{x}\left(Cp_I - P_L\right) + P_L$$

and

$$\sum_{products} n_i Cp_i = U\left(Cp_f - Cp_I\right) + \frac{U}{x}\left(Cp_I - P_U\right) + P_U \tag{A21}$$

where

$$P_U = 0.21 q_U Cp_{CO} + 0.21 g_U Cp_{CO_2} + 0.21 t_U Cp_{H_2} + 0.21 h_U Cp_{H_2O} + 0.79 Cp_{N_2} - 0.21 f_U Cp_f \tag{A22}$$

And

$$= \sum_{products} n_i RT = \left[0.79 + 0.21\frac{U}{x} + 0.21\left(1 - \frac{U}{x}\right)\left(q_U + g_U + t_U + h_U - f_U\right)\right]$$

$$RT = \left[0.79 + 0.21\frac{U}{x} + 0.21\left(1 - \frac{U}{x}\right)P_{PU}\right]RT \tag{A23}$$

Where,

$$P_{PU} = q_U + g_U + t_U + h_U - f_U \tag{A24}$$

Inserting Equations (A10), (A13), (A18), (A20), (A21), and (A23) into Equation (A2):

$$0.21 f_U \left(1 - \frac{U}{x}\right)(\Delta h_C^0)_U + \int_{298}^{T}\left[U\left(Cp_f - Cp_I\right) + \frac{U}{x}\left(Cp_I - P_U\right) + P_U\right]dT$$
$$+ \int_{T_0}^{298}\left[U\left(Cp_f - Cp_I\right) + \frac{U}{x}\left(Cp_I - P_L\right) + P_L\right]dT +$$
$$RT_0 - \left[0.79 + 0.21\frac{U}{x} + 0.21\left(1 - \frac{U}{x}\right)P_{PU}\right]RT = -\alpha e \underline{A}_S \sigma \left(T^4 - T_0^4\right)\Delta t \tag{A25}$$

The following equation is obtained

$$U = -\frac{\displaystyle\int_{298}^{T} P_U dT + \int_{T_0}^{298} P_L dT + 0.21 f_U \left(\Delta h_c\right)_U}{\displaystyle\int_{T_0}^{T} Cp_f - Cp_I)dT + \int_{298}^{T}\left(Cp_I - P_U\right)\frac{1}{x}dT +}{\displaystyle\int_{T_0}^{298}\left(Cp_I - P_L\right)\frac{1}{x}dT - 0.21 f_U \frac{1}{x}\left(\Delta h_C^0\right)_U - \frac{0.21}{x}\left(1 - P_{PU}\right)RT} \tag{A26}$$

A.3.2 Estimation of xU

At the point that $x = xU$, where the ratio of CO_2/CO in the burned gas increases with $1 - x$ (Liaw et al., 2012).

Substituting Equation (A26) into the above equation resulted in:

$$X_U = -\frac{0.21 f_U \left[\displaystyle\int_{T_0}^{T} Cp_I dT + R\left(T_0 - T\right) + \alpha e \underline{A}_S \sigma \left(T^4 - T_0^4\right)\Delta t\right]}{\displaystyle\int_{298}^{T} P_U dT + \int_{T_0}^{298} P_L dT + RT_0 - \left(0.79 + 0.21 P_{PU}\right)RT}{+\alpha e \underline{A}_S \sigma \left(T^4 - T_0^4\right)\Delta t + 0.21 f_U \left[\displaystyle\int_{T_0}^{T} Cp_f - Cp_I\right)dT + \left(\Delta h_c^0\right)_U\right]} \tag{A27}$$

A.3.3 UFL Estimate for 1 − x > 1 − x_U

In the region of $1 - x > 1 - x_U$, the heat of reaction can be estimated as (Liaw et al., 2012):

$$\Delta H_{298}^0 = U\left(\Delta h_C^0\right)_U \tag{A28}$$

The heat capacity before combustion is given as Equation (A20). The heat capacity following combustion is (Liaw et al., 2012):

$$\sum_{products} n_i Cp_i = UQ_U + a_1 U\left(Cp_{CO} - Cp_{CO_2}\right) + \frac{U}{x}\left(Cp_I - P_L\right) + P_L - \frac{U}{f_U}Cp_{O_2} \quad (A29)$$

where

$$Q_U \equiv aCp_{CO_2} + \frac{t_U}{f_U}Cp_{H_2} + \frac{h_U}{f_U}Cp_{H_2O} - Cp_I \quad (A30)$$

and

$$\sum_{products} n_i = 1 - U\left[1 + \frac{1}{f_U} - (a + 0.5b)\right] = 1 - P_{P2U}U - \frac{U}{f_U} \quad (A31)$$

Where

$$P_{P2U} = 1 - (a + 0.5b)$$

Inserting Equations (A10), (A13), (A20), (A28), (A29), and (A31) into Equation (A2), and replacing $(\Delta h^\circ_c)_U$ with Equation (A19), the following equation is derived:

$$U\left(\Delta h_C^0 - a_1\left(\underset{CO_2}{\Delta h_{CO}^0}\right) - \frac{b_1}{2}\left(\Delta h_{H_2O}^0\right)\right) +$$
$$\int_{298}^{T}\left[UQ_U + a_1 U\left(Cp_{CO} - Cp_{CO_2}\right) + \frac{U}{x}\left(Cp_I - P_L\right) + P_L - \frac{U}{f_U}Cp_{O_2}\right]dT$$
$$+\int_{T_0}^{298}\left[U\left(Cp_f - Cp_I\right) + \frac{U}{x}\left(Cp_I - P_L\right) + P_L\right]dT + RT_0 - \left(1 - P_{P2U}U - \frac{U}{f_U}\right)RT_0 =$$
$$-\alpha e\underline{A}_s\sigma\left(T^4 - T_0^4\right)\Delta t$$

$$(A32)$$

As derived in a previous study (Liaw et al., 2012):

$$a_1 = p - 0.42\left(\frac{1}{U} - \frac{1}{x}\right) \quad (A33)$$

where

$$p = 2a + \frac{b - b_1}{2} - c \quad (A34)$$

Inserting Equation (A33) into Equation (A32) results in:

$$U = -\frac{\begin{aligned}&\int_{298}^{T}\left[P_L - 0.42\left(Cp_{CO} - Cp_{CO_2}\right) - 0.21Cp_{O_2}\right]dT + \\ &\int_{T_0}^{298} P_L dT + 0.42\left(\underset{CO_2}{\Delta h_{CO}^0}\right) + R\left(T_0 - 0.79T\right) + \alpha e \underline{A}_S \sigma\left(T^4 - T_0^4\right)\Delta t\end{aligned}}{\begin{aligned}&\int_{298}^{T}\left[Q_U + \frac{1}{x}\left(Cp_I - P_L\right) + \left(Cp_{CO} - Cp_{CO_2}\right)\left(p + \frac{0.42}{x}\right) + \frac{0.21}{x}Cp_{O_2}\right]dT + \\ &\int_{T_0}^{298}\left[\left(Cp_f - Cp_I\right) + \frac{1}{x}\left(Cp_I - P_L\right)\right]dT + \left(\Delta h_p\right) + \left(P_{P2U} - \frac{0.21}{x}\right)RT\end{aligned}}$$

(A35)

$$\left(\Delta h_p\right) = \Delta h_C^0 - \frac{b_1}{2}\left(\Delta h_{H_2O}^0\right) - \left(p + \frac{0.42}{x}\right)\left(\underset{CO_2}{\Delta h_{CO}^0}\right)$$ (A36)

The flame temperature is assumed to be linear with the quantity of CO, a1 (Liaw et al., 2012):

$$T = \frac{a_1}{a_{1U}}T_U + \left(1 - \frac{a_1}{a_{1U}}\right)T_L$$ (A37)

Where a_1 is equivalent to a_{1U} at the point that $x = x_U$

A.4 Flame Temperature
For a fuel–air mixture without any inert gas:

$x = 1$
$L = L_0$
$U = U_0$
$f_U = f_{U0}$
$P_U = P_{U0}$

Equation (A14) can be reduced to:

$$L_0\left(\Delta h_C^0\right) + \int_{298}^{T}\left[L_0\left(Q_L - P_L\right) + P_L\right]dT + $$
$$\int_{T_0}^{298}\left[L_0\left(Cp_f - Cp_I\right) + P_L\right]dT + RT_0 - \left(1 + L_0 P_{PL}\right)$$
$$RT = -\alpha e \underline{A}_S \sigma\left(T^4 - T_0^4\right)\Delta t$$

(A38)

Equation (A25) can be reduced to:

$$0.21 f_U \left(1 - U_0\right)\left(\Delta h_C\right)_U + \int_{298}^{T} \left[U_0 \left(Cp_f - P_{U0}\right) + P_{U0}\right] dT$$
$$+ \int_{T_0}^{298} \left[U_0 \left(Cp_f - P_L\right) + P_L\right] dT + RT_0$$
$$- \left[0.79 + 0.21 U_0 + 0.21 \left(1 - U_0\right) P_{PU0}\right] RT = -\alpha e \underline{A}_S \sigma \left(T^4 - T_0^4\right) \Delta t \qquad \text{(A39)}$$

Equations. (A38) and (A39) are applied to calculate the flame temperatures at the LFL and UFL, respectively.

REFERENCES

American Society for Testing and Materials (ASTM) E1226-05, 2005. *Standard Test Method for Pressure and Rate of Pressure Rise for Combustible Dusts*. ASTM, West Conshohocken, PA.

American Society for Testing and Materials (ASTM) E2079-07, 2007. *Standard Test Methods for Limiting Oxygen (Oxidant) Concentration in Gases and Vapors*. ASTM, West Conshohocken, PA.

American Society for Testing and Materials (ASTM) E681, 1994.*Standard Test Method for Concentration Limits of Flammability of Chemicals (Vapors and Gases)*. ASTM, West Conshohocken, PA.

American Society for Testing and Materials (ASTM) E918-83, 2005. *Standard Practice for Determining Limits of Flammability of Chemicals at Elevated Temperatures and Pressures*. ASTM, West Conshohocken, PA.

Britton, L.G., 2002. Two hundred years of flammable limits. *Process Saf. Prog.* 21, 1–11.

Brooks, M.R., Crowl, D.A., 2007. Flammability envelopes for methanol, ethanol, acetonitrile and toluene. *J. Loss Prev. Proc.* 20, 144–150.

Buckmaster, J.D., 1976. The quenching of deflagration waves. *Combust. Flame* 26, 151–162.

Chen, C.C., Liaw, H.J., Wang, T.C., Lin, C.Y., 2009a. Carbon dioxide dilution effect on flammability limits for hydrocarbons. *J. Hazard. Mater.* 163, 795–803.

Chen, J.R., Tsai, H.Y., Pan, H.J., Huang, Y.P., 2008. *Characterization of upper flammability limits of propane/air mixtures at elevated pressures*. In: *Seventh International Symposium on Hazards, Prevention, and Mitigation of Industrial Explosions*, St. Petersburg, Russia, pp. 166–174.

Chen, C.C., Wang, T.C., Liaw, H.J., Chen, H.C., 2009b. Nitrogen dilution effect on the flammability limits for hydrocarbons. *J. Hazard. Mater.* 166, 880–890.

Chong, C.T., Hochgreb, S., 2011. Measurements of laminar flame speeds of acetone/methane/air mixtures. *Combust. Flame* 158, 490–500.

Coward, H.F., Jones, G.W., 1952. *Limits of Flammability of Gases and Vapors*. Bureau of Mines, United States Government Printing Office, Washington, DC.

Crowl, D.A., Louvar, J.F., 2011. *Chemical Process Safety: Fundamentals with Applications*, third ed. Pearson Education, Inc., Boston, MA.

Design Institute for Physical Properties (DIPPR), 2012. *DIPPR ®Data Compilation of Pure Chemical Properties*. American Institute of Chemical Engineers, New York,

NY. Dooley, S., Burke, M.P., Chaos, M., Stein, Y., Dryer, F.L., Zhukov, V.P., Finch, O., Simmie, J.M., Curran, H.J., 2010. Methyl formate oxidation: speciation data, laminar burning velocities, ignition delay times, and a validated chemical kinetic model. *Int. J. Chem. Kinet.* 42, 527–549.

Egolfopoulos, F.N., Du, D.X., Law, C.K., 1992. A comprehensive study of methanol kinetics in freely-propagating and burner-stabilized flames, flow and static reactors, and shock tubes. *Combust. Sci. Technol.* 83, 33–75.

Gérard, E., Götz, H., Pellegrini, S., Castanet, Y., Mortreux, A., 1998. Epoxide–tertiary amine combinations as efficient catalysts for methanol carbonylation into methyl formate in the presence of carbon dioxide. *Appl. Catal., A: Gen.* 170, 297–306.

Hansen, T.J., Crowl, D.A., 2010. Estimation of the flammability zone boundaries for flammable gases. *Process Saf. Prog.* 29, 209–215.

Joulin, G., Clavin, P., 1976. Analyse asymptotique des conditions d'extinction des flammes laminaires. *Acta Astronaut.* 3, 223–240.

Ju, Y., Maruta, K., Niioka, T., 2001. Combustion limits. *Appl. Mech. Rev.* 54, 257–277.

Kondo, S., Takizawa, K., Takahashi, A., Tokuhashi, K., 2006a. Extended Le Chatelier's formula and nitrogen dilution effect on the flammability limits. *Fire Saf. J.* 41, 406–417.

Kondo, S., Takizawa, K., Takahashi, A., Tokuhashi, K., 2006b. Extended Le Chatelier's formula for carbon dioxide dilutioneffect on flammability limits. *J. Hazard. Mater.* A138, 1–8.

Kondo, S., Takizawa, K., Takahashi, A., Tokuhashi, K., Sekiya, A., 2007. Flammability limits of isobutane and its mixtures with various gases. *J. Hazard. Mater.* 148, 640–647.

Kuchta, J.M., 1985. *Investigation of Fire and Explosion Accidents in the Chemical, Mining, and Fuel-Related Industries—A Manual, Bureau of Mines. United States* Government Printing Office, Washington, DC.

Liaw, H.J., Chen, C.C., Chang, C.H., Lin, N.K., Shu, C.M., 2012. Model to estimate the flammability limits of fuel–air–diluent mixtures tested in a constant pressure vessel. *Ind. Eng. Chem. Res.* 51, 2747–2761.

Liaw, H.J., Chen, C.C., Lin, N. K., Shu, C.M., Shen, S.Y., 2016. Flammability limits estimation for fuel–air–diluents mixtures tested in a constant volume vessel. *Proc. Safety Environ. Protec.* 100, 150-162.

Liaw, H.J., Chiu, Y.Y., 2003. The prediction of the flash point for binary aqueous–organic solutions. *J. Hazard. Mater.* 101, 83–106.

Maček, A., 1979. Flammability limits: a re-examination. *Combust. Sci. Technol.* 21, 43–52.

Mannan, S., 2005. *Lees' Loss Prevention in the Process Industries, vol. 1.*, third ed. Elsevier Butterworth-Heinemann, Oxford, UK.

Melhem, G.A., 1997. A detailed method for estimating mixture flammability limits using chemical equilibrium. *Process Saf. Prog.* 16, 203–218.

Pekalski, A.A., Zevenbergen, J.F., Pasman, H.J., Lemkowitz, S.M., Dahoe, A.E., Scarlett, B., 2002. The relation of cool flames and auto-ignition phenomena to process safety at elevated pressure and temperature. *J. Hazard. Mater.* 93, 93–105.

Ronney, P.D., Wachman, H.Y., 1985. Effect of gravity on laminar premixed gas combustion, I: Flammability limits and burning velocities. *Combust. Flame* 62, 107–119.

Rowley, J.R., Rowley, R.L., Wilding, W.V., 2010. Experimental determination and re-examination of the effect of initial temperature on the lower flammability limit of pure liquids. *J. Chem. Eng. Data* 55, 3063–3067.

Shebeko, Y.N., Fan, W., Bolodian, I.A., Navzenya, V.Y., 2002. An analytical evaluation of flammability limits of gaseous mixtures of combustible–oxidizer–diluents. *Fire Saf. J.* 37, 549–568.

SKC, 2015. *World leader in sampling technologies,* Available at:_http://www.skcinc.com/catalog/index.php?cPath=20000000020200000202000300_ (accessed in 2015).

Spalding, D.B., 1957. A theory of inflammability limits and flame quenching. *Proc. R. Soc. London, Ser. A* 240, 83–100.

Veloo, P.S., Egolfopoulos, F.N., 2011. Studies of n-propanol, iso-propanol, and propane flames. *Combust. Flame* 158, 501–510.

Vidal, M., Wong, W., Rogers, W.J., Mannan, M.S., 2006. Evaluation of lower flammability limits of fuel–air–diluents mixtures using calculated adiabatic flame temperatures. *J. Hazard. Mater.* 130, 21–27.

Welty, J.R., Wicks, C.E., Wilson, R.E., 1976. *Fundamentals of Momentum, Heat, and Mass Transfer*, second ed. John Wiley & Sons, New York, NY, USA.

Zabetakis, M.G., 1965. *Flammability Characteristics of Combustible Gases and Vapors.* Bureau of Mines, United States Government Printing Office, Washington, DC.

11 Advanced Calorimetric Technology for the Kinetic and Thermal Safety Analysis of Tert-butylperoxy-3,5, 5-trimethylhexanoate

11.1 INTRODUCTION

Over the years, there have been many casualties around the world as the result of accidents in chemical industries. Such accidents resulted in losses of life, property damage, health hazards to workers, and the environment. The main reason for accidental fire hazards and explosions is thought to be the thermal runaway reactions of free-radical initiators. More than 250 severe fire accidents have occurred worldwide, from 1979 to 1999 as a result of just such reactions (Talouba et al., 2011). Important parameters of safety (IPSs) regarding the physicochemical properties of all chemicals must be established before their use on an industrial scale, since it can prevent fire-related accidents and promote industrial sustainability. But the main concern is that such reports of IPSs for predicting the kinetic behavior of all free-radical initiators are not clearly mentioned in past studies. By conducting experiments or by using well-known kinetic models, the kinetic behavior of such free-radical initiators can be predicted (Talouba et al., 2011). Mostly in the polymer and rubber industries, organic peroxides (OPs) among the free-radical initiators are widely employed due to their free radicals (RO·) and energy supply efficiency (Hou et al. 2011). Presently, in many chemical industries the thermal sensitivity and runaway characteristics of OPs are managed with precautions for safety measures. Due to any thermal abnormalities caused by a sudden rise in temperature, mechanical shock, or the presence of any contaminants, violent exothermic reactions of OPs can get triggered, which might lead to fire hazards (Yan et al., 2014). OPs are well-known for their thermal instability and exothermic reaction is the fundamental characteristic of their thermal decomposition. OPs are sensitive to temperature and have high values of activation energy because of the weak bond energy of O–O linkage (80 to 200 kJ mol^{-1}). Their incompatibility with other materials (acids, bases, metal and ions) means that it is crucial to assess the exothermic reactions and kinetic behavior in order to understand their potential thermal hazards during handling and storage. Further such understandings can be used in determining suitable safety evaluation methods which could prevent thermal risks (Duh et al. 2008).

11.2 THERMAL SENSITIVITY AND RUNAWAY CHARACTERISTICS OF TERT-BUTYL PEROXY-3,5,5-TRIMETHYLHEXANOATE

Mostly for polymerization of styrene, and acrylates, organic peroxide tert-butyl peroxy-3,5,5-trimethylhexanoate (TBPMH) is used widely as an initiator or curing agent at 90–170 °C. Due to the presence of the O–O bond in TBPMH, it releases free radicals spontaneously, which initiates the free radical polymerization of monomer–polymer interconnections. However, a vast amount of heat and flammable gases were also released during decomposition due to the O–O bond. Such circumstances can result in severe damage by fire or explosion. The results of the kinetic and thermal analysis of TBPMH are very scant in the literature, and thus little is known about their thermal stability/kinetic parameters, their kinetic model, or suitable methods of analysis. However, at low onset temperature, TBPMH has high heat of reaction. It is thus recommended that the thermal characteristics of TBPMH should be analysed before its usage, transportation, or storage, in order to prevent serious thermal accidents. It is therefore recommended that such thermal characteristics must be determined beforehand in order to ensure safe manufacturing and transportation of such unstable compounds (Liu et al., 2013). Literature on the use of calorimetric approach-coupled kinetic models were explored to determine the exothermic runaway reactions of several hazardous chemicals such as tert-butyl peroxy-2-ethyl hexanoate, di-tert-butyl peroxide, and cumene hydroperoxide. Such methods can be used to reliably and precisely predict the innate thermal safety of TBPMH, particularly under process upsets (You et al., 2009; Lin et al., 2010).

In this chapter, the apparent exothermic onset temperature (T_o), peak temperature (T_p), final temperature (T_f), apparent activation energy (E_a), pre-exponential factor (A), reaction order (n), time to maximum rate under adiabatic conditions (TMR$_{ad}$), adiabatic temperature rise (ΔT_{ad}), and self-accelerating decomposition temperature (SADT), were determined through non-isothermal differential scanning calorimetry (DSC) combined with kinetic models. The findings can provide valuable information about thermal safety for storing, transporting, and manufacturing TBPMH, thereby preventing thermal accidents.

11.3 SAMPLE PREPARATION

11.3.1 SAMPLE

The TBPMH (100 mL, 97 mass% colorless liquid) was purchased from Acros Organics Corp (Geel, Belgium) and was stored in a refrigerator at temperatures < 4 °C due to its thermal sensitivity (Safety Data Sheet, 2015, 2016).

11.3.2 DIFFERENTIAL SCANNING CALORIMETRY

The influence of heat flow on temperature was assessed using DSC to evaluate the thermal stability of TBPMH. DSC experiments were performed in stainless steel high-pressure crucibles using a PerkinElmer DSC 8500 instrument. DSC experiments were performed at 0.5, 1, 2, and 4 °C min^{-1} heating rates in the temperature range of 30 to 300 °C with approximately 1–4 mg of TBPMH (Tong et al., 2014).

11.4 DETERMINATION OF THE KINETIC MODEL

The basic steps involved in model fitting methods were: evaluating the values of E_a and A, choosing a kinetic model, and performing a numerical simulation (Vyazovkin et al., 2011). Montserrat et al. (Montserrat et al., 1998) proposed a model function $y(\alpha)$ as:

$$y(\alpha) = Af(\alpha) = \frac{d\alpha}{dt} \exp\left(\frac{E_a}{RT}\right) \qquad (11.1)$$

where $d\alpha/dt$ and E_a can be determined through the Friedman method. The suitable kinetic model is determined by the shape of the plot of function $y(\alpha)$ vs α. The plot of the reaction function $y(\alpha)$ at different heating rates and α_{max} (maximum point) was presented in Figure 11.1. The different kinetic models and the nature of the curves of $y(\alpha)$ vs α was presented in Table 11.1. The kinetic models used were Johnson–Mehl–Avrami (JMA) and autocatalysis model at $0 < \alpha_{max} < \alpha_p$, where α_p is the heat flow at a maximum degree of conversion (Yoo et al., 2010). According to the literature, the value of α_p was 0.632 for JMA model with order of the reaction > unity (Málek, 1995) and is different from the experimental values. The kinetic behavior of TBPMH was simulated by the autocatalytic model to assess thermokinetic parameters such as ΔH_d, n, E_a, and A.

11.5 TIME FOR MAXIMUM RATE AT ADIABATIC CONDITIONS

At adiabatic conditions, the changes in time and temperature during thermal decomposition was evaluated at TMR_{ad} for a failure in cooling/stirring systems. This could

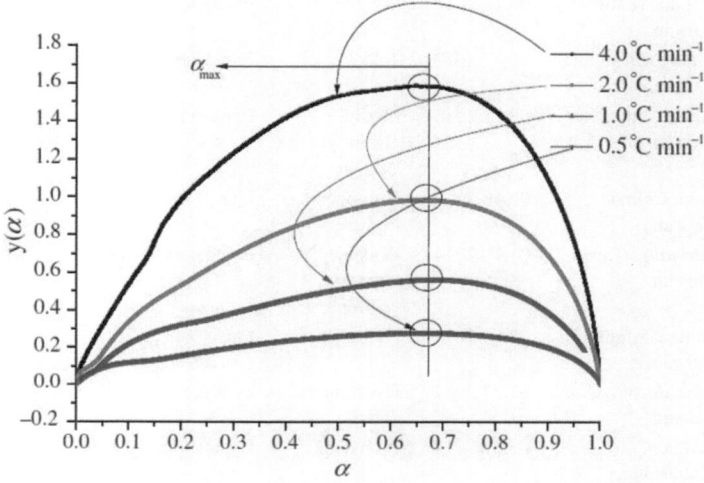

FIGURE 11.1 Prediction of function $y(\alpha)$ at heating rates of 0.5, 1.0, 2.0, and 4.0 °C min⁻¹ and various degrees of conversion (Reproduced with permission from Yang et al. Copyright © Springer, 2017)

TABLE 11.1
Major Accidents with Casualties in Taiwanese Chemical Plants, 2005–2015

Items	Plant Location	Date	Chemicals	Type	Injured/Death	Remarks
1	Tuku, Yunlin, Taiwan	04.02.2005	Trinitrotoluene	Explosion	3/2	Firecracker plant
2	Nantun, Taichung, Taiwan	07.03.2005	Sodium nitrite	Fire	22/0	Three firefighters injured
3	Hsitun, Taichung, Taiwan	02.20.2006	Hydrofluoric acid	Splash	2/0	
4	Mailiao, Yunlin, Taiwan	03.06.2006		Fire	5/1	Improper operation
5	Shulin, New Taipei City	05.21.2007	Nickel sulfate	Leakage	4/2	
6	Luchu, Taoyuan, Taiwan	06.25.2007	Propylene glycol methyl ether	Explosion	5/1	
7	Xinwu, Taoyuan, Taiwan	08.16.2007	Nitrite	Explosion	6/0	Improper operation
8	Guanyin, Taoyuan, Taiwan	09.22.2009	Calcium fluoride	Explosion	3/1	
9	Mailiao, Yunlin, Taiwan	11.18.2009	Phosgene	Leakage	12/0	
10	Nantun, Taichung, Taiwan	06.11.2010	Xylene	Explosion	5/2	
11	Nantun, Taichung, Taiwan	06.18.2011	Foam	Fire	1/1	Grinder sparks
12	Mailiao, Yunlin, Taiwan	09.14.2011	Styrene	Explosion	2/0	
13	Lukang, Changhua, Taiwan	05.17.2012	Toluene	Explosion	13/1	
14	Nantun, Taichung, Taiwan	08.01.2012	Methyl isobutyl ketone	Explosion	3/0	Static spark
15	Dalin, Chiayi, Taiwan	08.10.2012	Hydrogen	Fire	3/0	
16	Minsyong, Chiayi, Taiwan	01.10.2014	Dodecyl benzene sulfonic acid	Explosion	1/1	Welding spark
17	Mailiao, Yunlin, Taiwan	03.05.2014	Hydrogen	Fire	2/0	
18	Guishan, Taoyuan, Taiwan	03.27.2014	Hydrofluoric acid	Leakage	1/0	One firefighter injured
19	Cianjhen, Kaohsiung, Taiwan	03.31.2014	Propylene	Explosion	308/32	24 firefighters injured, 6 firefighters dead

(Data obtained with permission from Yang et al. Copyright © Springer, 2017)

help in the determination of maximum synthesis temperature and severity of a runaway reaction during the production process. TMR_{ad} was derived as follows (Singh et al., 2012 & Tsai et al., 2013). The effect of exothermic reaction on material characteristics was presented in Equation (11.2): where

$$\frac{dT}{dt} = \frac{M_s C_{p,s}}{M_c C_{p,c} + M_s C_{p,s}} \frac{-\Delta H_d d\alpha}{C_{p,s} dt}$$ (11.2)

where M_s/M_c and $C_{p,s}/C_{p,c}$ represents mass and specific heat capacity of sample and test cell, respectively. The ΔT_{ad} and phi (Φ) can be derived from Equations (11.3) and (11.4) (Saraf et al. 2003; Xiao et al. 2009)

$$\Phi = \frac{M_c C_{p,c} + M_s C_{p,s}}{M_s C_{p,s}}$$ (11.3)

$$\Delta T_{ad} = \frac{\Delta H_d}{C_{p,s}}$$ (11.4)

Substituting these equations in Equation (11.2) we get:

$$\frac{dT}{dt} = \frac{1}{\Phi} \Delta T_{ad} \frac{d\alpha}{dt}$$ (11.5)

where $d\alpha/dt$ was determined using Friedman method. The integration of dT/dt and Φ determines the value of TMR_{ad}.

11.6 SELF-ACCELERATING DECOMPOSITION TEMPERATURE

Self-accelerating decomposition reactions of OPs depend mainly on surrounding temperature and package configuration. For the transportation and storage of OPs, the determination of optimal temperature, package sizes, and SADT is essential. The SADT is defined as the lowest ambient temperature inducing the center that exceed 6 °C within 7 days or less for a commercial package and is usually measured when the center temperature of a package is 2 °C below the surrounding temperature (Kozlowski and Kurko, 2008; Fauske 2011). The heat generated from commercial packages of specific sizes of TBPMH were simulated by different kinetic models at boundary conditions as follows (Lu et al., 2015).

$$\frac{dT}{dt} = \frac{\lambda}{\rho C_{p,s}} \left(\frac{\partial^2 T}{\partial x^2} + \frac{g \partial T}{x \partial x} \right) + \frac{-\Delta H_d}{C_{p,s}} \frac{d\alpha}{dt}$$ (11.6)

where, $C_{p,s}$ is the sample heat capacity, λ is the thermal conductivity, ρ is the density of TBPMH, x is the package radius, and g is the geometry factor (2 for cylinder) (Malow & Wehrstedt, 2005).

11.7 RESULTS AND DISCUSSION

The safety of reactive chemicals and underlying thermal features can be analyzed by DSC. The determination of thermal stability characteristics and thermokinetic parameters is also facilitated by DSC. The heat flow curves of TBPMH at different heating rates is illustrated in Figure 11.2. The average values of ΔH_d was -924.0 J g^{-1} with T_o, T_p, T_f, approximately and 103, 132, and 147 °C for TBPMH at the four heating rates. From Table 11.2 it can be observed that, at higher heating rates, the maximum exothermic peak was higher and the T_0, T_p, and T_f was delayed, releasing a substantial amount of heat during the thermal decomposition of TBPMH. A chemical's ΔH_d (250.0 and 300.0 J g^{-1}) level can increase adiabatic temperatures by 100 to 200 °C and is considered potentially hazardous, according to the Center for Chemical Process Safety of AIChE. In a chemical reaction, deflagration and detonation can occur when ΔH_d exceeds from 1000 to 3000.0 Jg^{-1} (Sato et al., 2011). The thermal decomposition during preparation, processing, and storage of TBPMH may result in thermal runaway (exothermic reaction induced by detonation/deflagration) causing fire and explosion. Friedman method (differential approach) can be used to calculate E_a and determine the reaction hazard during various decomposition stages of TBPMH and is derived as follows (Tsai et al., 2012):

$$\ln \frac{d\alpha}{dt} = \ln\left[A(\alpha)f(\alpha) \right] - \left(\frac{E(\alpha)}{RT(t)} \right) \tag{11.7}$$

A plot of ln($d\alpha/dt$) vs $1/T(t)$ represents straight line with slope and intercept $E(\alpha)$, and ln[$A(\alpha)f(\alpha)$], respectively at a constant α. Figures 11.3 and 11.4 represents the plot of ln($d\alpha/dt$) vs $1/T$ and the values of E_a and ln[$A(\alpha)f(\alpha)$], respectively.

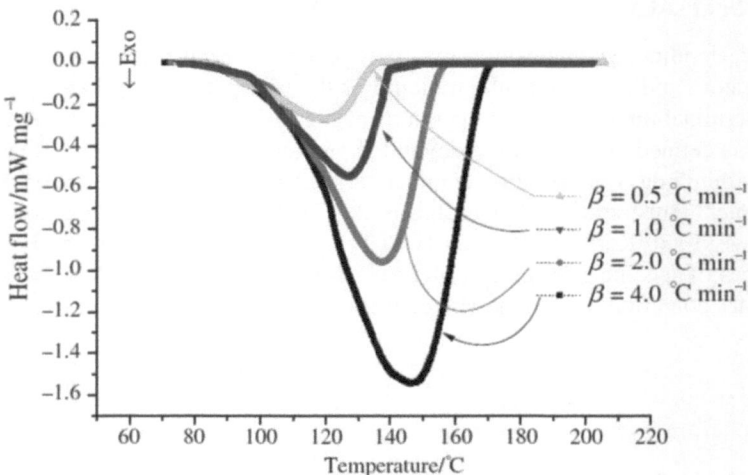

FIGURE 11.2 DSC thermal curves for TBPMH (97.0 mass%) at heating rates of 0.5, 1.0, 2.0, and 4.0 °C min^{-1} (Reproduced with permission from Yang et al. Copyright © Springer, 2017)

TABLE 11.2
Major Accidents with Casualties in Chemical Plants Globally, 2005–2015

Items	Plants Location	Date	Chemicals	Type	Injured/Death	Remark
1	Texas, USA	03.23.2005	Light oil	Explosion	180/15	
2	Jilin, China	11.13.2005	Nitrobenzene	Explosion	60/8	
3	Buncefield, UK	12.11.2005	Gasoline	Explosion	43/0	
4	Osaka, Japan	12.19.2006	Aluminum	Explosion	2/0	
5	Hebei, China	05.11.2007		Fire	80/4	Improper operation
6	Florida, USA	12.19.2007	Trinitrotoluene	Explosion	14/4	
7	Istanbul, Turkey	01.31.2008	Trinitrotoluene	Explosion	68/17	
8	Markazi Province, Iran	05.25.2008		Explosion	38/30	Welding spark
9	Georgia, USA	07.02.2008	Dust	Explosion	42/13	
10	Liaoning, China	09.14.2008		Fire	2/3	
11	Penang, Malaysia	04.24.2009	Gas	Explosion	5/1	
12	Ahmedabad, India	07.06.2009	Trinitrotoluene	Explosion	100/30	
13	Kharg Island, Iran	07.24.2010		Explosion	0/4	Boiler pressure was too high
14	Liaoning, China	01.19.2011	Heavy oil	Explosion	30/0	
15	Guangxi, China	11.23.2011	Trinitrotoluene	Explosion	11/4	Fire cracker plant
16	Amuay, Venezuela	08.25.2012	Propane	Explosion	130/50	
17	Hyogo, Japan	09.29.2012	Acrylic acid	Explosion	30/1	One firefighter dead
18	Jeollanam, Korea	03.14.2013	High density polyethylene	Explosion	11/6	Welding spark
19	Jiangsu, China	04.16.2014	Dust	Explosion	9/8	
20	Moerdijk, Netherlands	06.03.2014	Ethylbenzene	Explosion	2/0	

(Data obtained with permission from Yang et al. Copyright © Springer, 2017)

It was observed that the values of E_a decreased from 0.1 and 0.9 and is dependent on the reaction characteristics/profile (Naranjo et al., 2012). The decrease in values of E_a with an increase in α is characteristic of an autocatalytic reaction. The decomposition of TBPMH may have accelerated the reaction rate resulting in thermal runaway (Chi et al., 2012). The range of E_a values for TBPMH was approximately 80–200 kJ mol^{-1} with R^2 values of 0.99 indicating an excellent correlation, as signified by an error of less than 5% (Omrani et al., 2008). The values of E_a from literature was in the range 127–140 kJ mol^{-1} and is different from E_a (91 kJ mol^{-1}) derived using the Friedman method as represented in Table 11.3 (Montserrat et al. 1998; Yoo et al., 2010). This could be due to the fact that Friedman method considers the entire

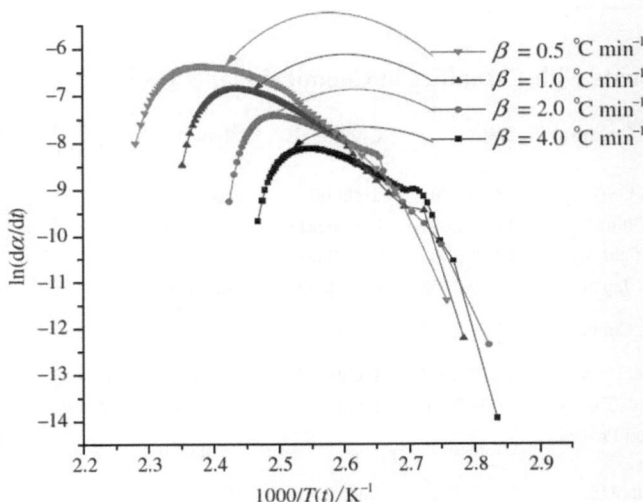

FIGURE 11.3 Evaluation of $E(\alpha)$ and $\ln A(\alpha)$ for TBPMH (97.0 mass%) by the Friedman method (Reproduced with permission from Yang et al. Copyright © Springer, 2017)

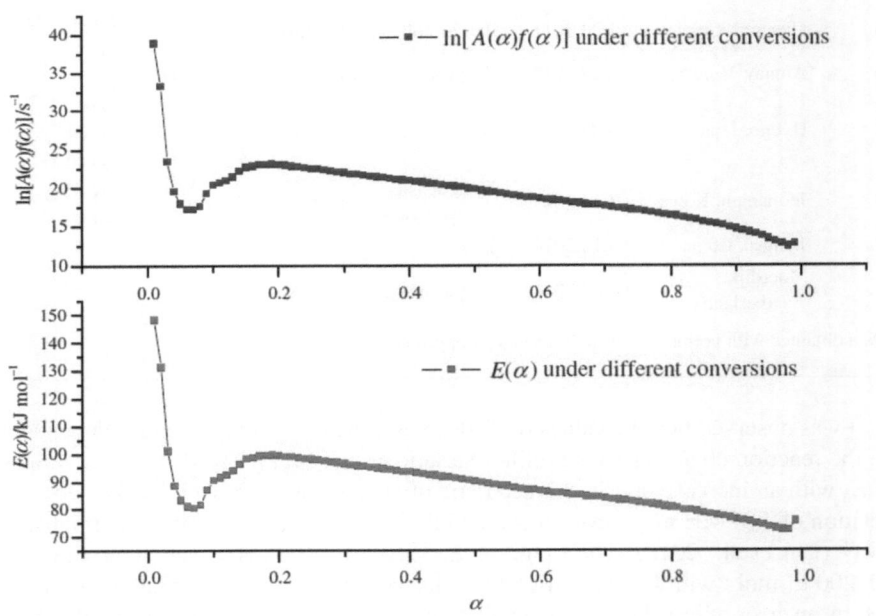

FIGURE 11.4 Calculation of $E(\alpha)$ and $\ln A(\alpha)$ as functions of conversion by the Friedman method (Reproduced with permission from Yang et al. Copyright © Springer, 2017)

TABLE 11.3
Data on Relevant Buildings

Company Name	Plant Type	Number of Floors	Floor Area (m²)	Structure	Application
Sin Hun	A raw material warehouse	One	1300	Iron sheet	Warehouse
Sin Hun	B raw material warehouse	One	1300	Iron sheet	Warehouse
Ever spring	Old plant	Tree	4300	Reinforced concrete, Iron roof	Plant, Office
Chun Yuan	An iron sheet plant	One	1650	Iron sheet	Plant

(Data obtained with permission from Yang et al. Copyright © Springer, 2017)

reaction process avoiding the error induced by the uncorrected f(α). The numerical evaluation was performed using the least squares method and the heat production vs time plots are shown in Figures 11.5 and 11.6. The Friedman method revealed that the TBPMH reaction was autocatalytic. However, a single kinetic model could not describe the thermal kinetic behavior of a reaction process. Accordingly, for TBPMH, a combination of nth order and autocatalytic reaction model represented an optimal curve fitting approach. It was observed that TBPMH reaction was a two-stage process and may advance directly to complex and multiple parallel reactions during thermal runaway. The model-fitting results of TBPMH predicted the exothermic curve at different heating rates and for TBPMH and Table 11.4 summarizes the values of kinetic parameters.

The simulation results of TMR_{ad} and ΔT_{ad} for TBPMH was presented in Figure 11.7. In order to calculate TMR_{ad}, a temperature of 90 °C was used as at this temperature, TBPMH is added as a reaction initiator for the polymerization reaction for acrylates, styrene, and methyl acrylate and the list of the parameters required for calculation is presented in Table 11.5. Assuming the reaction at adiabatic conditions, the value of U is considered as unity and is 1.0 and through U, ΔH_d, and C_p values ΔT_{ad} and TMR_{ad} are calculated (Tsai et al., 2012). The determined values of ΔH_d and C_p for TBPMH were approximately -924.0 J g^{-1} and 1.2 J g^{-1} K^{-1}, respectively. The simulation results of TMR_{ad} and ΔT_{ad} values reached 9.18 min and 552 °C, respectively, at an operating temperature of 90 °C for TBPMH. During the manufacturing of TBPMH, if the cooling or stirring system fails, the decomposition of TBPMH may be accompanied by a rise in the exothermic temperature and a rapid increase in the reaction rate, causing violent thermal runaway. The thermal stability of TBPMH stored in 10, 25, 35, and 50 kg commercial containers was investigated. The obtained value of SADT was 55 °C and is in agreement with literature (Safety Data Sheet, 2016).

The calculated SADT values for different package sizes of 10, 25, 35, and 50 kg were 58, 55, 54, and 530 °C, respectively. Results indicate that SADT values decrease with increase in mass. Therefore, to prevent the occurrence of abnormal self-heating

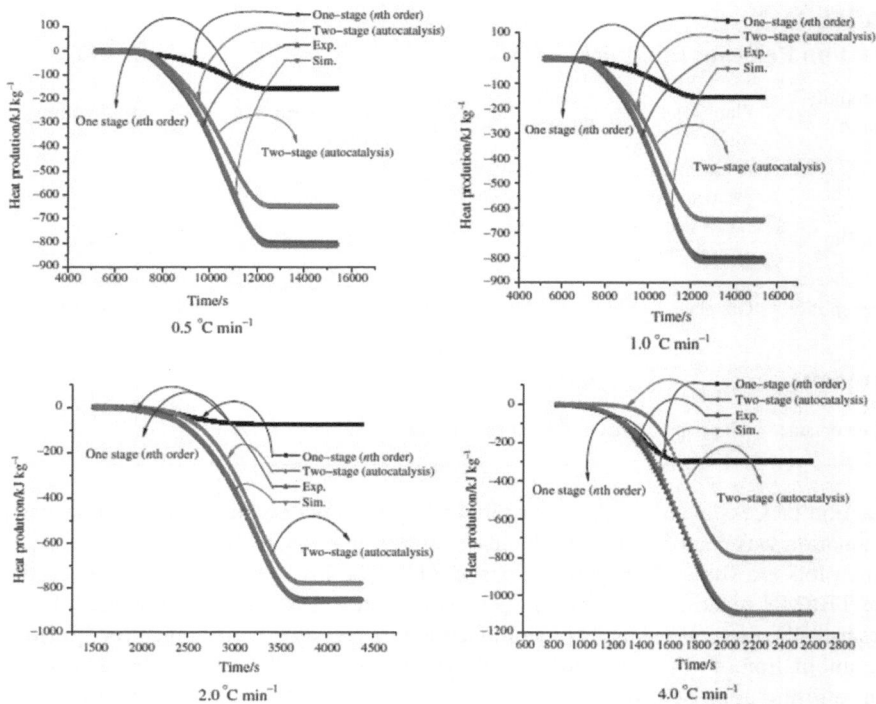

FIGURE 11.5 Heat production versus time curves derived through the model-fitting method at four heating rates, 0.5, 1.0, 2.0, and 4.0 °C min⁻¹, in the experiments and simulations (Reproduced with permission from Yang et al. Copyright © Springer, 2017)

decomposition behavior, the safer temperatures of different package sizes must be determined for storing or transporting of in TBPMH.

11.8 SUMMARY

This chapter reports the feasibility of DSC-based approach for conducting thermal runaway tests, instead of traditional large-scale tests. The potentially hazardous nature of TBPMH, related to thermal runaway, fire, explosion or toxic release was confirmed through the values of T_o, T_p, T_f, and ΔH_d of TBPMH were 103, 132, 147 °C, and -924.0 J g⁻¹, respectively. Such values revealed that at low temperatures, TBPMH could release high heat of decomposition. To determine the reaction characteristics of TBPMH, the Friedman method was employed for evaluating E_a at different degrees of conversion (α); with R^2 value ~0.99, the correlation was found to be appropriate with error <5 %. Autocatalytic nature of TBPMH reaction was revealed from the change in E_a value. However, through this method the average value of E_a determined was 91 kJ mol⁻¹, which is different from values reported in past studies (between 127 and 140 kJ mol⁻¹). Such a discrepancy was attributed due to the fact that the Friedman method evaluated an entire reaction process, and it doesn't involve

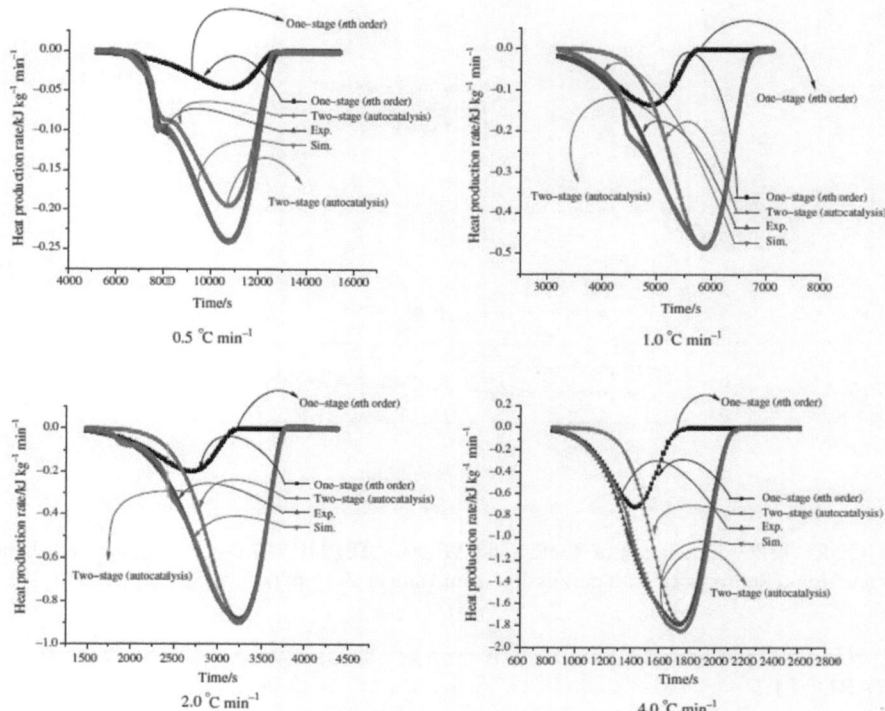

FIGURE 11.6 Heat production rate versus time curves derived through the model-fitting method for TBPMH (97.0 mass%) at four heating rates, 0.5, 1.0, 2.0, and 4.0 °C min^{-1}, in the experiments and simulations (Reproduced with permission from Yang et al. Copyright © Springer, 2017)

TABLE 11.4
Types and Amounts of Hazardous Material in Storage Tanks.

Number	Category	Chemical	Location	Capacity (kg)
1	Flammable liquid	Benzene	1st plant	19000
2	Flammable liquid	Methanol	2nd plant	12000
3	Flammable liquid	Isopropanol	2nd plant	3800
4	Flammable liquid	Acetonitrile	2nd plant	1700
5	Flammable liquid	N-heptane	2nd plant	3400

(Data obtained with permission from Yang et al. Copyright © Springer, 2017)

considering the kinetic function [f(α)]. Thus, the result obtained is more accurate. The TBPMH reaction comprises of a two-stage phenomenon. The first stage comprises of an nth order reaction, whereas the second stage was confirmed to be autocatalytic obtained by model fitting. Thus, the overall mechanism is found to be complicated. Complete kinetic parameters were also studied in this chapter. The

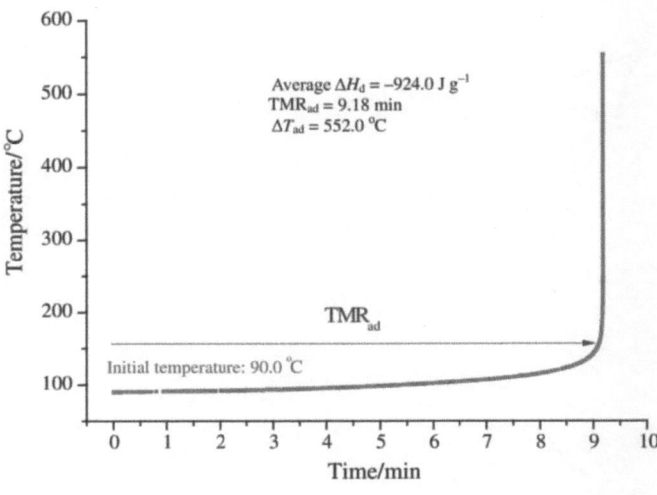

FIGURE 11.7 Evaluation of TMR$_{ad}$ and ΔT_{ad} for TBPMH (97.0 mass%) under adiabatic conditions (Reproduced with permission from Yang et al. Copyright © Springer, 2017)

TABLE 11.5
Types and Amounts of Hazardous Material in Storage Tanks

Number	Category	Chemical Stored	Capacity (kg)
1	Pyrophoric substances and substances which in contact with water emit flammable gas	Ammonia	4380
2	Flammable liquids	Hydrazine	5800

(Data obtained with permission from Yang et al. Copyright © Springer, 2017)

future perspective of this investigation is directed towards predicting exothermic curves associated with different heating rates. Such studies are important for preventing thermal accidents and for safety measures. The ΔT_{ad} and TMR$_{ad}$ values of TBPMH were recorded as 552 °C and 9.18 min, respectively. Such results indicated that at initial temperature ~90 °C, serious thermal runaway accidents may occur by TBPMH, if cooling systems don't operate properly. Maintaining the process temperature is very important during the reaction, whereas any failure situation of cooling and stirring failure should be prevented. The calculated SADT value of 55 °C was found to be consistent with reported values and conveyed that the simulation results are acceptable. The SADT values were found to be 58, 55, 54, and 53 °C for package size of 10, 25, 35, and 50 kg respectively. Such results represent a drop in SADT value with an increase in mass. These results suggest that for storing or transporting TBPMH, safety measures and guidelines should be followed to avoid any thermal hazards.

TABLE 11.6
Stock Objects in the Temporary Outdoor Storage Area

Chemical Stored	Raw Material/ Recycle	Quantity (Drum)	Public Dangerous Goods
Acetophenone	Raw material	27	Class III petroleum
Benzoate	Raw material	48	Class III petroleum
Butanol	Raw material	6	Class II petroleum
Butanol	Recycled	43	Class II petroleum
Cyclohexane	Recycled	8	Class I petroleum
Dimethylsebacate	Raw material	400	Class III petroleum
Dipropylamine	Recycled	58	Class I petroleum
Dipropylamine	Raw material	93	Class I petroleum
Formicacid	Raw material	8	Class II petroleum
N-heptane	Recycled	112	Class I petroleum
Isopropanol	Raw material	13	Alcoholic
Isopropanol	Recycled	23	Alcoholic
Methanol	Recycled	27	Alcoholic
Octylamine	Raw material	234	Class II petroleum
Xylene	Raw material	46	Class I petroleum
Xylene	Recycled	82	Class I petroleum

(Data obtained with permission from Yang et al. Copyright © Springer, 2017)

REFERENCES

Chi, J.H., Wu, S.H., Charpentier, J.C., Yet-Pole, I., Shu, C.M., 2012. Thermal hazard accident investigation of hydrogen peroxide mixing with propanone employing calorimetric approaches. *J. Loss Prevent. Proc. Ind.* 25, 142–147.

Duh, Y.S., Wu, X.H., Kao, C.S. 2008. Hazard ratings for organic peroxides. *Process. Saf. Prog.* 27, 89–99.

Fauske, H.K. 2011. *Gassy system vent sizing the role of two-phase flow.* Burr Ridge: Fauske and Associates.

Hou, H.Y., Shu, C.M., Duh, Y.S., 2011. Exothermic decomposition of cumene hydroperoxide at low temperature conditions. *AIChE J.* 47, 1893–1906.

Kozlowski, C., Kurko, K. 2008. *Consideration of autocatalytic behavior in determination of self-accelerating decomposition temperature.* Burr Ridge: Fauske and Associates.

Lin, C.P., Chang, C.P., Chou, Y.C., Shu, C.M., 2010. Modeling solid thermal explosion containment on reactor HNIW and HMX. *J. Hazard. Mater.* 176, 549–558.

Liu, S.H., Hou, H.Y., Shu, C.M., 2013. Effects of thermal runaway hazard for three organic peroxides conducted by acids and alkalines with DSC, VSP2, and TAM III. *Thermochim. Acta.* 566, 226–232.

Lu, G., Zhang, C., Chen, L., Chen, W., Yang, T., Zhou, Y., 2015. Kinetic analysis and self-accelerating decomposition temperature (SADT) of b-nitroso-a-naphthol. *Process. Saf. Environ. Prot.* 96, 69–76.

Málek J. 1995. The applicability of Johnson-Mehl-Avrami model in the thermal analysis of the crystallization kinetics of glasses. *Thermochim. Acta.* 267, 61–73.

Malow, M., Wehrstedt, K.D. 2005. Prediction of the self–accelerating decomposition temperature (SADT) for liquid organic peroxides from differential scanning calorimetry (DSC) measurements. *J. Hazard. Mater.* 120, 21–24.

Montserrat, S., Málek, J., Colomer, P., 1998. Thermal degradation kinetics of epoxy-anhydride resins: I. Influence of a silica filler. *Thermochim Acta.* 313, 83–95.

Naranjo, R.A., Conesa, J.A., Pedretti, E.F., Romero, O.R., 2012. Kinetic analysis: simultaneous modelling of pyrolysis and combustion processes of dichrostachys cinerea. *Biomass Bioenerg.* 36, 170–175.

Omrani, A., Simon, L.C., Rostami, A.A., Ghaemy, M., 2008. Cure kinetics, dynamic mechanical and morphological properties of epoxy resin–Im6NiBr 2 system. *Eur. Polym. J.* 44, 769–779.

Safety Data Sheet, Akzo Nobel base chemicals BV, The Netherlands (2015). http://www.akzonobel.com/.

Safety Data Sheet, Alibaba Group. China (2016). http://www.alibaba.com/showroom/tert–butyl-peroxy–3-5-5–trimethylhexanoatetbpmh.html.

Saraf, S.R., Rogers, W.J., Mannan, M.S., 2003. Prediction of reactive hazards based on molecular structure. *J. Hazard. Mater.* 99, 15–29.

Sato, Y., Okada, K., Akiyoshi, M., Murayama, S., Matsunaga, T., 2011. Diphenylmethane diisocyanate self-polymerization: thermal hazard evaluation and proof of runaway reaction in gram scale. *J. Loss Prevent. Proc. Ind.* 24, 558–562.

Singh, H., Chavda, A., Nandula, S., Jasra, R.V., Maiti, M., 2012. Kinetic study on stereospecific polymerization of 1,3-butadiene using a nickel based catalyst system in environmentally friendly solvent. *Ind. Eng. Chem. Res.* 51, 11066–11071.

Talouba, I.B., Balland, L., Mouhab, N., Chang, C.T., Abdelghani-Idrissi, M.A., 2011. Kinetic parameter estimation for decomposition of organic peroxides by means of DSC measurements. *J Loss Prevent Proc Ind.* 24, 391–406.

Tong, J.W., Chen, W.C., Tsai, Y.T., Cao, Y., Chen, J.R., Shu, C.M., 2014.Incompatible reaction for (3-4-epoxycyclohexane) methyl-30-40-epoxycyclohexyl-carboxylate (EEC) by calorimetric technology and theoretical kinetic model. *J. Therm. Anal. Calorim.* 116, 1445–1452.

Tsai, L.C., Tsai, Y.T., Lin, C.P., Liu, S.L., Wu, T.C., Shu, C.M., 2012. Isothermal versus non-isothermal calorimetric technique to evaluate thermokinetic parameters and thermal hazard of tert-butyl peroxy-2-ethyl hexanoate. *J. Therm. Anal. Calorim.* 109, 1291–1296.

Tsai, Y.T., You, M.L., Qian, X.M., Shu, C.M., 2013. Calorimetric techniques combined with various thermokinetic models to evaluate incompatible hazard of tert-butyl peroxy-2-ethyl hexanoate mixed with metal ions. *Ind. Eng. Chem. Res.* 52, 8206–8215.

Vyazovkin, S., Burnham, A.K., Criado, J.M., Perez-Maqueda, L.A., Popescu, C., Sbirrazzuoli, N., 2011. ICTAC kinetics committee recommendations for performing kinetic computations on thermal analysis data. *Thermochim Acta.* 520, 1–19.

Xiao, H.M., Ma, X.Q., Lai, Z.Y., 2009. Isoconversional kinetic analysis of co-combustion of sewage sludge with straw and coal. *Appl. Ener.* 86, 1741–1745.

Yan, Q.L., Zeman, S., Jiménez, P. S., Zhao, F. Q., Pérez-Maqueda, L. A., & Málek, J., 2014. The effect of polymer matrices on the thermal hazard properties of RDX-based PBXs by using model-free and combined kinetic analysis. *J. Hazard. Mater.* 271, 185–195.

Yang, Y., Tsai, Y.-T., Cao, C.-R., Shu, C.-M., 2017. Kinetic and thermal safety analysis for tert-butyl peroxy-3,5,5-trimethylhexanoate by advanced calorimetric technology. *J. Therm. Anal. Calorim.* 127, 2253–2262.

Yoo, M.J., Kim, S.H., Park, S.D., Lee, W.S., Sun, J.W., Choi, J.H., Nahm, S., 2010. Investigation of curing kinetics of various cycloaliphatic epoxy resins using dynamic thermal analysis. *Eur Polym J.* 46, 1158–1162.

You, M.L., Liu, M.Y., Wu, S.H., Chi, J.H., Shu, C.M., 2009. Thermal explosion and runaway reaction simulation of lauroyl peroxide by DSC tests. *J Therm. Anal. Calorim.* 96, 777–782.

12 Thermal Hazard Analysis and Its Application on Process Safety Assessments

12.1 INTRODUCTION

The intensive use of widely diverse chemicals are being used nowadays in huge quantities due to rapid industrial expansion. Two of the major factors in the cause of various fire accidents are known to be the reactivity and variability of these chemicals. In modern chemical production, the evaluation of the hazards and chemical reactivity of such compounds has become the major focus in attempts to avoid industrial fire and explosion accidents. On one occasion an explosion followed by a fire broke out at the INEOS phenol plant in Mobile, Alabama, USA in 2002. Failure in the immediate emergency shutdown system caused the disaster since the feed valve of cumene hydroperoxide (CHP) remained open for some considerable time. Further, runaway reactions have occurred due to the mixing of excessive amount of CHP with acid (Hsu et al., 2012; Iwata et al., 2006).

In China an explosion and fire occurred at the FuXin chemical plant in Heilongjiang province during 2011. The incident occurred during the manufacturing process of an initiator 2,2′-azodi-isobutyronitrile (AIBN) and caused the immediate death of 9 of the 14 workers who were on site at the time. Decomposition of AIBN occurs instantly at 100 ºC and results in both explosions and fires. This reaction was accompanied by the release of organic cyanide and nitrogen, which causes harmful health hazards to humans. Moreover, hydrogen peroxide (H_2O_2) finds profound usage as an oxidant/bleach. For improving the semiconductor wet cleaning technology, often high-purity H_2O_2 solutions were mixed in various proportions with ammonia solution (NH_4OH), sulfuric acid (H_2SO_4), and hydrochloric acid (HCl). But it has to kept in mind that H_2O_2, being a strong oxidant, poses the risks of fire and explosions (Casson et al., 2012). Spontaneous reactions occurs easily if the mass concentration of the H_2O_2 solution is greater than 65%. Moreover, the smooth decomposition and the release of huge amount of heat leading to fire outbreaks occurs as the solution mass concentration exceeds 90%, it is particularly found to happen in the presence of catalysts or cross-linking agents (Lu et al., 2006). Extended thermal hazards can be easily generated due to the high thermal sensitiveness of organic peroxides and azo compounds. Due to the generation of huge heat and gaseous pollutants such compounds pose a great threat for explosion and fire outbreak hazards (Maschio et al., 1992). Organic peroxides have been classified as explosive substances due to their exothermic decomposition and thermal explosion by the 27th Globally Harmonized System of Classification and Labeling of Chemicals (GHS) hazard classes. Azo compounds bear a functional group R–N=N–R', where R and R' denotes either an aryl or alkyl group.

Azo compounds have found widespread application in blowing agents, pigments, dyes, and initiators. Due to the presence of a bivalent –N–N– composition, which poses the risk of breaking easily under high ambient temperatures and thus undergoes self-decomposition. Such spontaneous decomposition causes a runaway reaction and ultimately leads to fire or explosion because of the failure of the cooling processes (Dubikhin et al., 2012). Azo compounds, including unstable liquids, solids, and mixed materials, are termed as self-reactive substances, which can undergo random exothermic decomposition, even in the absence of air and/or oxygen. At 120 °C or even nearly at room temperature, most of the organic peroxides undergo exothermic decomposition. Whereas the exothermic decomposition of azo compounds were noticed at a temperature higher than 40 °C. Hence it was recommended by the United Nations (UN) transportation regulations that during the transportation of such samples the temperature should not exceed its self-accelerating decomposition temperature (SADT). In a specified commercial package, SADT is described as the lowest ambient air temperature at which a self-reactive substance undergoes an exothermic reaction in a period of seven days. It was also suggested to obtain the SADT values of such packaged samples through reliable experimental methods so that no fire risk hazards can occur during the storage and transportation processes (Kotoyori, 1999; U.N., 2003).

12.2 ORGANIC PEROXIDES AND ITS ASSOCIATED THERMAL HAZARDS

In the assessment of thermal hazards and reaction incompatibilities for organic peroxides a series of studies have been investigated (Hou et al., 2012; Miyake et al., 2005; Yeh et al., 2003; Duh et al., 2008; Li and Koseki, 2005; Hou et al., 2001) and mainly for azo compounds, thermal analysis was performed (Li and Koseki, 2004). It was observed that both the macroscopic and microscopic methods were required for understanding the diverse reactivity phenomenon of organic peroxides and to handle them optimal process conditions should be obtained. It was necessary to form material safety parameters, and to take preventive and safety measures for handling such materials (Casson and Maschio, 2012; Maschio et al., 2010). The incompatible chemical interactions of organic peroxides and azo compounds with bases and acids, which give rise to runaway behaviors were studied in this chapter through thermal analysis by DSC and TAM III, as well as through VSP2 adiabatic calorimetry. Thermokinetic data such as the apparent exothermic onset temperature, the heat of decomposition, and the self-heating rate were obtained from calorimetric studies, and were further compared with organic peroxides and azo compounds. In this chapter, the thermal decomposition behaviors were verified, and a thermal analytic model was proposed through a hazard analytic case of organic peroxides and azo compounds. Such test results were found to be helpful in developing the safety measures of such compounds during handling, storage, transportation, and disposal procedures.

12.3 THERMAL HAZARD ANALYSIS

The commonly used thermal hazard analysis techniques and equipment can be divided into five categories: thermal analysis technology (DSC, DTA), iso-thermal calorimetric technology (TAM III), adiabatic calorimetric technology (VSP2, ARC), reaction calorimetric technology (RC1, C80), and emergency relief control technology (VSP2, RSST). The intrinsic hazardous data of unknown substances were measured through such techniques, which includes thermal stability, incompatibility, the lowest exothermic temperature, and hazardous consequence analysis. Such analysis determines the hazardous properties and ensures the provision of process safety measures. In this chapter the thermal stability analysis techniques, such as thermal analysis technology, and isothermal calorimetric technology, along with adiabatic calorimetry, were investigated in carrying out the hazard characteristics study (Maschio et al., 1999).

12.3.1 THERMAL ANALYSIS TECHNOLOGY

The heat flow across a sample was measured by DSC when it is heated, cooled, or held isothermally at constant temperature. By utilizing few sample amounts in mg, DSC can detect the endothermic and exothermic effects. Such a phenomenon shows the heat flow during the thermal decomposition of any materials. In order to obtain the thermokinetics, thermal screening tests for nth-order reaction, autocatalytic reaction, and single thermal curve were performed additionally. In order to determine the properties such as heat of reaction, apparent activation energy, and other kinetic parameters such as rate of reaction, DSC measurement was recommended by ASTM specification for evaluating specific harmfulness during operation of any substance (ASTM E698, 2011).

12.3.1.1 Experimental Setup

Mettler DSC governed by a TA8000 thermal analyzer was utilized in conducting the DSC experiments. In an attempt to withstand the pressures up to 100 bar and temperatures up to 600 °C, a high pressure crucible was employed. High-purity indium and zinc (Mettler, 2014) were used for calibrating the DSC for obtaining better temperature and heat flow measurements. In order to achieve a better state of thermal equilibrium, such instruments were utilized, which provides a dynamic scanning rate of 1–10 °C min^{-1}. For the dynamic scanning process, a series of high pressure crucibles were employed at 4°C min^{-1}. The hazardous nature and the thermal characteristics of reactive chemicals can be obtained using DSC using small amount of samples. From the slope of exothermic peak and the baseline, the extrapolated exothermic onset temperature of chemicals can be determined (Hofelich and Labarge, 2002).

The evaluated values of endothermic heat at a melting point of 50 °C was 129 J g^{-1}. The exothermic onset temperature was 67 °C and the heat of reaction of 1260 J g^{-1} with the maximum thermal peak of 3.4 mWg^{-1} at 118 °C for 2,2′-azodi-2-methylbutyronitrile (AMBN; Figure 12.1a) indicating the explosive nature of organic peroxide (heat of reaction > 250 J g^{-1}) (Gibson et al., 1987). As illustrated in Figure 12.1b, it was found that the relationship between the thermal decomposition reaction and the sample melting point is not clear, when solid-state samples undergo thermal scanning. Due to the continuous parallel phenomenon of the melting and thermal decomposition of solid samples, such a relation was assumed to happen, since the two obtained peaks may overlap and affect each other. For example, the endothermic melting of solid samples for AIBN and lauroyl peroxide (LPO) is accompanied by the exothermic chemical decomposition reaction during physical phase change upon heating (Figures 12.1b and 12.1c). In the petrochemical industries, during both the upstream and downstream processes, incompatible hazards such as alkaline solutions, metal ions, and rust, along with mixed acid, often create major problems.

The presence of such components, in terms of their concentration and their chemical properties, not only severely affect the product yield during the production process but also pose a relative threat to the design of the safety processes. Components with hazardous properties related to thermal decomposition may lead to a high heat of reaction, and ultimately release large amounts of toxic gaseous pollutants which may cause fire-related hazard scenarios. During the process, catastrophes can occur due to any specific instance of mismanagement during storage, transport, and manufacturing. Such mishaps will cause the process from the normal operating range, this may generate substantial reaction heat and, if not released in a timely fashion, such a phenomenon can lead to a disaster. Therefore, in scanning thermal hazards and testing the potentially incompatible substances during the production process, the utilization of DSC can prove very important. Such precautionary steps can therefore be useful in avoiding hazardous runaway reactions.

In the presence of air or oxygen, through the oxidation of cumene CHP is basically produced. Hou et al. (2012) reported that for more than 94.5% of the phenol production processes CHP and acid are used to react and produce phenol and acetone. In the case of unsaturated polymeric resins, CHP acts as a repairing agent. Furthermore, in a polymerization reaction, CHP is utilized as an initiator; in addition, in the alkaline process it is used as an intermediate to yield the cross-linking agent dicumyl peroxide (DCPO). However, in the case of organic peroxides the auto-thermal decomposition pathway differs from the acids and base catalyzed decomposition. For catalyzed reactions, the rate is fast, and the overall decomposition process is altered (Figure 12.2). On September 26, 2003, in the Changhua Coastal Industrial Area of Changhua County, Taiwan, one chemical plant produced CHP through the use of the alkaline process (Chen et al., 2008). The process includes the association of alkaline catalyst sodium hydroxide (45 mass%) with highly concentrated CHP (88 mass%). Since the set temperature was around 70–80 °C, a decomposition reaction was generated. Failing to deal with the uncertainty in order to completely control the temperature and reduce the heat of reaction in the absence of control measures/

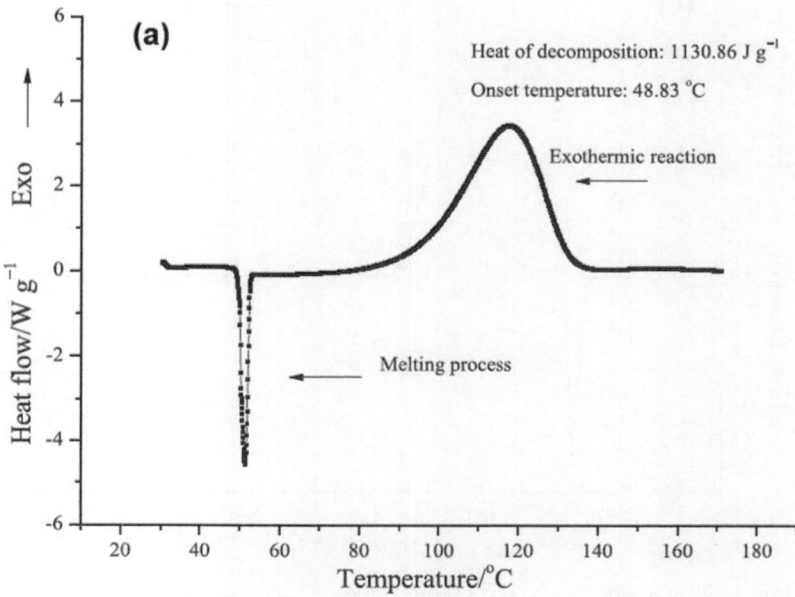

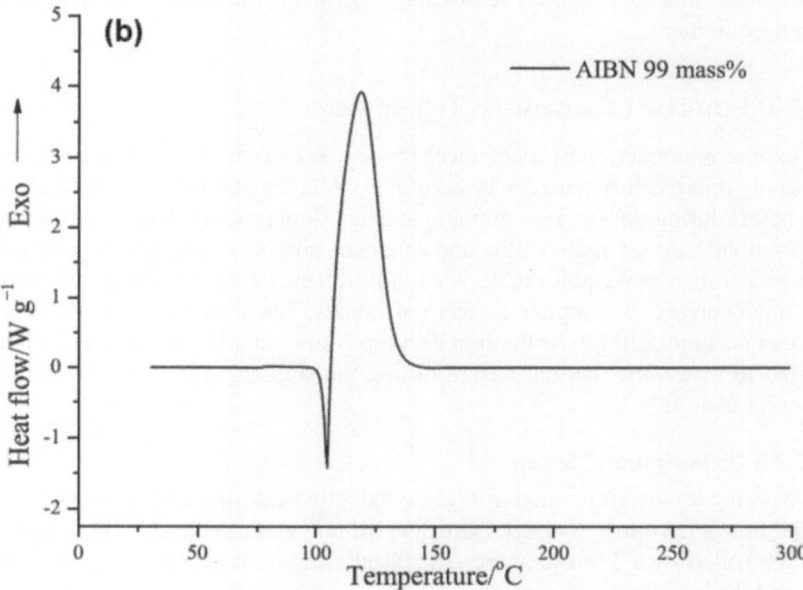

FIGURE 12.1 (a) DSC curve for AMBN (b) DSC curve for AIBN (c) DSC curve for LPO(Reproduced with permission from Liu et al. Copyright © 2015, Elsevier).

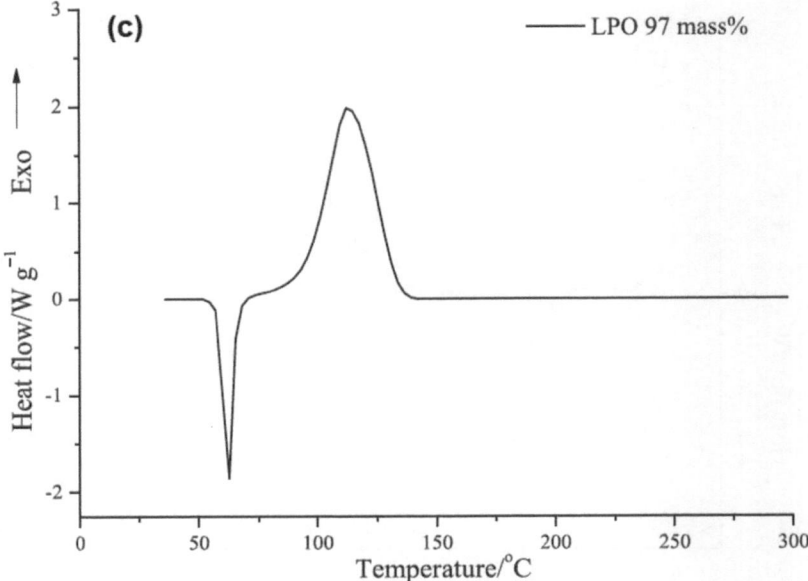

FIGURE 12.1 (a) DSC curve for AMBN (b) DSC curve for AIBN (c) DSC curve for LPO(Reproduced with permission from Liu et al. Copyright © 2015, Elsevier).

devices leads to huge economic losses and a thermal explosion that levelled the plant and the equipment.

12.3.2 Isothermal Calorimetry Technology

The exothermic reaction of a chemical process is extremely complex and a highly sensitivity micro calorimeter can be helpful in evaluating thermodynamic and kinetic parameters during the storage, shipping and handling process. The isothermal test is often simple, has vast applicability and generates reliable kinetic data as they involve fewer experimental variables (Chervin and Bodman, 1997). A *n*-order and autocatalytic model predictions applied isothermal studies. The accuracy of the process hazards can be simulated by the thermal decomposition kinetics. The research data can be used to assess the thermal hazard during the shipping and handling of thermal explosive materials.

12.3.2.1 Experimental Setup

The working temperature range for heat conduction calorimeter is around 15 to 150 °C and this is designed to check the extensive range of chemicals. The temperature was controlled by a Thermometric AB, Jarfalla, Sweden heating plate using an oil bath and the heat flow in fractions of 10 nW and can sustain within a range of ±0.01 °C. The heat flow and incompatibility during the thermal decomposition of organic peroxides and azo compounds placed in sealed glass tubes can be observed in this case. The autocatalytic and *n*-order reaction models are the most probable models

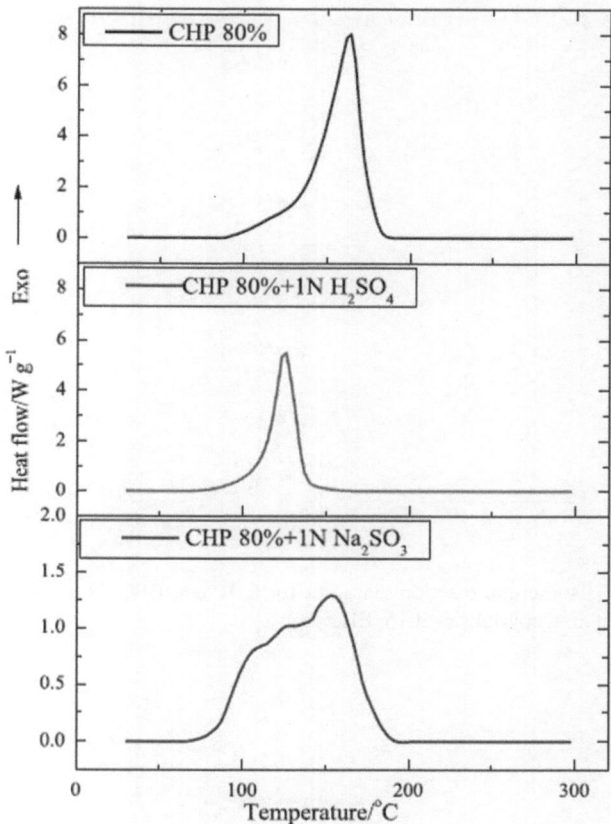

FIGURE 12.2 The DSC curves for CHP alone (top) and with acidic (middle) and basic (bottom) catalysts (Reproduced with permission from Liu et al. Copyright © 2015, Elsevier).

following isothermal reactions. Low reaction rates for organic peroxides, such as CHP, DCPO, BPO, MEKPO, was observed at the initial stages of the reaction. Maximum reaction rates were obtained as the reaction accelerated, generating auto-catalytic products as shown in Figure 12.3. The n-order reaction model was suggested by the LPO and TBPB reactions, where the maximum heating rate is achieved during the initial stages (Figure 12.4). An autocatalytic or n-order reaction has been described through thermal curve with the isothermal tests. LPO and TBPB decomposition temperature were chosen under isothermal circumstances individually. Table 12.1 and Figure 12.5 recapitulate the isothermal reactions performed at different temperatures as 85, 90, and 95 °C for BPO.

From the figure it is clear that, after the induced decomposition, the reaction rates begin to increase. The appearance of steep peaks were observed at higher temperatures and maximum heat flux decreased with a decrease in temperature. At lower temperatures, the reaction peak appears later and are relatively flat, which is the

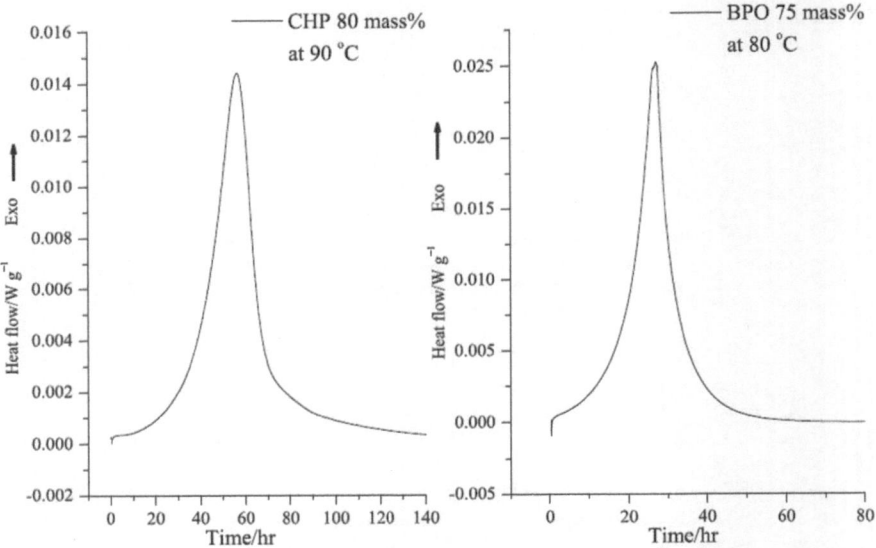

FIGURE 12.3 Isothermal reaction diagrams for CHP and BPO (Reproduced with permission from Liu et al. Copyright © 2015, Elsevier).

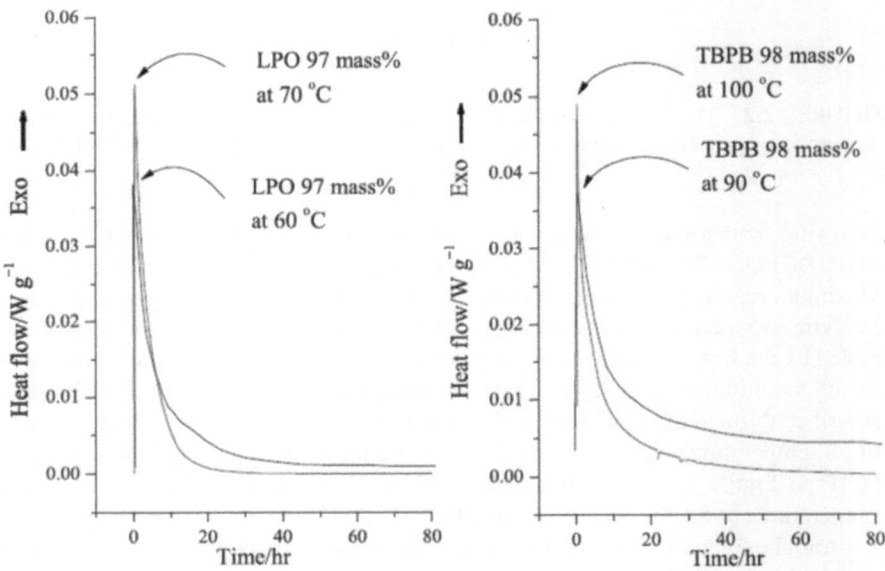

FIGURE 12.4 Isothermal reaction diagrams for LPO and TBPB (Reproduced with permission from Liu et al. Copyright © 2015, Elsevier).

TABLE 12.1

Data From the Isothermal Reaction of BPO 75 Mass% Isothermal Experiment as Measured by TAM III

Sample	Temperature (°C)	Sample Mass, Mg	Test Cell	Time to Peak (h)	Maximum Peak Power (Wg⁻¹)	Peak Area (J)	Reaction Type
BPO (75 mass%)	85	50.1	Glass	14.8	0.0579	99.7	Autocatalytic reaction
	90	50.8	Glass	4.71	0.1661	106.9	
	95	50.1	Glass	1.31	0.5509	103.4	

(Data obtained with permission from Liu et al. Copyright © 2015, Elsevier)

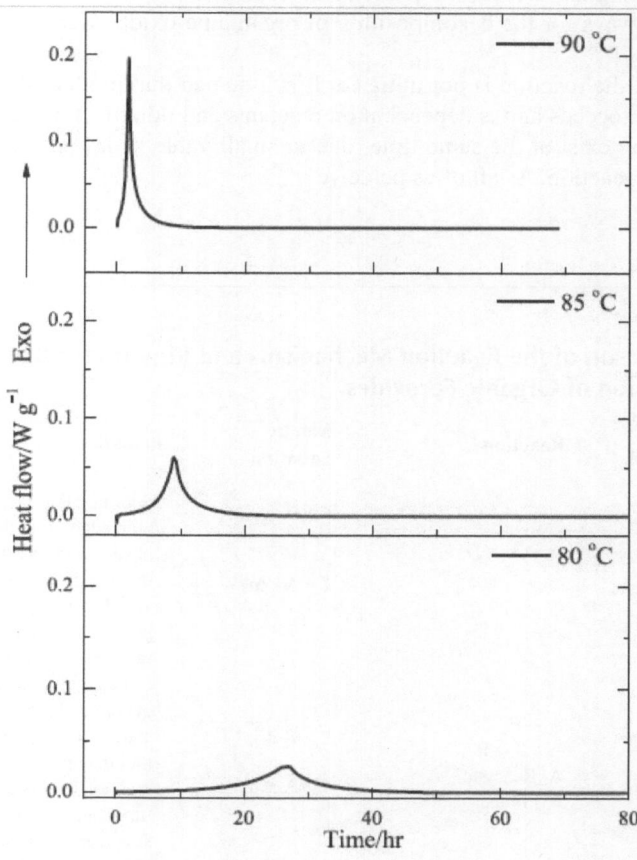

FIGURE 12.5 A comparison of the thermal power of BPO 75 mass% under different isothermal conditions (Reproduced with permission from Liu et al. Copyright © 2015, Elsevier).

characteristic feature of autocatalytic reactions. Published literature on autocatalytic reactions suggests that the hazardous temperature of organic peroxides is lower than the runaway temperature. During the manufacturing and transportation, the cumulative thermal effect will undergo a runaway reaction excursion which increases the danger of organic peroxides. Through the use of the isothermal mode, frequent reactions have been observed: polymerization, decomposition, and oxidation and even changes in crystallization (Chiu et al., 1983; Maschio and Moutier, 1989; Soh and Sundberg, 1982). Isothermal scans are often used to validate the kinetic parameters, are more elaborate, and provide better interpretation of the data when compared to the power control mode, which measures fewer variables at single temperatures. The order and pathways are unknown while examining new reactions and the isothermal scans expand the possibility of thermokinetic parameters and also require the use of limited materials. For a selected sample, isothermal aging tests assist in an understanding of the exothermic behavior, and the time needed to achieve maximum exothermic power at different temperatures that can be used to determine the kinetic parameters using the Arrhenius equation (Liu et al., 2011). The kinetic equations and reaction pathways for the decomposition of organic peroxides were summarized in Table 12.2.

At $[P]_0=0$, the reaction is not initiated. It is assumed that product P is formed as the reaction proceeds and is dependent on reactants and other factors. In preference, initial reaction exist at the same time, due to small value of k_1, and it initiates the autocatalytic reaction. As all of us perceive

$$k = Ae^{-E_a/RT} \tag{12.1}$$

TABLE 12.2
The Comparison of the Reaction Mechanisms and Kinetics for the Decomposition of Organic Peroxides

Isothermal Reaction Model	Reaction	Kinetic Equation	Remark
n-order reaction	$A \rightarrow B + D$	$r = kC_A^n$ $k = A \exp\left(-\dfrac{E}{RT}\right)$	Assuming that $[P]_0 = 0$ at the beginning, the reaction never starts. We thus assume that the product P exists at the very beginning, that is, the product is contained in the reactants or other factors causing this product to be generated before the reaction starts. Alternatively, we assume that the initial decomposition reaction occurs at the same time, but its k_1 value is very small, and its effect is to initiate the autocatalytic reaction.
Autocatalytic reaction	$A \xrightarrow{k_1} P$ $A + P \xrightarrow{k_1} R$ A: Reactant P: Product	$r = k[A]_t^m[P]_t^n$ $k = A \exp\left(-\dfrac{E}{RT}\right)$	

(Data obtained with permission from Liu et al. Copyright © 2015, Elsevier)

At individual temperatures the general logarithm of the reaction rate has been taken for the outcome of the expression.

$$\ln\left(\frac{k_2}{k_1}\right) = \left(E_a / R\right)\left(\frac{1}{T_1} - \frac{1}{T_2}\right)$$ (12.2)

At this temperature, the peak power Q is proportional to the reaction rate k, Equation (12.2) can be modified as:

$$\ln\left(\dot{Q}_2 / \dot{Q}_2\right) = \left(E_a / RT\right)\left(\frac{1}{T_1} - \frac{1}{T_2}\right)$$ (12.3)

The kinetic parameters can be achieved or by taking the natural algorithm of Equation (12.1) (shown in Equation (12.4)) and resolving the slope.

$$\ln k = \ln A - \left(E_a / R\right)\left(1 / T\right)$$ (12.4)

At the isothermal conditions of 70, 80, 90, 100, and 110 °C, the simulated values of ln(k) vs 1/T are represented in Figure 12.6a. The calculated value of E_a and ln(A) of CHP were 101 kJ mol^{-1} and 34.6 min^{-1}, respectively, with a correlation coefficient of 0.996. For LOP, at isothermal conditions at 50, 60, 70, and 80 °C, the values of E_a and ln(A) were 103 kJ mol^{-1} and 14.29 min^{-1} with a correlation coefficient of 0.993, as shown in Figure 12.6b.

12.3.3 ADIABATIC CALORIMETRY TECHNOLOGY

In analysing the safer design of process reactors, the VSP2 calorimeter that works on the principle of automatic heat-wait-search (HWS) mode was developed by the Design Institute for Emergency Relief Systems (DIERS) under the American Institute of Chemical Engineers (AIChE), as shown in Figure 12.7. VSP2 analysis was performed in a test cell of 120 mL with low phi-factor (F) and the experimental data can be scaled up without computational techniques. Test data consists of adiabatic rates of temperature and pressure change, such as initial exothermic temperature (T_o), pressure-temperature plots, the self-heating rate (dT dt^{-1}), and the pressure rise rate (dP dt^{-1}). The incompatibility reaction and the thermal hazard of a runaway reaction can be determined by using the thermokinetic parameters, and the essential data have been used for the control and prevention of the runaway reactions.

12.3.3.1 Experimental Setup

The thermal hazard data was measured by the VSP2 (Fauke & Associates, Inc.) in an adiabatic environment representative of actual process conditions (FAI, 2002). In the VSP2 experiments, the reactants (organic peroxides and azo compounds) were loaded into an evacuated test cell and were allowed to decompose via autocatalysis. The recommended ratio of impurities such as acid, base, and rust is less than 5% to simulate incompatibility hazards in the process. The need for improvement in purity and changes in the applications of chemicals such as the blending of high-purity H_2O_2 with NH_4OH, H_2SO_4, and HCl in various proportions for semiconductor wet

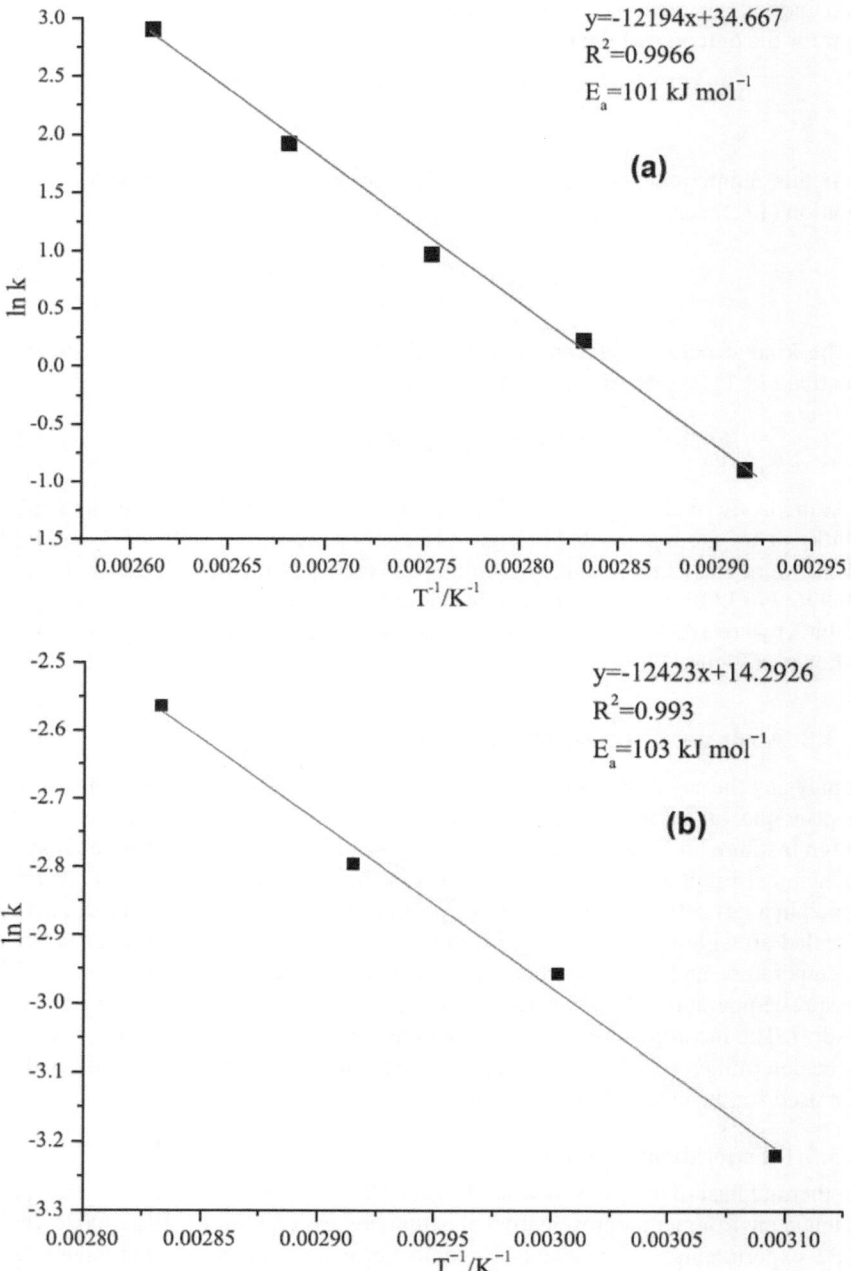

FIGURE 12.6 (a) CHP of activation energy analysis graph for Arrhenius plot at different isothermal temperatures at 70, 80, 90, 100, and 110 °C (b) LPO of activation energy analysis graph for Arrhenius plot at different isothermal temperatures at 50, 60, 70, and 80 °C(Reproduced with permission from Liu et al. Copyright © 2015, Elsevier).

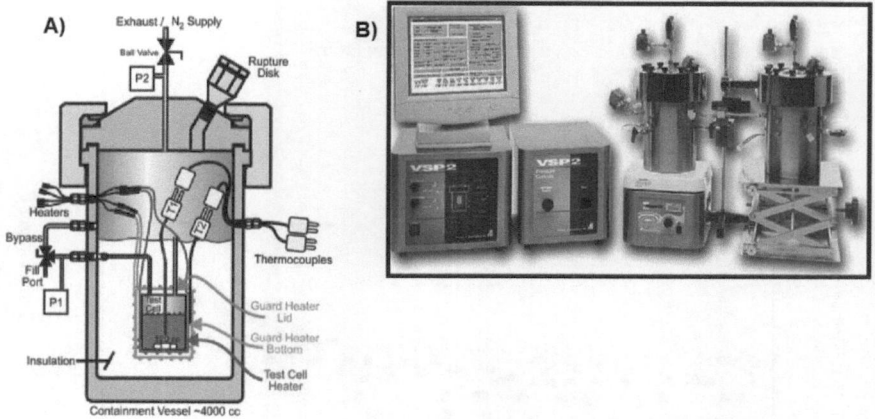

FIGURE 12.7 (A). The structure of test cell setting and detection during H–W–S model in the vessel. (B). All units of VSP2 under adiabatic condition (Reproduced with permission from Liu et al. Copyright © 2015, Elsevier).

cleaning technology resulted in the use of hydrogen peroxide as an oxidant/bleaching agent. However, a spontaneous reaction of H_2O_2 solution with concentration > 65 mass% could result in fire and explosion. H_2O_2 is a typical peroxide with the molecular formula H-O-O-H. The oxygen released during thermal decomposition supports fire and explosion when it comes in contact with other combustible materials. The mixture of H_2O_2 solution with concentration > 90mass% with catalysts/cross-linking agents are susceptible to thermal decomposition and releases an excess amount of heat, creating a fire and explosion hazard. Therefore, according to the National Fire Protection Association (NFPA 432), H_2O_2 is defined as an explosive substance. Exothermic phenomena were observed when 65 mass% H_2O_2 underwent a thermal reaction at 74 °C. At a reaction pressure of 4192 kPa, 5g of the sample can result in a reaction temperature up to 235 °C under adiabatic conditions. The temperature increases at a rate of 18,233 °C min⁻¹ and the pressure increases at a rate of 524,819 kPa min⁻¹ during runaway reactions and Table 12.3 summarizes the physical data obtained from the thermal analysis. From the above, it is clear that the rates of the temperature and pressure increase cannot be resolved by an external cooling systems when the reaction extents to a non-reversible runaway reaction.

An explosion in the Taoyuan Yongxing resin plant containing a H_2O_2 concentration of 50 mass% (approximately 10 tons) resulted in the complete rupture of the storage tank. During a runaway reaction, the volume of water can expand up to 1700 times generating large amounts of gaseous substances if adequate heat is provided to vaporize water. The rapid decomposition, improper operation, temperature deviations, and the mixing of incompatible materials during the production process can damage the reactor/storage tank, causing a fire and explosion hazard. At the self-propagating exothermic reaction temperature of 74 °C, the time required for the reaction to reach the maximum rate of temperature raise is 290 as shown from Figure 12.8a.

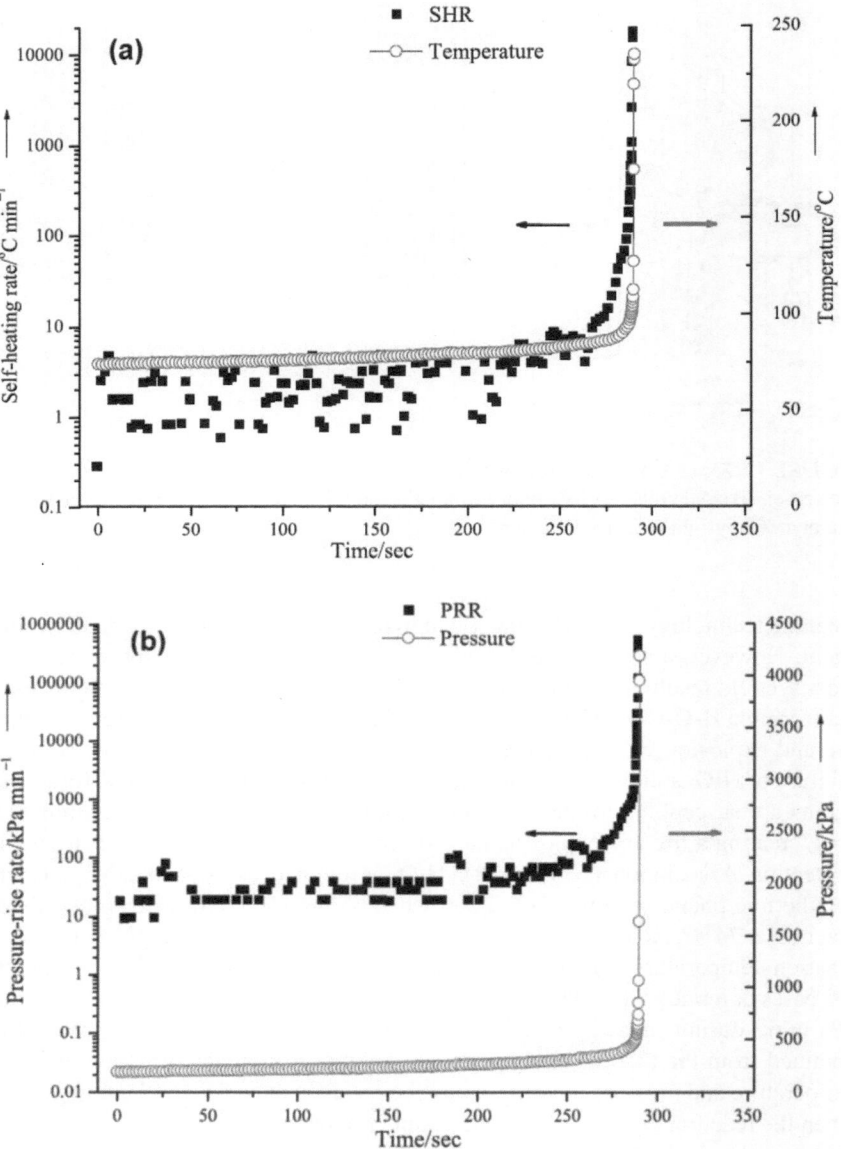

FIGURE 12.8   (a) Self-heating rate and temperature of H_2O_2 65 mass% vs. time for an adiabatic runaway reaction (b) Pressure-rise rate and pressure of H_2O_2 65 mass% vs. time for an adiabatic runaway reaction(Reproduced with permission from Liu et al. Copyright © 2015, Elsevier).

The rate of the temperature increase has grown exponentially from 11 °C min^{-1} to the maximum heating rate of 18233 °C min^{-1} within 20 s when the temperature reaches 85 °C. The time required for the pressure of the test tank to rise from 345 kPa to the maximum pressure of 4192 kPa was only 20 s, as presented in Figure 12.8b. Literature suggest that the presence of heat, metal ions, or acids in the environment can

TABLE 12.3
The Thermal Hazard Data of H_2O_2 and MEKPO as Analyzed by VSP2

Sample	Incompatibility		T_0, K	T_{max}, K	P_{max}, kPa	$(dT\ dt^{-1})_{max}$, K min⁻¹	$(dT\ dt^{-1})_{max}$, kPa min⁻¹
	Substance	Mass, kg					
H_2O_2 65 mass%	0.005 kg	–	–	347	508	4,192	18,233	524,819
MEKPO 31 mass%	0.005 kg	–	–	355	562	1,820	176	1,606
	0.005 kg	6N NaOH	0.00025	RT	570	3,157	10,386	333,531
	0.005 kg	6N H_2SO_4	0.00025	324	413	896	731	6,474

(Data obtained with permission from Liu et al. Copyright © 2015, Elsevier)

intensify the rapid exothermic decomposition which could be due to the presence of three peroxy-O-O- bond isomers in MEKPO. At room temperatures, the exothermic onset temperature of MEKPO is significantly reduced under the catalytic effect of acid and alkaline solutions. The rate of the pressure rise is increased to more than 200 times that of the MEKPO thermal runaway reaction shown from the Figures 12.9a, 12.9b and Table 12.3.

Accidents such as the Yongxing resin plant explosion resulted in 10 fatalities and injuries to 47 people due to the insufficient cooling capacity of the MEKPO reaction tank and improper material loading that led to a runaway reaction. The presence of catalysis that advances the reaction and the subsequent stages were triggered due to the thermal decomposition reaction releasing heat that causes the rise in temperature. The hazard's onset conditions, characteristics of the temperature/pressure changes, as well as the severity of the accident can be explored by the thermal hazard analysis techniques in advance. To obtain the kinetic parameters of the reaction, such as the reaction order, the apparent activation energy, and the frequency factor, the adiabatic experimental data and the theoretical model analysis were used. The adiabatic decomposition kinetics of 31 mass% MEKPO was deduced through the self-heating rate and the relationship between temperature and time (Townsend and Tou, 1980). The reaction rate equation used was:

$$r = \frac{dC}{dt} - kC^n \tag{12.5}$$

where n is the reaction order, k is the reaction rate constant, and C is the concentration of the reactant. To obtain the self-heating rate, m_t, at time t, the concentration changes can be exchanged into temperature changes, if the conversion rate for reactant A, $C = C_0(T_f\text{-}T)/(T_f\text{-}T_0) = C_0(T_f\text{-}T)/\Delta T_{ad}$, is replaced into Equation (12.5). The reaction rate formula can be reworked as.

$$\frac{dT}{dt} = k\left(\left(T_f - T\right)/\Delta T_{ad}\right)^n \Delta T_{ad} C_0^{n-1} \tag{12.6}$$

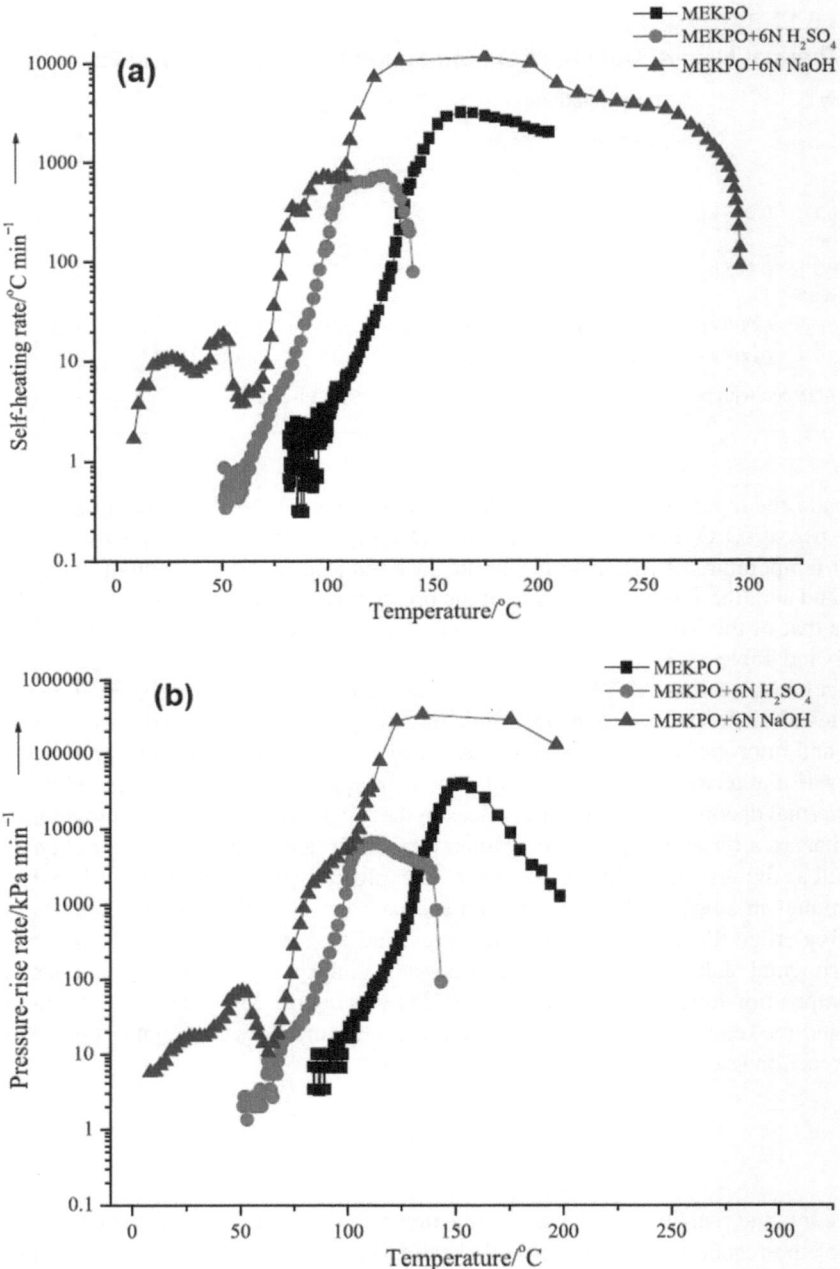

FIGURE 12.9 (a) Self-heating rate vs. the adiabatic runaway reaction temperature for MEKPO 31 mass% in acid and alkaline solutions (b) Pressure-rise rate vs. the adiabatic runaway reaction temperature vs. for MEKPO 31 mass% in acid and alkaline solution(Reproduced with permission from Liu et al. Copyright © 2015, Elsevier).

We defined the pseudo-zero order rate constant, k* for simplifying the Equation (12.2)

$$k^* = C_0^{n-1} k = m_t \, / \left(\left(T_f - T \right) / \Delta T_{ad} \right)^n \Delta T_{ad} \qquad (12.7)$$

Logarithm is obtained when Equation (12.7) is granted in the form of the Arrhenius equation.

$$\ln\left(k^*\right) = \ln\left(C_0^{n-1} A\right) - \frac{E_a}{RT} = \ln\left(A^*\right) - E_a / RT \qquad (12.8)$$

Assuming $n = n_1$ and substituting the experimentally calculated ΔT_{ad}, T_f, and dT dt^{-1} at different temperatures, the corresponding (k^*) is obtained. The frequency factor (A^*) and the activation energy E_a, were calculated from the slope of the plot of $\ln(k^*)$ against $(1/T)$ as shown in Figure 12.10. The determined Arrhenius parameter were $\ln(A^*)$ 39.3 s^{-1} and E_a 28.2 kcal mol^{-1} and are in accordance with literature $(E_a = 28.7$ kcal mol$^{-1})$ (Lee, 1969).

The necessity of using the controls during the storage and transport processes are determined by the value of SADT. The minimum storage climate temperature reaches the SADT, when the self-heating rate is observed 6 °C in a week (UN, 2003). It is essential to provide controls when the self-accelerating decomposition is at 50 °C during the use, storage and transport of chemical. Improvement of the shape,

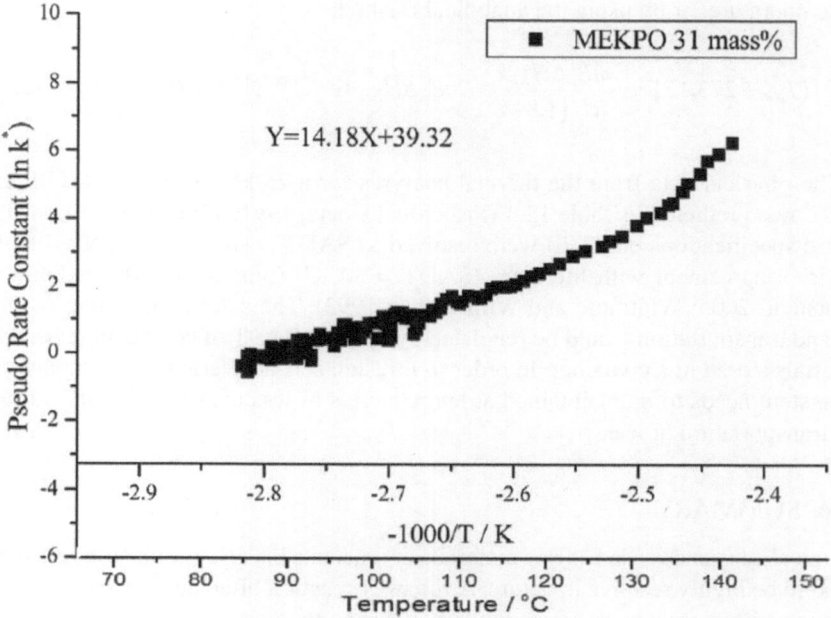

FIGURE 12.10 The pseudo-zero order rate constant vs. temperature for the thermal decomposition reaction of MEKPO 31 mass % (Reproduced with permission from Liu et al. Copyright © 2015, Elsevier).

TABLE 12.4
The Calculation Data of SADT for DTPB

Item	Value
Heat of reaction (J g^{-1})	1250
Frequency factor	2.5×10^{16}
Activation energy (J mol^{-1})	163030
Heat transfer coefficient (J min^{-1} m^{-2} K^{-1})	0.307
Amount (25 kg package) (g)	24948
Vessel surface area (m^2)	0.48
Flammable limit (%)	7.50–100

(Data obtained with permission from Liu et al. Copyright © 2015, Elsevier)

material, or volume of the container, the addition of inhibitors, air conditioning equipment and the installation of refrigerating are incorporated by the appropriate controls. SADT can be achieved based on the Wilberforce theory:

$$SADT = T_{NR} - R\left(T_{NR} + 273.15\right)^2 / E_a \qquad (12.9)$$

By using the chart method, the $T_{NR,}$ irreversible temperature can be achieved (Wilberforce method), which plots the time required to the higher reaction rate across the temperature, or by using the analytical method:

$$\left(T_{NR} + 273.15\right)^2 = \frac{mE_a \Delta H_{AB}k}{RU\left(1.8\right)a} = mE_a \Delta H_{AB}Ae^{-\frac{E_a}{R\left(T_{NR} + 273.15\right)}} / RU\left(1.8\right)a \quad (12.10)$$

The physical data from the thermal analyses for a 25 kg package of DTBP and SADT was predicted in Table 12.4 (Fisher and Goetz, 1991; Wang et al., 2006). The hazard specifications of DTPB were resolved as SADT = 80.3 °C and TNR = 87 °C and is in agreement with literature (SADT = 80 °C) (Sun et al., 2001, Malow & Wehrstedt, 2005; Whitmore and Wilberforce, 1993). The safer temperature of storage and transportation would be regulated by T_{NR} and SADT of combustible harmful materials stored in a container. In order to forestall self-accelerating decomposition, the system needs to be maintained at temperatures of less than 55 °C in the storage and transportation process.

12.4 SUMMARY

Chemical substances used as raw materials or reactants in various industries are often found to be highly reactive in nature. A runaway reaction often takes place in case of unexpected factors which causes hindrance during the normal operation of the system, or during mishandling and transportation. Hence the appropriate approach was established for organic peroxides and azo compounds to ensure safety from such potential hazard compounds. For process designing and transportation, the widely applied SADT and E_a values were estimated. It was recommended by the United Nations (UN) that the temperature should not exceed the SADT value of the reactive chemicals, during transportation. The UN also recommends reliable experimental methods for the manufacturers to determine the SADT of actual commodities by utilizing thermal hazard analysis techniques and calorimetric equipment, which are discussed in this chapter. The extensive analysis of the hazards caused by reactive chemicals in producing thermal runaway reactions, together with their incompatibility hazard during the manufacturing process, could be more efficient in analysing the thermal hazards in large-scale plants. In addition, the product analyses could also help in predicting the possible reaction mechanisms and can ensure safety management measures to prevent thermal hazardous situations.

REFERENCES

American Society for Testing and Materials (ASTM) E 698-11, 2011. Standard Test Method for Arrhenius Kinetic Constants for Thermally Unstable Materials Using Differential Scanning Calorimetry and the Flynn/Wall/Ozawa Method (W. Conshohocken, PA, USA)

Casson, V., Maschio, G., 2012. Screening analysis for hazard assessment of peroxides decomposition. *Ind. Eng. Chem. Res.* 51, 7526–7535.

Casson, V., Salzano, E., Maschio, G., 2012. Hydrogen peroxide decomposition analysis by screening calorimetry technique. *Chem. Eng. Trans.* 26, 27–32.

Chen, K.Y., Wu, S.H., Wang, Y.W., Shu, C.M., 2008. Runaway reaction and thermal hazards simulation of cumene hydroperoxide by DSC. *J. Loss Prev. Process Ind.* 21, 101–109.

Chervin, S., Bodman, G.T., 1997. Mechanism and kinetics of decomposition from isothermal DSC data: development and application. *Process Saf. Prog.* 16, 94–100.

Chiu, W.Y., Carratt, G.M., Soong, D.S., 1983. A computer model for the gel effect in free-radical polymerization. *Macromolecules* 16, 348–357.

Dubikhin, V.V., Knerel'man, E.I., Manelis, G.B., Nazin, G.M., Prokudin, V.G., Stashina, G.A., Chukanov, N.V., Shastin, A.V., 2012. Thermal decomposition of azobis (isobutyronitrile) in the solid state. *Kinet. Catal.* 446, 171–175.

Duh, Y.S., Wu, X.H., Kao, C.S., 2008. Hazard ratings for organic peroxides. *Process Saf. Prog.* 27, 89–99.

FAI, 2002. *VSP2 Manual and Methodology*. Fauske & Associates, Inc., Burr Ridge, Illinois, USA.

Fisher, H.G., Goetz, D.D., 1991. Determination of self-accelerating decomposition temperatures using the accelerating rate calorimeter. *J. Loss Prev. Process Ind.* 4, 306–316.

Gibson, N., Rogers, R.L., Wright, T.K., 1987. Chemical reaction hazards: an integrated approach, hazards from pressure. In: *Institution of Chemical Engineers Symposium Series*, vol. 102, pp. 61–84.

Hofelich, T.C., Labarge, M.S., 2002. On the use and misuse of detected onset temperature of calorimetric experiments for reactive chemicals. *J. Loss Prev. Process Ind.* 15, 163–168.

Hou, H.Y., Shu, C.M., Duh, Y.S., 2001. Exothermic decomposition of cumene hydroperoxide at low temperature conditions. *AIChE J.* 47, 1893–1896.

Hou, H.Y., Su, C.H., Shu, C.M., 2012. Thermal risk analysis of cumene hydroperoxide in the presence of alkaline catalysts. *J. Loss Prev. Process Ind.* 25, 176–180.

Hsu, J.M., Su, M.S., Huang, C.Y., Duh, Y.S., 2012. Calorimetric studies and lessons on fires and explosions of a chemical plant producing CHP and DCPO. *J. Hazard. Mater.* 217–218, 19–28.

Iwata, Y., Momota, M., Koseki, H., 2006. Thermal risk evaluation of organic peroxides by automatic pressure tracking adiabatic calorimeter. *J. Therm. Anal. Calorim.* 85, 617–622.

Kotoyori, T., 1999. The self-accelerating decomposition temperature (SADT) of solids of the quasi-autocatalytic decomposition type. *J. Hazard. Mater.* 64, 1–19.

Lee, P.R., 1969. Safe storage and transportation of some potentially hazardous materials. *J. Appl. Chem.* 19, 345.

Li, X.R., Koseki, H., 2004. Interpretation of decomposition mechanisms of unstable substances near the SADT by an isothermal method. In: Loss Prevention and Safety Promotion in the Process Industries, 11th International Symposium Loss Prevention, Praha Congress Centre 31 May-3 June, pp. 2278–2285.

Li, X.R., Koseki, H., 2005. Thermal decomposition kinetic of liquid organic peroxides. *J. Loss Prev. Process Ind.* 18, 460–464.

Liu, S.H., Lin, C.P., Shu, C.M., 2011. Thermokinetic parameters and thermal hazard evaluation for three organic peroxides by DSC and TAM III. *J. Therm. Anal. Calorim.* 106, 165–172.

Liu, A.-H., Shu, C.-H., Hou, H.-Y., 2015. Applications of thermal hazard analyses on process safety assessments. *J. Loss Prev. Proc. Ind.* 33, 59–69.

Lu, K.T., Yang, C.C., Lin, P.C., 2006. The criteria of critical runaway and stable temperatures of catalytic decomposition of hydrogen peroxide in the presence of hydrochloric acid. *J. Hazard. Mater.* 135, 319–327.

Malow, M., Wehrstedt, K.D., 2005. Prediction of the self-accelerating decomposition temperature (SADT) for liquid organic peroxides from differential scanning calorimetry (DSC) measurements. *J. Hazard. Mater.* 120, 21–24.

Maschio, G., Bello, T., Scali, C., 1992. Optimization of batch polymerization reactors: modelling and experimental results for suspension polymerization of Methyl- MethAcrylate. *Chem. Eng. Sci.* 47, 2609–2614.

Maschio, G., Ferrara, I., Bassani, C., Nieman, H., 1999. An integrated calorimetric approach for the scale-up of polymerization reactors. *Chem. Eng. Sci.* 54, 3273–3282.

Maschio, G., Lister, D.G., Casson, V., 2010. Use of screening analysis calorimetry in the study of peroxides decomposition. *Chem. Eng. Trans.* 19, 347–352.

Maschio, G., Moutier, C., 1989. Polymerization reactor: the influence of "Gel Effect" in batch and continuous solution polymerization of methyl methacrylate. *J. Appl. Polym. Sci.* 37, 825–840.

Mettler Company, 2014. TA8000 Operation Instructions, Switzerland.

Miyake, A., Yamada, N., Ogawa, T., 2005. Mixing hazard evaluation of organic peroxides with other chemicals. *J. Loss Prev. Process Ind.* 18, 380–383.

Soh, S.K., Sundberg, D.C., 1982. Diffusion-controlled vinyl polymerization. I. The gel effect. *J. Polym. Sci. Polym. Chem. Ed.* 20, 1299–1313.

Sun, J., Li, Y., Hasegawa, K., 2001. A study of self-accelerating decomposition temperature (SADT) using reaction calorimetry. *J. Loss Prev. Process Ind.* 14, 331–336.

Townsend, D.I., Tou, J.C., 1980. Thermal hazard evaluation by an accelerating rate calorimeter. *Thermochim. Acta.* 37, 1–30.

UN, 2003. *Recommendations on the Transport of Dangerous Goods*, 13th ed, vol. 1. United Nations Publications, Geneva, Switzerland, p. 104.

Wang, Y.W., Duh, Y.S., Shu, C.M., 2006. Evaluation of adiabatic runaway reaction and vent sizing for emergency relief from DSC. *J. Therm. Anal. Calorim.* 85, 225–234.

Whitmore, M.W., Wilberforce, J.K., 1993. Use of the accelerating rate calorimeter and the thermal activity monitor to estimate stability temperature. *J. Loss Prev. Process Ind.* 6, 95–101.

Yeh, P.Y., Shu, C.M., Duh, Y.S., 2003. Thermal hazard analysis of methyl ethyl ketone peroxide. *Ind. Eng. Chem. Res.* 40, 1–5.

13 Safety of Flammable Liquid Mixtures

13.1 INTRODUCTION

The flash point for a given liquid known as the temperature, which is experimentally determined where the substance by emitting sufficient vapor forms a combustible mixture with air (CCPS/AIChE, 1993). It is hence found that there is a greater chance for fire hazard if the flash point value of a liquid mixture is low (Lees, 1996). Recently, after a series of fire hazards and explosions related to essential oils in Taiwan and Shengli, the importance of the flash point has today begun to be dramatically highlighted. During a series of such accidents, eight people were badly burned by six blasts which occurred from January through August 2003. Flash point primarily characterizes the fire and explosion hazard of the essential oils (Crowl and Louvar, 2002). Due to such casualties and hazards, after the Shengli event large quantities of waste organic solutions were temporarily moved in small amounts at various factory sites, to reduce the risk (Liaw & Chiu, 2003). Thus, in order to ensuring safety in storing huge volumes of such hazardous liquid mixtures, a knowledge of flash points became increasingly important. A particularly serious incident occurred on April 29, 2007 when a gasoline tanker crashed near the San Francisco–Oakland Bay Bridge, USA, and the blast produced such intense heat that this stretch of the highway melted and eventually collapsed. Flash point values are really important for ensuring the safe transportation of flammable liquids (DOT, Shippers – General Requirements for Shipments and Packagings, 2004). Hence, to characterize the fire- and explosion-related accidents for liquids, the flash point represents the most important variable, which could lead to safe usage, storage, or transportation. In 2008, the implementation of the Globally Harmonized System of Classification and Labeling of Chemicals (GHS) was encouraged by the United Nations (UN). The flash point of mixtures was considered to be the critical property in implementing the GHS for classifying the flammable liquids. Unfortunately, data on flash points are not generally available in the literature. In industrial processes, there can be considerable variation in the composition, and therefore the chemical properties, of specific mixtures. Deriving flash-point data utilizing test instruments is a time-consuming process. Thus, the classification of mixtures was delayed until 2015 by the European Union (EU) (Regulations, Regulation (EC) No. 1272/2008 of the European Parliament and of the Council, on Classification, Labeling and Packaging of Substances and Mixtures, 2008). In the case of partially miscible mixtures which are utilized in the liquid and liquid extraction processes (Kurihara et al., 2002 & Matsuda & Ochi, 2004) and heterogeneous distillation processes (Kosuge & Iwakabe, 2005), it was found that the flash point was one of the least studied aspects of the processes in many chemical plants. In order to facilitate the evaluation of fire and explosion hazards, flash point data are urgently needed for partially miscible mixtures. In order to ensure liquid phase equilibrium the data obtained were for completely stirred mixtures. In the real world, however, the complete stirring of liquid mixtures are not always considered during processes such as the collection or accumulation of waste solvents. Rather, the decantation process is exhibited by such mixtures depending on their varying density, with the lightest phase above.

13.2 FLASH POINT EVALUATION

In Taiwan it costs about NT\$ 20,000/US\$ 600 per sample to derive flash point data from test instruments, which is a very high cost. Hence, for various mixtures, especially for miscible ones several alternative models were formulated for predicting the flash points (Affens & McLaren, 1972; White et al., 1997; Liaw, Tang & Lai, 2004; Liaw & Chiu, 2006; Liaw & Wang, 2007; Catoire, Paulmier & Naudet, 2006; Gmehling & Rasmussen, 1982 & Lee & Ha, 2003). However, in the case of partially miscible mixtures, to date only three models have been proposed. Liaw et al. (2008a) first developed the model for flammable solvents forming binary partially miscible mixtures. Further, using the experimental data its accuracy was verified to observe the difference (Liaw et al., 2008a). Similarly, in the case of aqueous-organic mixtures forming partially miscible solutions the second model was developed and it was compared successfully and verified through experimental results (Liaw et al., 2008b). For flammable solvents forming ternary partially miscible mixtures, a third model was developed (Liaw et al. 2008b), and further verification of the predicted flash point was performed for both type-I and type-II mixtures (Liaw, Gerbaud & Chiu, 2010b). For the considered mixtures, liquid phase equilibrium with their compositions were the basic assumptions considered in all three of the developed models. For a given mixture the flash point value is basically relative to its vapor pressure, and this depends on the liquid phase composition. Since the liquid–liquid equilibrium assumption is not always valid, the flash point behavior in such a case is found to be quite different from that under liquid–liquid equilibrium (LLE). Thus, for binary partially miscible mixtures, the effect of stirring was investigated in both aqueous-organic and flammable solvent mixtures. In this chapter a mutual solubility region was investigated which existed for aqueous-organic mixtures, partially miscible in nature:

water+isobutanol, water+1-butanol, water+1-pentanol, and water+2-butanol. However, both octane and water are immiscible to each other. Minimum flash point behavior was exhibited by the mixtures of flammable solvents, such as methanol+2,2,4-trimethylpentane, and methanol+octane, whereas minimum value of flash point behavior was shown by methanol+decane, were also investigated in this chapter.

13.3 EXPERIMENTAL PROTOCOL

According to the requirement of ASTM D56 standard for partially miscible mixtures (ASTM D56, 1999), an HFP 362-Tag Flash Point Analyzer (Walter Herzog GmbH, Germany), was utilized for measuring flash points both with and without the stirring effect at different compositions (water+1-butanol, water+2-butanol, water+isobutanol, water+1-pentanol, water+octane, methanol+decane, methanol+2,2,4-trimethylpentane, and methanol+octane). A control device was installed within the apparatus to program a specific rate of heating the sample within a temperature range near about the expected flash point. By using an igniter at test intervals of specified temperature, the automatic testing of flash point was performed. Heat rate-1 was used along with igniter firing at test interval-1, when the expected flash point is lower than or equal to the change temperature. Similarly, heat rate-2 was adopted, if the expected flash point is higher, and at test interval-2 the igniter was fired. Expected flash point minus the start-test value is regarded as the equivalent temperature at which the first flash point test series is initiated. The experimental iteration is terminated when the test temperature exceeds the sum of the expected flash point plus the end-of-test value, and still the flash point is not being determined.

The test parameters selected for the analysis as per ASTM D56 were: start/end test 5 and 20 °C; heat rate-1 and 2 were 1 and 2 °C/min; test interval-1 and 2 were 0 5 and 1 °C and the change in temperature was 60 °C (ASTM D56, 1999). The mole fraction of the liquid mixture was measured using a Setra EL-410D digital balance. Two sets of mixtures, completely stirred and unstirred, were tested for comparison. The prepared mixtures of the former set were stirred (with a magnetic stirrer) for 30 min before the flash point test. The unstirred samples were analyzed using Flash Point Analyzer in a test cup. Methanol and isobutanol were HPLC/Spectro-grade reagents (Tedia Co. Inc., USA); 1-butanol, 1-pentanol, octane, and 2,2,4-trimethylpentane were also sourced from Tedia. 2-Butanol was purchased from Fisher Scientific International Inc. (USA). Decane was obtained from Alfa Aesar (Lancaster, England). A Milli-Q plus was used for water purification.

13.4 FLASH POINT MODEL PREDICTION FOR PARTIALLY MISCIBLE MIXTURES

At liquid–liquid equilibrium, for binary partially miscible aqueous–organic and flammable solvents mixtures, flash point was determined based on models proposed by Liaw et al., 2008a, and Liaw et al., 2008b.

13.4.1 MODEL FOR AQUEOUS–ORGANIC SOLUTIONS

For binary partially miscible aqueous-organic mixture, the flash point was evaluated (Liaw, Gerbaud & Chiu, 2010b) with water as component 1 and flammable material as component 2.

$$1 = \frac{\chi_2 \gamma_2 P_2^{sat}}{P_{2,fp}^{sat}} \tag{13.1}$$

$$\log P_2^{sat} = A_2 - \frac{B_2}{T + C_2} \tag{13.2}$$

Where $P_{i,fp}^{sat}$, in Equation (13.1), is the vapor pressure of the pure substance, i, at its flash point, and P_i^{sat} is the vapor pressure of substance, i, at the mixture's flash point.

The partially miscible region where two liquid phases are in equilibrium (tie line), the overall composition of both liquid phases is equal to the single vapor composition located on the vapor line (Van Ness & Abbott, 1982 & Renon & Prausnitz, 1968). Flash point in this region is constant regardless of the liquid composition on the liquid–liquid equilibrium tie line. The compositions between liquid phases in equilibrium was estimated from the compound fugacities in each phase (Liaw, Gerbaud & Chiu, 2010b),

$$\left(\chi_i \gamma_i \right)^\alpha = \left(\chi_i \gamma_i \right)^\beta ; i = 1, 2 \tag{13.3}$$

where α and β designate the two coexisting liquid phases.

The activity coefficients γ_i in Equations. (13.1) and (13.3), were estimated using thermodynamic activity coefficient models adequate for partially miscible mixtures, such as the NRTL (Renon & Prausnitz, 1968) or UNIQUAC equations (Abrams & Prausnitz, 1975).

13.4.2 MODEL FOR MIXTURES OF FLAMMABLE SOLVENTS

The flash point for binary partially miscible mixture of flammable solvents within the mutual-solubility region was calculated from Equation (13.4 and 13.5) (Liaw et al., 2008a)

$$1 = \frac{\chi_1 \gamma_1 P_1^{sat}}{P_{1,fp}^{sat}} + \frac{\chi_2 \gamma_2 P_2^{sat}}{P_{2,fp}^{sat}} \tag{13.4}$$

$$\log P_i^{sat} = A_i - \frac{B_i}{T + C_i} ; i = 1, 2 \tag{13.5}$$

Where $P_{i,fp}^{sat}$, in Equation (13.4), is the vapor pressure of the pure substance, i, at its flash point, and P_i^{sat} is the vapor pressure of substance, i, at the mixture's flash point. The flash point of the two liquid phase was calculated from the value of temperature derived from Equations (13.3)–(13.5). The partial miscibility of mixtures involves VLE and LLE where Equations (13.1, 13.2, 13.4, and 13.5) were used to evaluate flash point and Equation (13.3) is used to determine the tie line equilibrium liquid compositions. In the mutual solubility region, the flash point was estimated using VLE parameters. The flash point of flammable solvents and aqueous-organic mixtures were estimated based on the lower boiling point of pure compound and the span approaching flammable, respectively.

13.5 RESULTS AND DISCUSSION

13.5.1 PARAMETERS USED

The flash point prediction model for binary partially miscible aqueous-organic and flammable solvent mixtures, as described in section 13.3.2, were water+1-butanol, water+2-butanol, water+isobutanol, water+1-pentanol, water+octane, and methanol+decane, methanol+2,2,4-trimethylpentane, methanol+octane, respectively. The effect of stirring (with and without) on the flash point behavior was investigated and was compared with predicted data (two sets of measurements). The former set of data was reported by Liaw et al., 2008a & Liaw, Gerbaud & Chiu, 2010b and the latter one is listed in Tables 13.1 and 13.2. The average values of standard deviation of the unstirred (1.4 and 2.5 °C) was greater than stirred aqueous-organic solutions and mixtures of flammable solvents by 0.7 °C. Liquid-phase activity coefficients were estimated using the NRTL (Renon & Prausnitz, 1968) and/or UNIQUAC equations (Abrams & Prausnitz, 1975). Binary interaction parameters obtained either from the LLE or VLE data were used in this chapter, with parameters adopted from the literature (Tang, Li & Li, 1995; Klauck, Grenner & Schmelzer, 2006; Resa et al., 2006; Lu, Chiou & Chen, 2002; Tourino et al., 2003; Gramajo de Doz, et al., 2003; Gmehling, Onken & Arlt, 1981; Gmehling & Onken; 1977; & Gmehling, Onken & Arlt, 1982) (Tables 13.3 and 13.4). The parameters for relative van der Waals volume (r) and the surface area (q) for the pure components needed in the UNIQUAC equation were obtained from the literature (Gramajo de Doz, et al., 2003 & Poling, Prausnitz & O'Connell; 2001) are listed in Table 13.5 along with the Antoine coefficients sourced from the literature (Gmehling, Onken & Arlt, 1981; Gmehling & Onken; 1977; & Gmehling, Onken & Arlt, 1982).

The flash points for the pure substances used, were measured using the Flash Point Analyzer, and these values were comparable to their literature-derived analogues (NIOSH Pocket guide to chemical hazards, 2008; Merck, 2008; Fisher Scientific, 2008; Tedia, 2008; Univar USA, 2008; Mallinckrodt Baker, 2008; M. Bohnet, et al., 2007). (Table 13.6). There were between-source differences in the flash point data for 1-butanol, 2-butanol, isobutanol, 1-pentanol, octane, methanol, decane, and 2,2,4-trimethylpentane. However, these differences were acceptable except for the value of 1-butanol provided by NIOSH (NIOSH Pocket guide to chemical hazards, 2008), 2-butanol by Tedia (Tedia, 2008), 1-pentanol by Fisher (Fisher Scientific, 2008), decane by SFPE (SFPE, The SFPE Handbook of Fire Protection Engineering, second ed., Society of Fire Protection Engineers, Boston, 1995) and 2,2,4-trimethylpentane by Merck (Merck, 2008), and SFPE (SFPE, The SFPE Handbook of Fire Protection Engineering, second ed., Society of Fire Protection Engineers, Boston, 1995). The experimental flash points for these eight samples were

TABLE 13.1

Measured Flash Point for Unstirred Partially Miscible Aqueous-organic Mixtures

$x1$	Water (1) + 1-butanol (2) (oC)	Water (1) + 2-butanol (2) (oC)	Water (1) + isobutanol (2) (oC)	Water (1) + 1-pentanol (2) (oC)	Water (1) + octane (2) (oC)
0	36.9	22	28.5	49.5	14.5
0.01	–	21.7	–	49.6	–
0.02	–	21.9	28.5	49.7	–
0.03	–	22.1	28.5	–	–
0.05	36.8	–	–	–	14.3
0.1	36.9	22.5	28.4	50.4	15
0.2	37.6	22.5	28.7	50.6	14.9
0.3	37.2	22.3	28.9	50.5	14.6
0.4	38.1	22.4	29.1	50.8	14.7
0.5	37.8	22.5	29.3	50.7	15
0.6	38.1	22.9	30.1	50.7	14.4
0.7	37.8	23.5	30.4	50.9	14.1
0.8	38.3	24.1	31.6	51.2	14.5
0.9	38.3	23.8	31.2	51	13.8
0.95	38.6	25.3	–	53.4	14.7
0.97	–	26.6	32.4	–	–
0.98	39.8	27.3	33.7	–	–
0.99	41.6	44.1	43.9	56.2	14.6
0.992	51.3	–	–	–	–
0.993	54.4	51.6	–	–	–
0.994	57.6	–	52.9	–	–
0.995	62.9	58.1	57.5	–	–
0.996	69.3	63.6	60.5	59.5	–
0.997	–	–	66	64.5	–
0.998	–	–	–	71.5	–

(Reproduced with permission from Liaw et al. 2010a, Copyright © Elsevier)

close to the literature-derived values (Chevron Phillips Chemical Company; J. T Baker; ASTM D 93, 2000), except for the ones mentioned above, which had greater differences from other sources (Table 13.6).

13.5.2 PARTIALLY MISCIBLE AQUEOUS–ORGANIC MIXTURES

For water + 1-butanol mixture, the comparison of predicted and the experimental flash point values by the aforementioned models were presented in Figure 13.1. The VLLE based model predictions were in good agreement with the experimental data (NRTL or UNIQUAC model) for the completely stirred mixtures in the entire flammable range. The experimental flash point values of unstirred mixtures follow a similar trend and are at lower values when compared to completely stirred mixtures. The

TABLE 13.2
Measured Flash Point for Unstirred Partially Miscible Mixtures of Flammable Solvents

$x1$	Methanol (1) + decane (2) (oC)	Methanol (1) + 2,2,4- trimethylpentane (2) (oC)	Methanol (1) + octane (2) (oC)
0	51.8	−8.1	14.5
0.005	48		
0.01	30.4	−8.6	12.6
0.02	22.8	–	10.5
0.03	–	–	8.9
0.04	19.7	–	–
0.05	18.3	−8.4	7.3
0.06	15.5	–	–
0.1	15.7	−9.1	4.8
0.2	15.2	−8.7	5
0.3	15.9	−9	4.2
0.4	14.5	−9.6	4.3
0.5	16	−9.6	4.3
0.6	13.8	−9.5	4.2
0.7	12.9	−9.2	4.3
0.8	11.7	−9.6	3.9
0.9	11.4	−9.3	4.5
0.95	10.3	−9.6	3.8
0.97	–	−7.4	–
0.98	11.1	−4.7	4
0.985	–	–	5.1
0.99	–	−0.1	6.1
0.992	–	1.1	–
0.995	10.8	4.83	7.5
0.998	9.8	–	–
1	10	10	10

(Reproduced with permission from Liaw et al. 2010a, Copyright © Elsevier)

measured values of flash points were very close in the narrow mutual solubility region ($x_{water} > 0.95$). The flash point values decreased slightly with a decrease in water fraction and was greater than pure 1-butanol for unstirred mixtures ($x_{water} < 0.95$). Figures 13.2–13.4 represents similar behavior for other partially miscible aqueous-organic mixtures (water+2-butanol, water+isobutanol and water+1-pentanol). For unstirred immiscible water+octane mixture, the flash point values were in agreement with the predicted values and were almost equivalent to the completely stirred mixtures (near immiscibility of these compounds) (Maczynski et al. 2004) (Figure 13.5). However, for the unstirred aqueous-organic mixtures, two-phase liquid exists almost over the entire flammable range and stirring ensures complete miscibility in the flammable rich/lean regions. As the density of the flammable substances is

TABLE 13.3

LLE Parameters of the NRTL and UNIQUAC Equations for the Studied Systems

System	TC (K)	$\alpha12$	Parameters	ij		Reference
NRTL equation				12	21	
Water(1)+1-butanol(2)	–	0.45	a_{ij}	−2610.15	−3884.30	Kosuge et al. (2005)
			b_{ij}	19.4473	30.3191	
			c_{ij}	−0.0237040	−0.0527519	
Water(1)+2-butanol(2)	–	0.45	a_{ij}	−2744.73	−3871.43	Kosuge et al. (2005)
			b_{ij}	19.1484	25.0760	
			ci_j	−0.0228962	−0.0393948	
Water(1)+isobutanol(2)	–	0.3	τ_{ij}	3.770	0.025	Tang et al. (1995)
Water(1)+octane(2)	–	0.2	a_{ij}	−169.718	4197.06	Klauck et al. (2006)
			b_{ij}	12.5591	−7.5243	
			c_{ij}	0	0	
Methanol(1)+octane(2)	339.69	0.2	α'_{ij}	751.016	63.260	Kurihara et al. (2002)
			β'_{ij}	1.831	8.375	
			γ'_{ij}	−0.211	9.502×10^{-3}	
			δ'_{ij}	2.542×10^{-3}	-6.654×10^{-4}	
Methanol(1)+2,2,4-trimethyl pentane(2)	316.84	0.2	α'_{ij}	594.073	147.674	Kurihara et al. (2002)
			β'_{ij}	6.255	6.282	
			γ'_{ij}	−0.588	0.178	
			δ'_{ij}	1.070×10^{-2}	-5.702×10^{-3}	
UNIQUAC equation				12	21	
Water(1)+1-butanol(2)	–	–	a_{ij}	−1237.85	−4.72337	Kosuge et al. (2005)
			b_{ij}	7.12425	1.36693	
			c_{ij}	−0.0066927	−0.0047593	
Water(1)+2-butanol(2)	–	–	a_{ij}	−1276.11	−145.764	Kosuge et al. (2005)
			b_{ij}	7.59662	1.46978	
			c_{ij}	−0.0083095	−0.0038732	
Water(1)+1-pentanol(2)	–	–	a_{ij}	242.413	90.395	Resa t al. (2006)
			b_{ij}	0	0	
			c_{ij}	0	0	
Water(1)+octane(2)	–	–	a_{ij}	195.95	2446.88	Lu et al. (2002)
			b_{ij}	0	0	
			c_{ij}	0	0	
Methanol(1)+decane(2)	–	–	a_{ij}	8255.57	1472.06	Tourino et al. (2003)
			b_{ij}	−7.37400	−4.33899	
			c_{ij}	0	0	
Methanol(1) + 2,2,4-trimethylpentane(2)	–	–	A_{ij}	−30.557	738.15	Gramajo et al. (2003)

(Reproduced with permission from Liaw et al. 2010a, Copyright © Elsevier)

TABLE 13.4
VLE Parameters of the NRTL and UNIQUAC Equations for the Studied Systems

Mixtures	NRTL			UNIQUAC		Reference
	A12	A21	α12	A12	A21	
Water(1)+1-butanol(2)	1332.336	193.464	0.4056	193.397	129.827	Gmehling et al. (1981)
Water(1)+2-butanol(2)	891.640	133.786	0.4406	116.950	87.753	Gmehling et al. (1981)
Water(1)+isobutanol(2)	1109.011	114.185	0.3155	142.459	150.949	Gmehling et al. (1981)
Water(1)+1-pentanol(2)	1643.518	60.776	0.3309	252.687	77.061	Gmehling et al. (1981)
Methanol(1)+decane(2)	–	–	–	–58.522	933.899	Gmehling et al. (1977)
Methanol(1)+octane(2)						
Methanol (1) + 2,2,4-trimethylpentane(2)	728.279	697.771	0.4313	–30.042	793.817	Gmehling et al. (1982)

(Reproduced with permission from Liaw et al. 2010a, Copyright © Elsevier)

less than water, the upper layer of the aqueous-organic mixtures is an organic phase (Table 13.5).

The flash points of the unstirred aqueous-organic mixtures were dominated by the composition of the organic phase and were close to that of the pure flammable component which could be due to the low quantities of aqueous phase in the mixtures. The diffusion of water molecules into the organic phase amplifies the flash point values with the increase in the mole fraction of water. Only aqueous (liquid) phase was observed when the mole fraction of water approached unity (flammable mole fraction completely soluble in water). The non-ideal mixtures combine both liquid and vapor phase properties to evaluate the flash point. In unstirred mixtures, the mole fraction of flammable organic compounds were high as the liquid phases were not in equilibrium, reducing the flash point values of the organic phase when compared to completely stirred mixtures. The disturbances in the unstirred mixtures improves mixing and reaches the LLE value after complete stirring. The flash point values gradually increased with disturbances and remains limited by the LLE value. For the immiscible water+octane mixture, the flash point values were almost constant/similar to octane with complete stirring. For such mixtures, stirring has no significant influence on flash points. Overall, the flash point values of the aqueous–organic mixtures was never lower than the pure organic compound. However, it is suggested that such mixtures should be stirred completely before handling to reduce the fire and explosion hazard.

13.5.3 PARTIALLY MISCIBLE MIXTURES OF FLAMMABLE SOLVENTS

The predicted values of flash point for methanol + decane were evaluated using Equations (13.3)–(13.5) with the binary interaction parameters listed in Tables 13.3 and 13.4 and compared with experimental values in Figure 13.6. Equilibrium model

TABLE 13.5

Antoine Coefficients and Density for Solution Components, and Relative van der Waals Volumes (r) and Surface Areas (q) for the Pure Components for the UNIQUAC Model

Material	Antoine Coefficients				Relative van der Waals Volumes (r) and Surface Areas (q)				Density	
	A	B	C	Reference	r	q	Reference	ρ	Reference
1-Butanol	7.83800	1558.190	−76.119	Gmehling et al. (1981)	3.4543	3.052	Poling et al. (2001)	0.81	Merck, 2006
2-Butanol	7.47429	1314.188	−86.500	Gmehling et al. (1981)	3.4535	3.048	Poling et al. (2001)	0.808	Merck, 2006
Isobutanol	8.53516	1950.940	−35.853	Gmehling et al. (1981)	3.4535	3.048	Poling et al. (2001)	0.806	Merck, 2006
1-Pentanol	7.39824	1435.570	−93.202	Gmehling et al. (1981)	4.1287	3.592	Poling et al. (2001)	0.8110	Merck, 2006
Octane	6.93142	1358.800	−63.145	Gmehling et al. (1981)	5.8486	4.936	Poling et al. (2001)	0.7028	Merck, 2006
Methanol	7.97010	1521.230	233.970	Gmehling et al. (1982)	1.4311	1.432	Gramajo et al. (2003)	0.7960	Merck, 2006
Decane	7.44000	1843.120	230.220	Gmehling et al. (1977)	7.1974	6.016	Poling et al. (2001)	0.7365	Tourino et al. 2003
2,2,4-Trimethylpentane	6.80304	1252.590	220.119	Gmehling et al. (1982)	5.8463	5.008	Gramajo et al. (2003)	0.692	Oxford university, 2008

(Reproduced with permission from Liaw et al. 2010a, Copyright © Elsevier)

TABLE 13.6

Comparison of Flash-point Values Adopted From the Literature with Experimentally Derived Data for the Studied Solution Components

Component	Experimental Data (oC)	Literature (oC)
1-Butanol	36.9±2.8	35 (Tedia, 2008) 36 (Univar USA, 2008)
2-Butanol	22.0±2.4	24 (Merck, 2008) 26 (Oxford university, 2008) 28.88 (Tedia, 2008)
Isobutanol	28.5±0.9	28 (Merck, 2008) 29 (Tedia, 2008)
1-Pentanol	49.5±1.2	38 (Fisher Scientific, 2008) 50 (Bohnet et al., 2007)
Octane	14.5±1.4	13 (Merck, 2008) 15 (Oxford university, 2008)
Methanol	10.0±0.8	11 (http://www.alfa.com) 10 (Oxford university, 2008)
2,2,4-Trimethylpentane	−8.1±1.3	−7 (Oxford university, 2008) −8 (http://www.cpchem.com/enu/mscs.asp.)
Decane	51.8±1.0	44 (SFPE, 1995) 50.9±2.3 (ASTM 56, 1999) 52.8±2.3 (ASTM D 93,2000)

(Reproduced with permission from Liaw et al. 2010a, Copyright © Elsevier)

predictions were in agreement with experimental data of completely stirred mixtures over the entire composition. The difference in flash point behavior of the unstirred and completely stirred mixtures of methanol+decane were different from the aqueous-organic solutions. The flash point values of the unstirred mixtures were all greater than the completely stirred mixtures and less than that of decane (highest boiling pure compound). Similar behavior was observed for methanol+2,2,4-trimethylpentane and methanol+octane (Figure 13.7 and 13.8) The unstirred mixture of methanol+decane separated into two liquid phases because of the difference in density and the upper layer of this partially miscible mixture is the decane-rich phase (Table 13.5). The flash point values of unstirred mixture was evaluated by the composition of the decane-rich phase (lower quantity of methanol-rich phase) which is the highest boiling pure compound. The determined flash point values of the unstirred mixture lie between LLE and that of pure decane (Ellis, 1976 & Merck, 2006). With the increase in the mole fraction of methanol, the quantities in the methanol-rich phase increased and decane-rich phase decreased. For unstirred mixtures, the flash point of the methanol-rich phase is lower than that of the decane-rich phase. At a particular concentration, the upper layer volume of the decane-rich phase is not enough to cover the air-exposed surface area. The patches of the decane-rich phase and the some methanol-rich phase are in contact with the air in such regions.

For the unstirred methanol+2,2,4-trimethylpentane partially miscible mixture, flash point value was determined by the 2,2,4- trimethylpentane-rich phase and were slightly less than that of pure 2,2,4-trimethylpentane (Table 13.5). This could be due

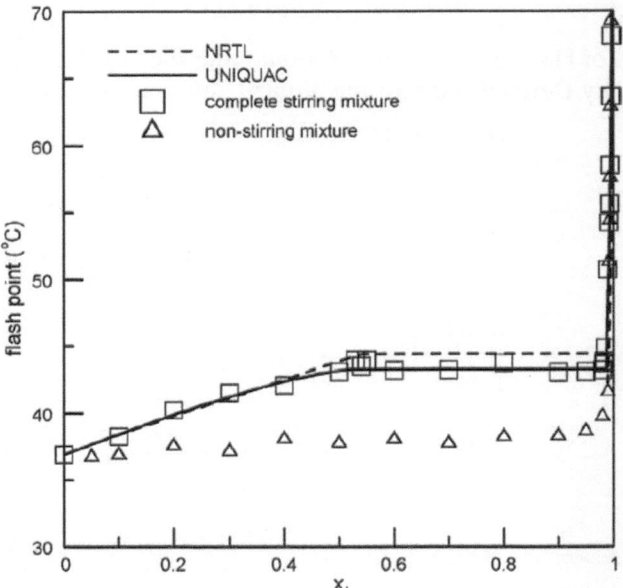

FIGURE 13.1 Comparison of predicted flash point and experimental data for completely stirred and unstirred water (1)+1-butanol (2) (Reproduced with permission from Liaw et al. 2010a, b Copyright © Elsevier).

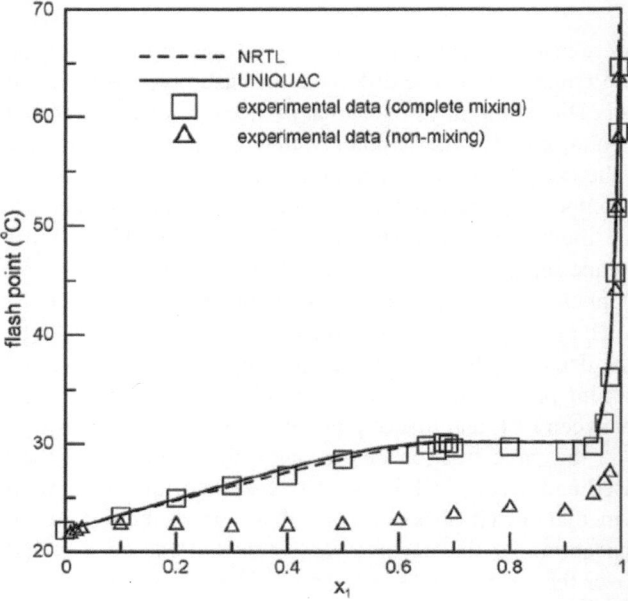

FIGURE 13.2 Comparison of predicted flash point and experimental data for completely stirred and unstirred water (1)+2-butanol (2) (Reproduced with permission from Liaw et al. 2010a, b Copyright © Elsevier).

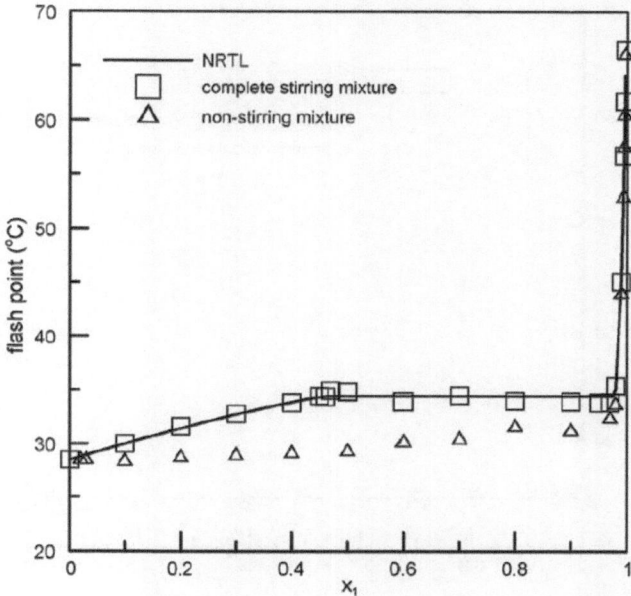

FIGURE 13.3 Comparison of predicted flash point and experimental data for completely stirred and unstirred water (1) + isobutanol (2) (Reproduced with permission from Liaw et al. 2010a, b Copyright © Elsevier).

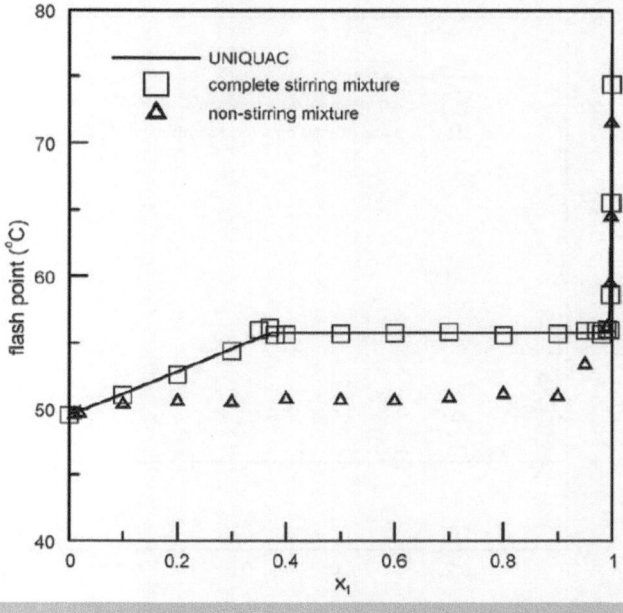

FIGURE 13.4 Comparison of predicted flash point and experimental data for completely stirred and unstirred wavter (1) + 1-pentanol (2) (Reproduced with permission from Liaw et al. 2010a, b Copyright © Elsevier).

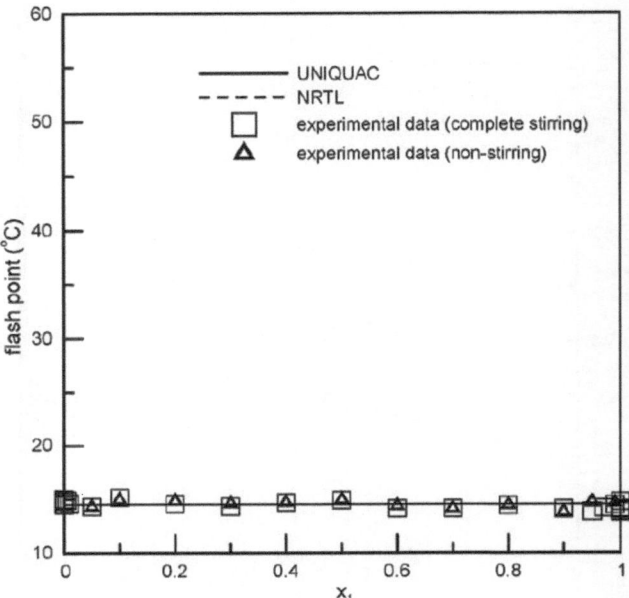

FIGURE 13.5 Comparison of predicted flash point and experimental data for completely stirred and unstirred water (1) + octane (2) (Reproduced with permission from Liaw et al. 2010a, b Copyright © Elsevier).

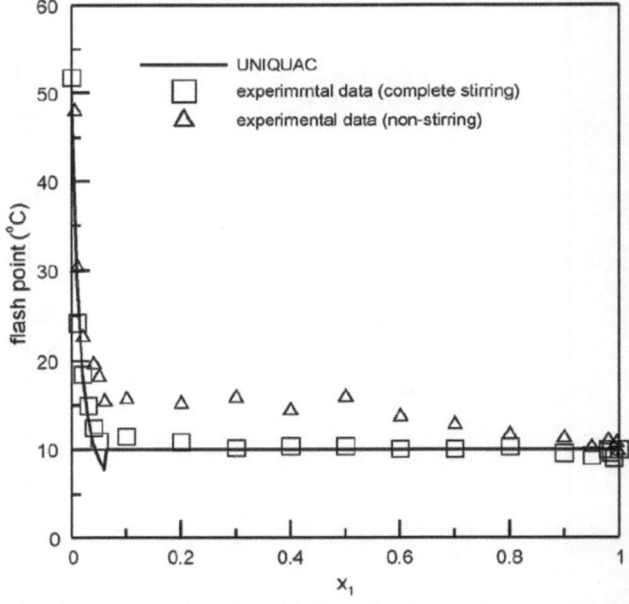

FIGURE 13.6 Comparison of predicted flash point and experimental data for completely stirred and unstirred methanol (1) + decane (2) (Reproduced with permission from Liaw et al. 2010a, b Copyright © Elsevier).

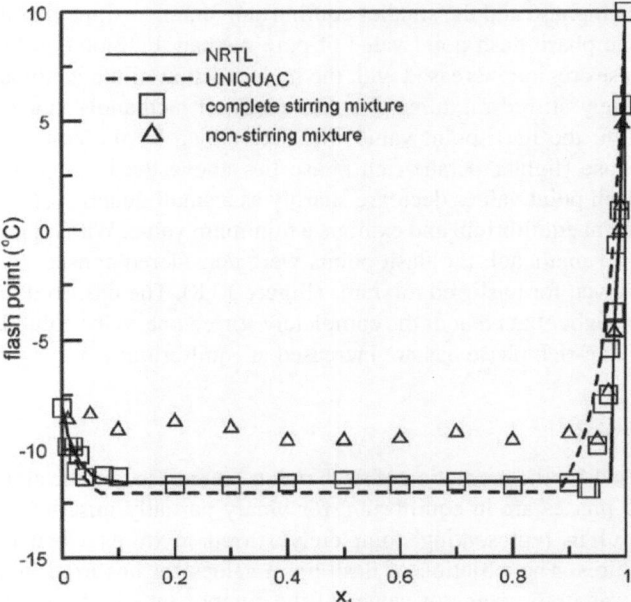

FIGURE 13.7 Comparison of predicted flash point and experimental data for completely stirred and unstirred methanol (1) + 2,2,4-trimethylpentane (2) (Reproduced with permission from Liaw et al. 2010a, b Copyright © Elsevier).

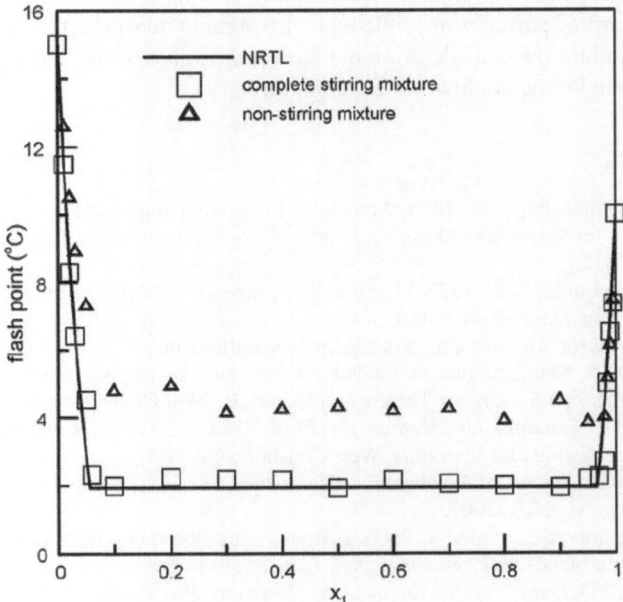

FIGURE 13.8 Comparison of predicted flash point and experimental data for completely stirred and unstirred methanol (1) + octane (2) (Reproduced with permission from Liaw et al. 2010a, b Copyright © Elsevier).

to the heavy rich phase and the smaller equilibrium quantity of methanol in the mixture. The single phase flash point value of pure methanol at non-equilibrium in the two liquid phases region increased with the mole fraction of methanol and was similar to completely stirred mixtures. For the unstirred methanol+octane mixture, at non-equilibrium, the flash point value was determined by the composition of the octane-rich phase (lighter octane-rich phase lies above the heavier methanol-rich phase). The flash point values decrease sharply as a small quantity of methanol was added to octane at equilibrium and exhibits a minimum value. With an increase in the mole fraction of methanol, the flash points were considered constant and a similar trend was observed for unstirred mixtures (Figure 13.8). The unstirred mole fraction of methanol is higher than that of the completely stirred one as the solubility of methanol in the octane-rich phase has not increased at equilibrium.

13.6 SUMMARY

The models used for the prediction of flash points were based on the assumption that the two liquid phases are in equilibrium for binary partially miscible mixtures and were successful in representing completely stirred mixtures when compared to unstirred mixtures. The variation of flash point values for unstirred flammable solvent mixture was in the range of values of the component's highest flash point and the value of the completely stirred mixtures. As per the recommendations of GHS, all the samples on the industrial sites should be completely stirred before testing to ensure safety against fire and explosion hazard. The range of the flash point was between the values of the completely stirred equilibrium mixture and the pure flammable solvent for aqueous-organic partial miscible mixtures. To assess the fire and explosion hazard of partially miscible aqueous-organic mixture, it is always recommended to take into the consideration the flash point value of the flammable component as it is found to be the lowest of the mixture.

REFERENCES

Abrams, D.S., Prausnitz, J.M. 1975. Statistical thermodynamics of liquid mixtures: new expression for the excess Gibbs energy of partly or completely miscible systems, *AIChE J.* 21, 116–128.

Affens, W.A., McLaren, G.W. 1972. Flammability properties of hydrocarbon solutions in air, *J. Chem. Eng. Data* 17, 482–488.

Alfa Aesar, A *Johnson Matthey Company*, http://www.alfa.com.

ASTM D 93, 2000. Standard Test Methods for Flash-point by Pensky–Martens Closed Cup Tester, American Society for Testing and Materials, West Conshohocken, PA.

ASTMD56, 1999. *Standard test Method for Flash Point by Tag Closed Tester, American Society for Testing and Materials*, West Conshohocken, PA.

Bohnet, M., et al., *Ullmann's Encyclopedia of Industrial Chemistry, Wiley Inter-Science* (John Wiley & Sons), USA, 2007.

Catoire, L., Paulmier, S., Naudet, V. 2006. Estimation of closed cup flash points of combustible solvent blends, *J. Phys. Chem. Ref. Data* 35, 9–14.

CCPS/AIChE, 1993, *Guidelines for Engineering Design for Process Safety*, American Institute of Chemical Engineers, New York .

Chevron Phillips Chemical Company, http://www.cpchem.com/enu/msds.asp.

Crowl, D.A., Louvar, J.F. 2002. *Chemical Process Safety: Fundamentals with Applications*, 2nd ed., Prentice Hall PTR, New Jersey.

DOT, Shippers-General Requirements for Shipments and Packagings, 2004. Class 3-Assignment of Packing Group, 49CFR173.121, National Archives and Records Administration, p. 488.

Ellis, W.H. 1976. Solvent flash points–expected and unexpected, *J. Coat. Technol.* 48, 44–57.

Fisher Scientific, 2008, https://www.fishersci.com/wps/portal/CMSTATIC?pagename=msds.

Gmehling, J., Onken, U. 1977. *Vapor–Liquid Equilibrium Data Collection*, vol. 1, Part2a, DECHEMA, Frankfurt, Germany pp. 72, 275.

Gmehling, J., Onken, U., Arlt, W. 1981. *Vapor–Liquid Equilibrium Data Collection, Part1a*, DECHEMA, Frankfurt, Germany.

Gmehling, J., Onken, U., Arlt, W. 1982. *Vapor–Liquid Equilibrium Data Collection*, vol. 1, Part 2c, DECHEMA, Frankfurt, Germany pp. 229–250.

Gmehling, J., Rasmussen, P. 1982. Flash points of flammable liquid mixtures using UNIFAC, *Ind. Eng. Chem. Fundam.* 21, 186–188.

Gramajo de Doz, M.B., Bonatti, C.M., Solimo, H.N. 2003. Liquid–liquid equilibria of ternary and quaternary systems with two hydrocarbons, an alcohol, and water at 303.15K Systems containing 2,2,4-trimethylpentane, toluene, methanol, and water, or 2,2,4-trimethylpentane, toluene, ethanol, and water, *Fluid Phase Equilib.* 205, 53–67.

J.T Baker, http://www.jtbaker.com/msds/englishhtml/T3913.htm.

Klauck, M., Grenner, A., Schmelzer, J. 2006. Liquid-liquid(-liquid) equilibria in ternary systems of water + cyclohexylamine + aromatic hydrocarbon (toluene or propylbenzene) or aliphatic hydrocarbon (heptane or octane), *J. Chem. Eng. Data* 51, 1043–1050.

Kosuge, H., Iwakabe, K. 2005. Estimation of isobaric vapor–liquid–liquid equilibria for partially miscible mixture of ternary system, *Fluid Phase Equilib.* 233, 47–55.

Kurihara, K., Midorikawa, T., Hashimoto, T., Kojima, K., Ochi, K. 2002. Liquid–liquid solubilities for the binary system of methanol with octane and 2,2,4- trimethylpentane, *J. Chem. Eng. Jpn.* 35, 360–364.

Lee, S.-J., Ha, D.-M. 2003. The lower flash points of binary systems containing nonflammable component, *Korean J. Chem. Eng.* 20, 799–802.

Lees, F.P. 1996. *Loss Prevention in the Process Industries, vol. 1*, 2nd ed., Butterworth-Heinemann, Oxford, UK.

Liaw, H.-J., Chen, C.-T., Gerbaud, V. 2008a. Flash-point prediction for binary partially miscible aqueous–organic mixtures, *Chem. Eng. Sci.* 63, 4543–4554.

Liaw, H.-J., Chiu, Y.-Y. 2003. The prediction of the flash point for binary aqueous organic solutions, *J. Hazard. Mater.* 101, 83–106.

Liaw, H.-J., Chiu, Y.-Y. 2006. A general model for predicting the flash point of miscible mixture, *J. Hazard. Mater.* 137, 38–46.

Liaw, H.-J., Gerbaud, V., Chen, C.-C., Shu. 2010a. Effect of stirring on the safety of flammable liquid mixtures. *J. Hazard. Mat.* 177(1–3), 1093–1101.

Liaw, H.-J., Gerbaud, V., Chiu, C.-Y. 2010b. Flash point for ternary partially miscible mixtures of flammable solvents, *J. Chem. Eng. Data* 55, 134–146.

Liaw, H.-J., Lee, Y.-H., Tang, C.-L., Hsu, H.-H., Liu, J.-H. 2002. A mathematical model for predicting the flash point of binary solutions, *J. Loss Prev. Proc.* 15, 429–438.

Liaw, H.-J., Lu, W.-H., Gerbaud, V., Chen, C.-C. 2008b. Flash-point prediction for binary partially miscible mixtures of flammable solvents, *J. Hazard. Mater.* 153, 1165–1175.

Liaw, H.-J., Tang, C.-L., Lai, J.-S. 2004. A model for predicting the flash point of ternary flammable solutions of liquid, *Combust. Flame* 138, 308–319.

Liaw, H.-J., Wang, T.-A. 2007. A non-ideal model for predicting the effect of dissolved salt on the flash point of solvent mixtures, *J. Hazard. Mater.* 141, 193–201.

Lu, Y.L., Chiou, D.R., Chen, L.J. 2002. Liquid–liquid equilibria for the ternary system water + octane + diethylene glycol monobutyl ether, *J. Chem. Eng. Data* 47, 310–312.

Mączyński, A., Wiśniewska-Gocłowska, B., & Góral, M. 2004. Recommended liquid–liquid equilibrium data. Part 1. Binary alkane–water systems, *J. Phys. Chem. Ref. Data* 33, 549–577.

Mallinckrodt Baker, 2008, http://www.mallbaker.com/Americas/catalog/default.asp?searchfor =msds.

Matsuda, H., Ochi, K. 2004. Liquid–liquid equilibrium data for binary alcohol + nalkane (C10–C16) systems: methanol + decane, ethanol + tetradecane, and ethanol + hexadecane, *Fluid Phase Equilib.* 224, 31–37.

Merck (Ed.), 2006. *The Merck Index*, 14th ed., Merck & CO., NJ

Merck, 2008, http://www.chemdat.info/mda/inten/index.html.

NIOSH Pocket Guide to Chemical Hazards, 2008, http://www.cdc.gov/noish/npg/npgname-o. html.

Poling, B.E., Prausnitz, J.M., O'Connell, J.P. 2001. *The Properties of Gases and Liquids*, fifth ed., McGraw-Hill, New York.

Regulations, Regulation (EC) No. 1272/2008 of the European Parliament and of the Council, on Classification, Labeling and Packaging of Substances and Mixtures, 2008, Amending and Repealing Directives 67/548/EEC and 1999/45/EC, and Amending Regulation (EC) No. 1907/2006, Official *J. Eur. Union*, L353.

Renon, H., Prausnitz, J.M. 1968. Local compositions in thermodynamic excess functions for liquid mixtures, *AIChE J.* 14, 135–144.

Resa, J.M., Goenaga, J.M., Iglesias, M., Gonzalez-Olmos, R., Pozuelo, D. 2006. Liquid–liquid equilibrium diagrams of ethanol + water + (ethyl acetate or 1-pentanol) at several temperatures, *J. Chem. Eng. Data* 51, 1300–1305.

SFPE, The SFPE Handbook of Fire Protection Engineering, second ed., Society of Fire Protection Engineers, Boston, 1995.

Tang, Y., Li, Z., Li, Y. 1995. Salting effect in partially miscible systems of n-butanol-water and butanone-water. 2. An extended Setschenow equation and its application, *Fluid Phase Equilib.* 105, 241–258.

Tedia, 2008, http://www.tedia.com/products.php3.

Tourino, A., Casas, L.M., Marino, G., Iglesias, M., Orge, B., Tojo, J. 2003. Liquid phase behaviour and thermodynamics of acetone +methanol + n-alkane (C9–C12) mixtures, *Fluid Phase Equilib.* 206, 61–85.

Univar USA, 2008, http://www.univarusa.com/assistmsds.htm.

Van Ness, H.C., Abbott, M.M. 1982. *Classical Thermodynamics of Nonelectrolyte Solutions: With Applications to Phase Equilibria*, McGraw-Hill, New York.

White, D., Beyler, C.L., Fulper, C., Leonard, J. 1997. Flame spread on aviation fuels, *Fire Saf. J.* 28, 1–31.

14 Calorimetric Approach on the Thermal Hazard Assessment of Cumene Hydroperoxide

14.1 INTRODUCTION

Chemical processing industries require the use of extensively reactive chemicals for preparation, processing, sampling, storage, transportation, and even during disposal. The various perturbed conditions, such as mischarging, overdosing, or exposure to external fire at any stage, could lead to the generation of heat caused by the reactive material decomposition or polymerization (NFPA 43B, 1999). The peroxy group (-O-O) present in the organic peroxides is the reason for thermal runaways due to its intrinsic unstable and reactive nature, which is similar to cumene hydroperoxide (CHP) referred to in this chapter (Hou et al., 2000). When the rate of heat generated is more than the rate of heat removed in a reactive system, thermal runaway occurs. The peroxy group in CHP is most elusive to heat and the exothermic threshold is as low as ambient temperatures and sometimes below 120 °C. Since the 1970s, due to the highly unstable nature of CHP it has caused several fires and explosions over industries in Taiwan. In common with all other alkyl-peroxides, hydroperoxides contains a hydroxy group/acidic hydrogen, which makes it more reactive than others. Moreover, hydroperoxides are extremely sensitive towards impurities, metal ions, acids, and bases and thus hydrogen peroxide (H_2O_2), CHP, and tert-butyl hydroperoxide (TBHP), are handled carefully (Shu et al., 2000 & Boundy and Boyer, 1952). For peroxide systems during a runaway reaction, solvents with a higher boiling point will have no effect on the reaction kinetics and pressure relief systems. The different concentration of 30, 60, and 81 mass% of commercial CHP is produced industrially. In the production of dicumyl peroxide, CHP is used as a catalyst and is sensitive to pH. The oxygen balance, functional groups of explosives, and incompatibility charts are used to evaluate the hazardous storage of different materials that estimate the amount of oxygen released during oxidation. In past cases, the instability property of CHP has led to a number of fire hazards explosions in various industries.

14.2 THERMAL RUNAWAY OF CUMENE HYDROPEROXIDE

CHP is known to be widely used in polymerization reaction, where it acts as an initiator. It has been used extensively for synthesizing acrylonitrile-butadiene-styrene (ABS) co-polymer in various industries across Taiwan. Its usage is also known in synthesizing phenol and acetone through the catalytic cleavage reaction mechanism (Boundy and Boyer, 1952). The National Fire Protection Association (NFPA) has classified CHP to be a class III type flammable (fire hazard). The emphasis in research studies has been on understanding the pressure relief for organic peroxides by the members of the Design Institute for Emergency Relief System (DIERS) under the American Institute of Chemical Engineers (AIChE) (Wang et al., 2001). In order to design the measures for preventing such hazardous reaction by peroxides and also to lower the generated pressure through adequate vent sizing, various aspects need to be addressed. The most important perspectives such as thermal decomposition, reactive incompatibility, the external fire in process as well as the storage area should be taken into account for handling such hazards, along with other credible scenarios. Many previous works have been reported of various process conditions regarding the runaway hazards, reaction mechanisms and decomposition kinetics of CHP (Wang et al., 2001; Ho, Duh and Chen, 1998 & Duh et al., 1997).

In the production of dicumyl hydroperoxide (DCPO) and phenol, CHP has been widely used in Taiwan. During the polymerization synthesis of the ABS, it is applied as an initiator. The after-reaction process includes the release of toxic elements with runaway reactions uncontrolled in nature. Such reaction leads toward fires and explosion which further results in severe casualties and damages. In many cases, thermal runaways are unavoidable since the high calorific capacity of gaseous substances and their decomposed materials incur different hazards due to their specific physical and chemical properties. In order to mitigate the related hazard and damages incurred due to the runaway reactions it is important to analyze the hazardous chemicals with adequate instruments, such as differential scanning calorimetry (DSC) and a thermal activity monitor (TAM). Such procedures could proactively help in preventing possible disaster, or, if it happened, could reduce it to an acceptable level.

In previous researches, investigation was carried out for auto-catalytic nature utilizing DSC and TAM test at low temperature conditions, but with nth order reaction behaviors under high temperature ranges (Duh et al., 1997). From the preventive measure point of view, during thermal runaway reactions, it was suggested that apart from the related thermokinetic parameters of the peroxides and isothermal conditions, safety alertness during manufacture and transportation should be viewed as a top priority before the design process stage. DSC and TAM testing primarily evaluates the data from thermokinetic parameters utilizing experimental thermal curves. The thermal hazards of reactive chemicals could be determined with the help of these two calorimeters, under normal or perturbed operating conditions. Utilizing such experimental results, along with data from several other sources, CHP's thermal decomposition phenomenon, along with other contaminants, could be easily calculated using a micro amount of the sample. Although DSC and TAM utilizes a sample quantity of milligrams or grams, such obtained experimental curves could be extrapolated through the adoption of a technological approach for commercial applications.

This information could be utilized for safety information during process design and could provide guidance for any specific reactive chemical during storage and transportation.

14.3 EXPERIMENTAL STUDIES

14.3.1 SAMPLES

The 80 mass% solution of CHP in cumene was purchased directly from Merck Co. The density and concentration was measured and then stored in a refrigerator at 4°C for subsequent use. The contaminants that could be easily encountered in process or storage such as H_2SO_4, NaOH, and $FeCl_3$, were chosen to be from 0.1 to 0.5 mass% for DSC experiments. The CHP samples of 35 mass% was diluted in cumene and 80 mass% were used in DSC and TAM experiments, respectively.

14.3.2 DSC (DIFFERENTIAL SCANNING CALORIMETER)

DSC is the heat-flux type, which could measure as well as acquire relationships between micro-heat transform and temperature difference of materials of interest. Practically speaking, it is placed with sample and reference in two small crucibles in a heat flow furnace. By the configuration, a heater around the furnace is employed to control the temperature of the crucibles (Duh et al., 1997). Typically, there is a heat flow sensor between sample and reference, which has a pair of thermocouple wires measuring the temperature difference between these two. As the sample temperature rises to the conversion point, such as the glass transition point, the fusion point, the boiling point or the heat changes of decomposition reaction, the sample absorbs the energy supplied by the heater (an endothermic reaction) or releases energy (an exothermic reaction) to indicate the temperature difference between the sample and the reference. Without this step, the balance resulting from generating or liberating heat flow could not be suitably maintained.

14.3.3 TAM (THERMAL ACTIVITY MONITOR)

Structurally, TAM possesses two parts: a 25-liter thermostat water bath, and an external water circulator. Normally, although not always, water is continuously circulated by pump upwards into the thermostated water bath and connected to the external water circulator, which keeps the sample under an excellent constant temperature (or isothermal) environment (The Isothermal Calorimetric Manual for Thermometric, 1999). The ampoules containing sample and reference were inserted into a measuring cup between a pair of Peltier thermopile sensors. These sensors are in contact with a pair of metal heat sinks, where it is indicated that a pair of metal heat sinks of all cylinders are immersed into the thermostated water bath. Meanwhile, a pair of metal heat sinks carry messages to the Digital Voltmeter (DVM) of the heat difference between the sample and reference. DVM can continuously provide all messages to users by showing the related information, along with the situation of the experiments. As for data acquisition, the data, as shown in DVM, are simple heat powers from the sample, which illustrates the typical thermal curves of simple heat power (μW/g) versus time (day) from samples.

14.3.4 APPLICATIONS

In essence, DSC and TAM are two independent calorimeters with different purposes. Although both have different approaches in terms of design principles, they can continuously detect small heat flow (μW) transform from a sample. If working suitably, both DSC and TAM could be smoothly coordinated to complement each other. Because, sample by TAM could not release all of its own calorific value under an isothermal experiment, there is typically 20% calorific value which could be detected by the follow-up DSC scanning experiment to determine the remaining calorific value. In terms of the principles of instrumental structure for both DSC and TAM, either one has both its advantages and its disadvantages. Through a comprehensive understanding of these experimental principles and operating procedures, optimal usages or control by both DSC and TAM can be accomplished to fit a specific application, such as pressure relief designs by DIERS technology for the safer venting of runaway reactions incurred by organic peroxides in the incipient stages of a runaway reaction (Wang et al., 1999 & Leung & Fisher, 1998).

14.4 RESULTS AND DISCUSSION

The DSC experiments were performed to determine the acquired exothermal phenomenon of CHP, associated with its catalytic decomposition reaction with contaminants, such as H_2SO_4, NaOH, and $FeCl_3$ was selected to be as contaminants. Table 14.1 and 14.2 indicates the DSC experimental data for the decomposition of 35 mass% CHP with incompatible chemicals and 93 mass% of CHP with NaOH, respectively. The heat flow vs temperature curves of 35 mass% CHP with and without contaminants were presented in Figure 14.1.

The addition of incompatible chemicals to CHP decreased the exothermic onset temperature when compared to 35 mass% CHP without contaminants increasing the exothermal heat of reaction. The results demonstrate the substantial impacts of CHP with potentially incompatible contaminants encountered during preparation,

---

TABLE 14.1

Experimental Data of Decomposition for CHP 35 Mass% and Contaminants Conducted by DSC Tests

Sample	Contaminants		T_0	$\triangle H$	TMR
Mass of CHP 35 mass% (mg)	Chemical	Dosage (mg)	(°C)	(J g^{-1})	(min)
6.15	–	–	135	607.3	192.6
4.33	H_2SO_4 (1 N)	1.05	90	667.3	167.3
5.09	NaOH (1 N)	0.40	55	768.7	147.9
5.20	$FeCl_3$ (1 M)	0.81	40	768.7	144.6

(Obtained from Lin et al. (2007) © National Institute of Occupational Safety and Health)

TABLE 14.2

Experimental Data of Decomposition for CHP 93 Mass% and CHP 93 Mass% with NaOH 99 Mass% by DSC Tests

Sample	T_0 (°C)	T_{max} (°C)	$\triangle H_d$ (Jg^{-1})
93 mass% CHP	80	163	1399
93 mass% CHP + 99 mass% NaOH (0.5 mg) (first peak)	40	98	1153
93 mass% CHP + 99 mass% NaOH (0.5 mg) (second peak)	40	96	986

(Obtained from Lin et al. (2007) © National Institute of Occupational Safety and Health)

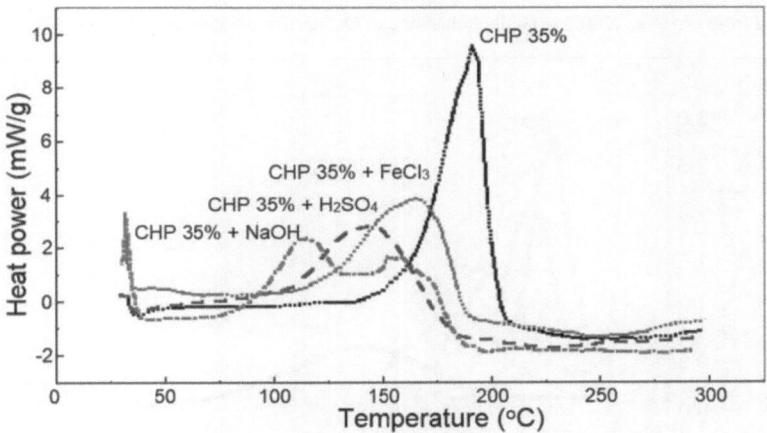

FIGURE 14.1 Comparisons with heat power vs. temperature for CHP 35 mass% and contaminants by DSC tests (Obtained from Lin et al. (2007) © National Institute of Occupational Safety and Health).

processing, storage, transportation, or even disposal, in which runaway reactions and accompanying hazards cannot be overlooked. The isothermal TAM experimental data of CHP demonstrated the autocatalytic runaway reaction features at low temperatures, as shown in Table 14.3.

During the induction period, the runaway reaction was accelerated after reaching the autocatalytic runaway onset temperature, reducing the possibility of controlling the catastrophic fire and explosion. The thermograms of 80 mass% CHP at five different isothermal temperatures of 75, 80, 83, 88, and 90 °C are shown in Figure 14.2. The results indicated that the runaway reaction temperature of CHP was less than 75 °C, which is due to the initial endothermicity and is much lower than the exothermic onset temperature of 110 °C by VSP2 (Vent Sizing Package2) (Hou, Shu & Duh, 2001 & VSP2 Manual and Methodology, 1997). The results demonstrated that the various stages in operation of CHP could accumulate a considerable amount of heat increasing the hazards that may result in unexpected thermal runaway reaction.

TABLE 14.3

Experimental Data of Autocatalytic Reaction for CHP 80 Mass% Conducted by TAM Tests

Isothermal Temperature (°C)	Sample Mass (g)	Reaction Course (Days)	Time (Hours)		Time (Days)	
			1st Peak	$\triangle$H (J g^{-1})	2nd Peak	$\triangle$H (J g^{-1})
75	0.505	42.7	27.60	5.28	20.9	1082.51
80	1.020	20.0	20.80	4.05	13.0	980.98
83	0.510	22.0	13.30	6.29	9.7	1128.86
88	0.504	16.5	8.30	4.46	6.5	1181.65
90	0.506	14.1	7.55	2.61	4.0	1248.45

(Obtained from Lin et al. (2007) © National Institute of Occupational Safety and Health)

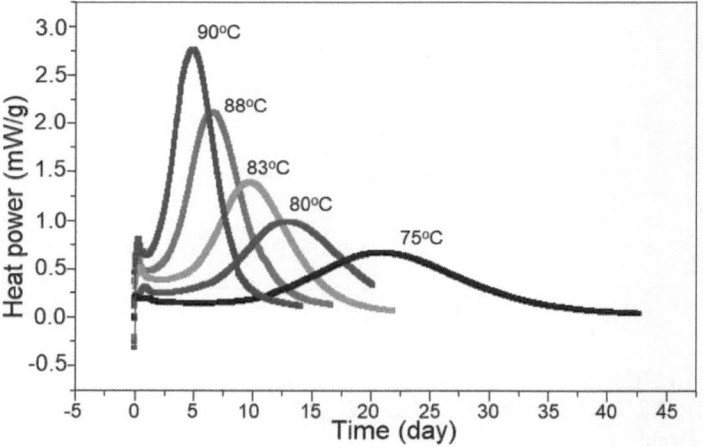

FIGURE 14.2 Comparisons with heat power vs. time of CHP 80 mass% under different isothermal temperatures by TAM tests (Obtained from Lin et al. (2007) © National Institute of Occupational Safety and Health).

14.4.1 SIGNIFICANCE AND APPLICATIONS OF CHP DERIVED BY DSC AND TAM

The use of various calorimetric techniques to determine the reactive hazardous characteristics of CHP during the various process stages could help in understanding the potential unexpected hazards from thermal runaway reactions. The significant observation drawn from both the DSC and TAM experiments of CHP in estimating the thermal characteristics due to its unstable nature and reactive structure were:

1. Addition of NaOH or FeCl$_3$ with CHP resulted in a higher heat of reaction with a considerable decrease in exothermic onset temperatures. The formation of OH$^-$ or Fe^{3+} suggests the possible different reaction pathways for the increase in the reaction hazard.

2. The contamination of CHP with incompatible substances such as acids/ bases, metal, ions and dusts during the sampling, preparation, and storage/ disposal could result in runaway reactions. Critical care should be taken to avoid contamination with reactive chemicals during each stage

3. The prolonged storage of CHP at ambient temperatures results in thermal runaway owing to an autocatalytic reaction. The TAM experiments at 75, 80, 83, 88 and 90 °C proved that over a period of time, CHP could initiate a thermal runaway reaction at high temperatures. Therefore, it is crucial to avoid direct heat accumulation and external fire exposure to all the CHP storage vessels and warehouses. However, CHP hazards could not be detected by the conventional adiabatic VSP2 calorimeter.

4. The DSC could adequately simulate the thermal runaway reaction of CHP as the heat liberated during the thermal decomposition is less than the exothermic heat liberated by the self-heating rate of CHP. It would be very difficult for the emergency rescue team to evacuate the personnel without any knowledge of that temperature of no return (TNR).

14.4.2 COMPARISON OF THERMOKINETIC PARAMETERS FOR CHP DERIVED FROM DSC AND TAM

The kinetic models could simulate, and predict, the degree of hazards, physical property changes, and reactions, including oxidation, crystallization, polymerization, and decomposition, involved in a chemical process. It could also evaluate the potential reaction hazards of materials during various stages of operation. The fire and explosion from CHP due to quick pressure rise during thermal decomposition (by itself/ contaminants) has resulted in fatalities, injuries and property loss. Table 14.4 represents the thermokinetic parameters in both isothermal and non-isothermal conditions of TAM and DSC. The experimental results were adequate to depict the real-time conditions during processing, transportation or storage. In addition, the application of DSC and TAM is expected to predict the CHP runaway reaction proactively during the early stages of process development.

14.5 SUMMARY

To evaluate the degree of hazards of reactive materials, one normally cannot be only dependent on a single instrumental analysis. Therefore, assessment of the related thermal hazard using different technologies will play a crucial role in chemical

TABLE 14.4

Comparisons on Thermokinetic Parameters for CHP 80 Mass% Derived from DSC and TAM Tests

Calorimeters	T^0 (°C)	n	E_a (kJ mol^{-1})	A (min^{-1})	k (min^{-1})
DSC	100	0.45	141.13	9.26×10^{13}	4.5×10^{-7}
TAM	70	0.50	95.02	9.064×10^{9}	1.618×10^{-4}

(Obtained from Lin et al. (2007) © National Institute of Occupational Safety and Health)

processing industries. In theory, physicochemical properties of hazardous materials such as thermal stability, sensitivity, amount of test material, range of operating temperature and materials of test cell could be analyzed by calorimeters. The thermal hazard nature of CHP, along with its in incompatibilities, could be evidently performed by DSC and TAM. The results indicated that DSC and TAM can be used appropriately based on the advantages, disadvantages and applications corresponding to specific conditions with this new approach. The thermokinetics and potential hazards of reactive chemicals could be readily understood based on the accurate data provided by these calorimeters for the process industries that demonstrate the hazardous behavior during handling and usage that could effectively lessen the probabilities and consequences of hazards generated during perturbed conditions among all the process stages. Through adequate testing, the results could be employed to alleviate any chemical disasters by apprehending the inherent thermal hazard.

REFERENCES

Boundy, R. H., Boyer, R. F. 1952. *Styrene, Its Polymers, Copolymers, and Derivative.* Rinehold, New York.

Duh, Y.S., Kao, C.S., Hwang, H.H., Lee. W.L., 1998. Thermal Decomposition Kinetics of Cumene Hydroperoxide. *Trans IChemE.* 76 (Part B), 272–276.

Duh, Y.S., Kao, C.S., Lee, C., Yu, S.W., 1997. Runaway Hazard Assessment of Cumene Hydroperoxide from the Cumene Oxidation Process. *Proc. Safety Environ. Protec.* 75(2), 73-80

Ho, T.C., Duh, Y.S., Chen, J.R., 1998. Case Studies of Incidents in Runaway Reactions and Emergency Relief. *Proc. Safety Prog.* 17, 259.

Hou, H.Y., Shu, C.M., Duh, Y.S., 2001. Exothermic Decomposition of Cumene Hydroperoxide at Low Temperature Conditions. *AIChE J* 47, 1893–1896.

Hou, H.Y., Shu, C.M., Duh, Y.S., Yeh, P.Y., Peng, D.J., 2000. *North American Thermal Analysis Society (NATAS),* Ottawa, Canada.

Leung, J.C., Fisher, H.G., 1998. Runaway Reaction Characterization: A Round-Robin Study on Three Additional Systems. Process Integration on Runaway Reactions, Pressure Relief Design, and Effluent Handling, New Orleans, LA, USA 1998; 109.

Lin, S.-H., Li, J.-C., Chang, C.-M., Jr Peng, D., Shu, C.M., 2007. Basic Thermal Hazard Assessment on Cumene Hydroperoxide by Calorimetric Approaches. *J. Occup. Safety Heal.* 15, 297–307.

Lin, W.H., Shu, C.M., 2005. Reactive Hazards of Cumene Hydroperoxide Incompatible with Sodium Hydroxide. National Yunlin University of Science and Technology, 5–22.

NFPA 43B, 1999. Code for the Storage of Organic Peroxide Formulations, *National Fire Protection Association,* Quincy, MA, USA.

Shu, C.M., Hou, H.Y., Peng, D.J., Duh, Y.S., 2000. *The 2nd International Conference of EDUG,* Ludwigs, Germany.

The Isothermal Calorimetric Manual for Thermometric AB. Jarfalla, Sweden; 1999, Merck & CO., NJ, 2006.

VSP2 Manual and Methodology, 1997 Fauske and Associates, Inc., Burr Ridge, IL, USA.

Wang, Y.W., Shu, C.M., Duh, Y.S., Kao, C.S., 1999. Incompatibilities on Thermal Runaway Hazards of Cumene Hydroperoxide CHP. Methodology of Reaction Hazards Investigation and vent sizing, St Petersburg, Russia; 1999: 1–15.

Wang, Y.W., Shu, C.M., Duh, Y.S., Kao, C. S., 2001. Thermal Runaway Hazards of Cumene Hydroperoxide with Contaminants. *Ind Eng Chem Res.,* 40(4), 1125–1132.

15 Evaluation of the Information System of Maintenance Efficiency in Petrochemical Plants

15.1 INTRODUCTION

The frequent occurrence of accidents in process plants all across Taiwan has become a serious cause for concern. A large number of the accidents are attributed to human error, machine failure and faults in manufacturing processes, which caused severe casualties and deaths (Chang et al., 2000; Lorenzo, 1990; Errington and Bullemer, 1998). Computerized maintenance management systems (CMMS) have become such a medium which helps in maintaining the proper functioning of the plant facilities with the advancement of technology and the prevalent use of automation in plants for various processes. Process management also comes into consideration, which helps in minimizing human errors and thereby helps in increasing the efficiency of plant performance.

The detailed study of the operations and management helped in large-scale functions of the industry. No matter how many details on the plant performance were collected, like safety, equipment lifecycles and costs, yet a proper functioning analysis remained incomplete without the utilization of the historical data of the plant.

The computerized maintenance management system (CMMS) comes into the picture, which helps in recording the distribution of funds and the smooth running of operations with the timely maintenance of the assets. Thus, for decision-making, proper distribution costs and so on, it becomes an utmost necessity to establish a relationship between operation and maintenance (Waeyenbergh and Pintelon, 2002). Helping decision-makers with proper results and diagrams using software such as CMMS can prove very helpful. However, the current system only gives primary importance to maintenance in the manufacturing scenario, but lags behind in providing definite details on special tests and facilities of the petrochemical plant. The reason being that almost all the available software were designed as per locally available industries considering the local conditions and not specifically for a petrochemical plant.

One of the biggest challenges faced by the maintenance section is the absence of sufficient amount of data for the analysis of a particular condition. As per ideal requirements a very efficient maintenance management information system is the need of the hour with adequate data such that the maintenance staff do not have to be frustrated while conducting analysis on the cost and failures in a plant due to the unavailability of proper data. The assembly of such a detailed database again requires the use of the software, which has a statistical analysis of various organizations, facilities and factories, different processes etc. The management system would not only help in improving the organization and processing, but will also act as an information point for statistical, analytical and reliability analysis of the factory and the facilities available in it (Pintelon and Wassenfove, 1990). This, in turn, can help in improving the organization on a larger scale with focused reference to maintenance and managerial staff.

15.2 MAINTENANCE MANAGEMENT OF FACILITIES

It has been suggested that an integrated platform must be generated wherein all the information could be drawn, thereby enhancing both reliability and competitiveness. Having a good maintenance system is of the utmost necessity in bringing about improvements in a petrochemical plant. It has been found that in the past, unreliable data with traditional methods could not establish a good analysis platform. Although tools such as mean time between failure (MTBE) and mean time to failure (MTTF) were available, they did not comply well with the existing petrochemical plant.

As far as maintenance is considered, the system should be well equipped in such a way that it can record the job orders, records, resource utilized, and space to be used. In addition, the day-to-day work carried out in a plant must be converted into data files, which would make decision-making more effective. It was thus suggested that the below-mentioned tools should be made available:

Mean time to repair (MTTR)

Mean time to failure (MTTF)

Mean time between maintenance (MTBM)

Along with the above, a cost analysis of maintenance and material with Performance and failure analysis is also to be carried out (Chien et al., 2000).

With a view to improving the maintenance of the petrochemical plant, the executive information system (EIS) was established based on the CMMS). The best practices for good maintenance was done considering the key performance index (KPI), focusing on providing fast information in order to save time for data collection.

The data were collected by the logical and diagram selection after which the user can review the status of the asset and the facility available, arranged from a vertical selection of items and equipment, to proceed with the analytical approach:

1. A vertical axis of time period was designed which compared work unit, structure of assets and a certain time of the year.
2. A rectangular bar showing the first ten items was established. This bar basically pointed out the least useful work unit based on the occurrence of a serious condition.

3. The diagram was printed using a mouse function wherein these diagrams and their corresponding data could be transformed into MS Excel files.

15.3 PRELIMINARY DESIGN AND INDEX ESTABLISHMENT

With the progress of time, Taiwan has been investing into automation systems for its numerous industries. With the advent of such automation, a proper maintenance on production and the relevant cost analysis must be taken into consideration, bearing in mind the KPI. There also exists a further necessity in the investigation and solving of the system (Mather, 2002).

In order to obtain a minimum production a strategic planning must be put in place which keeps in mind the regular maintenance and service work during the mentioned period of production. If the need arises, the production could be maximized, resulting in the overall upgrading of period of production (Levitt, 2003). As described in Figure 15.1, the entire system could be arranged into different subheadings of management, such as asset, maintenance, report and system safety.

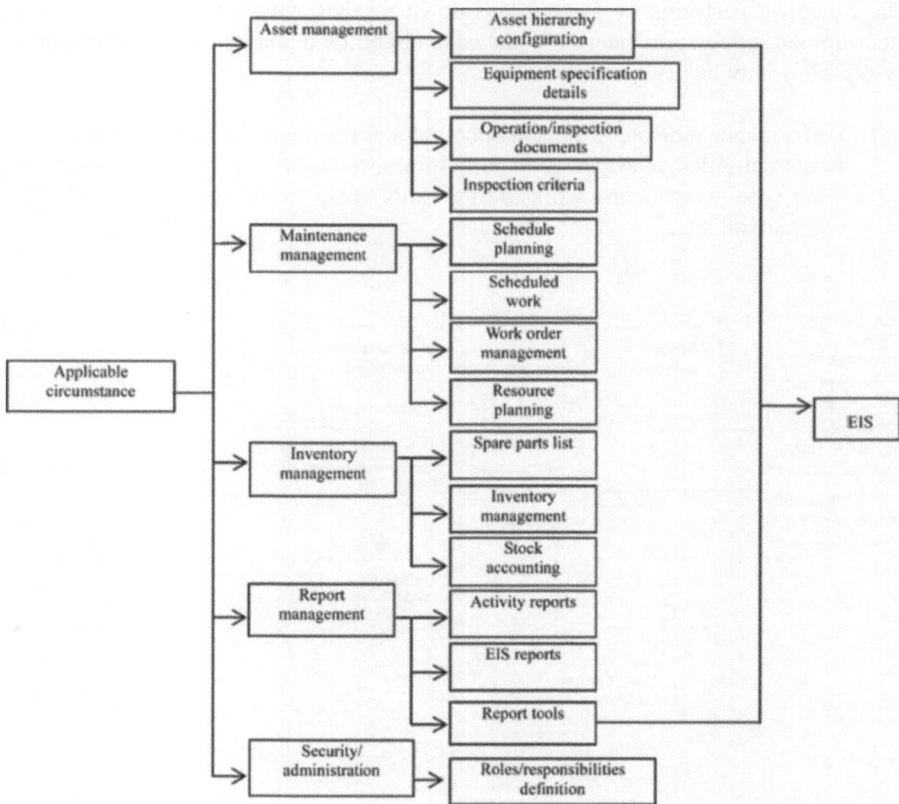

FIGURE 15.1 The applicable circumstances of system design (Reproduced with permission from Hwang et al. 2007 Copyright © Elsevier)

The main function of plant maintenance was to store information about understanding and analyzing the execution of maintenance status such that each unit has the facility for regular service and maintenance implementing logic and various figures to explain the database.

The status of the assets or the used equipment may be updated by the user based on the proper screening and calculation of data form a vertical arrangement, ranging from the lowest to the topmost index. Thus, large relevant information could be obtained, which could aid in the full understanding of the maintenance work and in achieving increased efficiency keeping under consideration factors such as the time period, the cost of materials, staff expenditure and so on.

The described maintenance scheme offers multiple perspectives in improving the management of maintenance and information in a petrochemical plant. It also played a key role in detecting emergencies and their corresponding solutions (Figure 15.2).

15.4 DESIGN OF THE SYSTEM

15.4.1 INDEX DESIGN

The executive performance system lays down the KP, which mainly includes the measurement of performance, cost analysis, reliability, management etc. (Wireman, 1999). They may be discussed as follows:

1. Performance indicators: This mainly includes the mean time needed for different activities or events such as failure, availability, repairs, turnaround, work type, work hours, work to time ratio, postpone of work, time period comparison, etc.

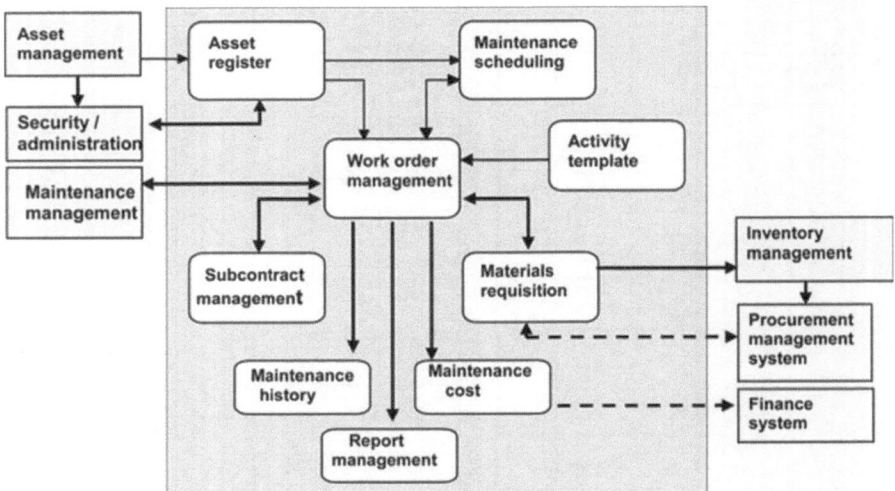

FIGURE 15.2 Information source of maintenance supporting system (Reproduced with permission from Hwang et al. 2007 Copyright © Elsevier)

2. Result of asset failure: To identify the frequency of failure for equipment and their components.
3. Summary of maintenance cost: To determine the maintenance cost, contract cost, material cost, comparing work unit, asset structure etc.
4. Summary of material cost: Gather a detailed plan on the asset structure, time period of work, work unit.

15.4.2 System Development

The present system developed by the research institute helps in extending flexibility to special needs. Thus, interest must be taken in view of addressing special needs as per availing conditions. To connect the real-life situation with the available data, the interaction behavior must be taken into consideration. ICAM Definition (IDEF) was used to plot the analysis diagram for maintenance in the execute information system (Huang and Chou, 2002) (Figures 15.3–15.5). The preliminary system was thus a multi-tiered system and operated using MS Windows with the application server being conducted on identical computers. Swift calls from client site or specific requests were placed using a component transaction server (CTS). The CTS acting as the server, demands could be placed via Internet Inter-ORB Protocol (IIOP) call.

Management of Security and the Associated User Privilege

Table 15.1 shows all the details of the users and staff right from the inventory to the managerial level, along with their roles and responsibilities.

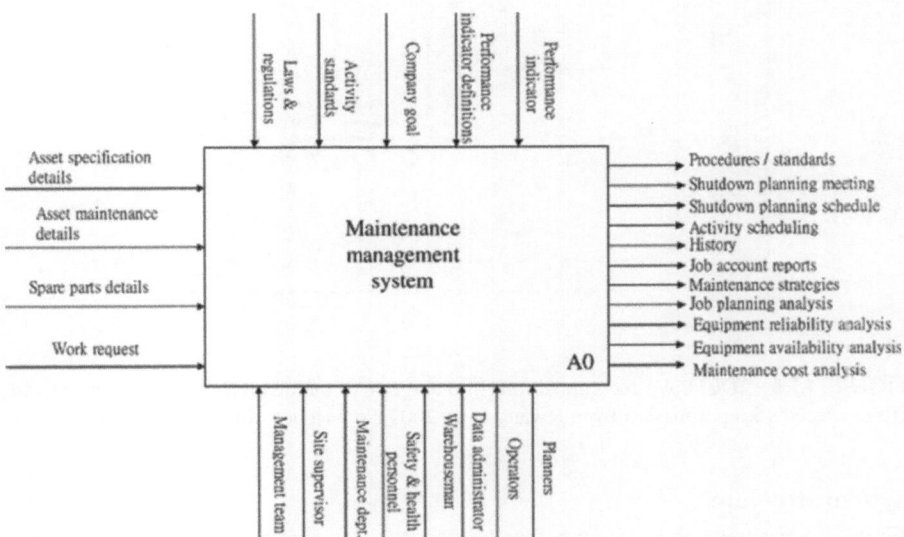

FIGURE 15.3 IDEF0 A0 level diagram of maintenance management system (Reproduced with permission from Hwang et al. 2007 Copyright © Elsevier)

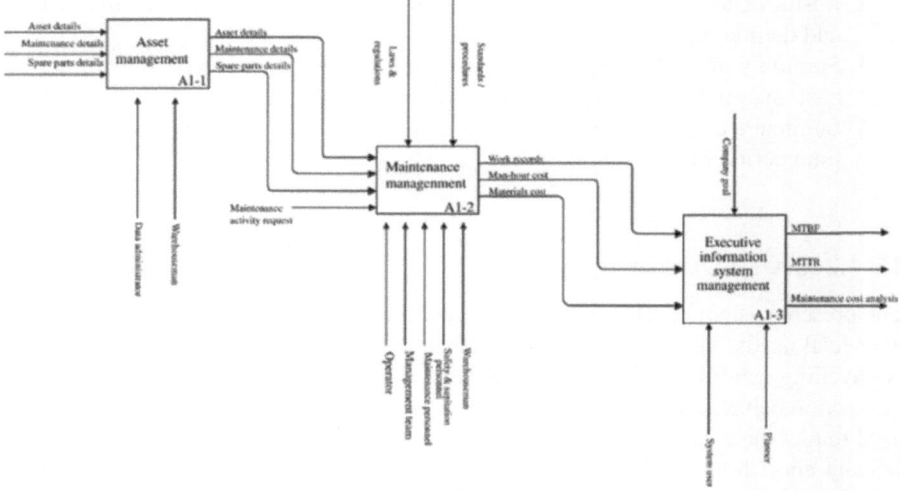

FIGURE 15.4   IDEF0 A0 child diagram of maintenance management system (Reproduced with permission from Hwang et al. 2007 Copyright © Elsevier)

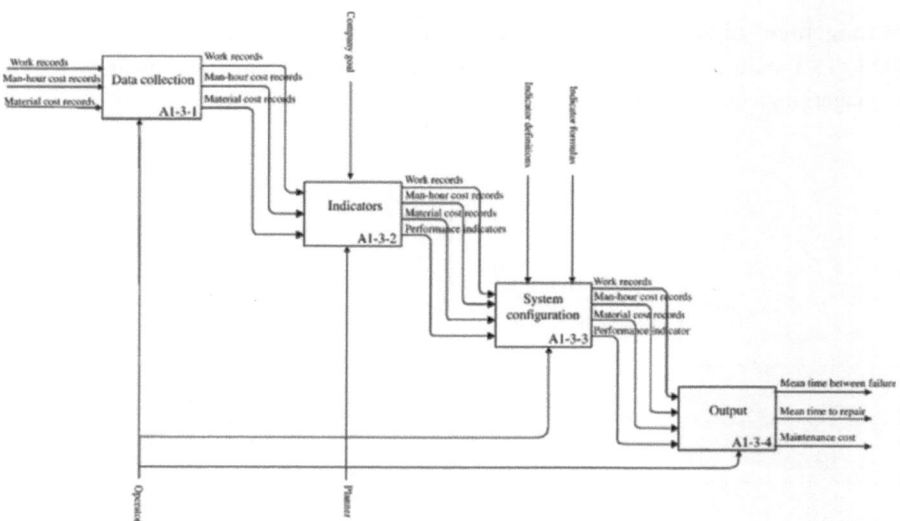

FIGURE 15.5   IDEF0 A0 decomposition child diagram of maintenance management system (Reproduced with permission from Hwang et al. 2007 Copyright © Elsevier)

System Structure

Figure 15.6 shows the software that has ample information for the management maintenance which has details of the experience of expert, rules carried out for analysis as well as graphical user interface.

TABLE 15.1
Summary of User Roles and Responsibilities

Users	Information and Functions
Management staff for information of common users	Request for maintenance demands • Maintain facility structure and renew facility information • Maintain the structure of organization, staff and roles • Maintain the contents and establishment of work flow • Maintain the contents and information of this system
Maintenance/service staff	• Conduct actual maintenance/ service work • Record actual work contents in the system
Stock management staff	• In charge of stock management of needed materials for maintenance (such as material distribution, return material, storage and checking)
Managerial staff	• Understand the contents of each relevant index via information supporting management mode

(Reproduced with permission from Hwang et al. 2007 Copyright © Elsevier)

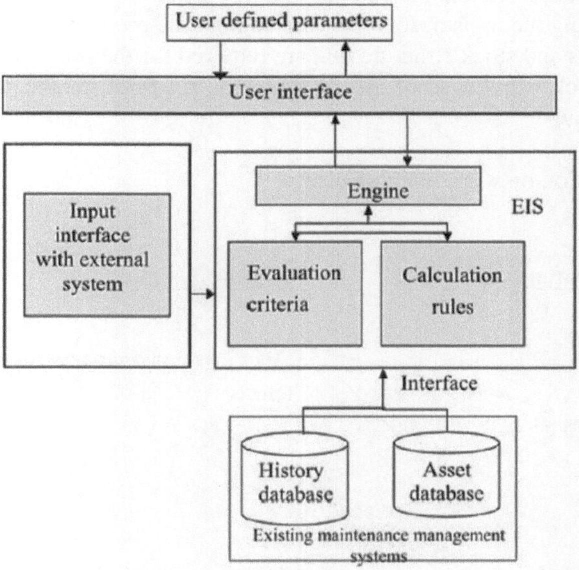

FIGURE 15.6 EIS software structure of this system (Reproduced with permission from Hwang et al. 2007 Copyright © Elsevier)

System Interface

Data mapping (data exchange, restoration etc.), synchronizing rules etc. needed to be performed for the development of the interface programme. Once, these were formulated, a plan was laid down to do the error reply and handling taking the information technology factors, such as IT security and data compactness followed by enhancing the overall efficiency.

15.4.3 Data Requirements

This part basically helps in recognizing the objects which were utilized in the data model. A system called template mapping was done to describe the object of the data model and which is illustrated in Figure 15.7.

The details, thus collected were again classified into three main categories which are discussed below:

1. Information about assets: This requires the identification of all the details pertaining to the focused asset:

   - Identification of the hierarchy for the initiation of production,
   - Define the type of equipment to be utilized.
   - Detailed forms must be recorded with respect of date of installation, size, class, type etc. of different assets.
   - Schedule of periodical maintenance must be prepared.
   - Determine the inspection required and the maintenance service type preferred.
   - Identify the service cycle, including the next scheduling and the final due dates.
   - Identify the work and the cost for the production.
   - Identifying the tools needed for carrying out the production.
   - Materials and stock structure that are required for the production.
   - Also stock warehouse of all the materials required for the process to be carried out.

2. Information on work maintenance:

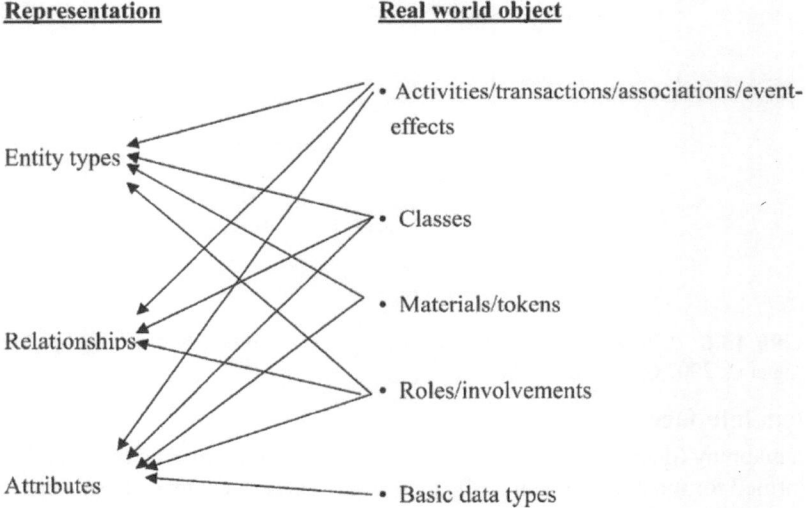

FIGURE 15.7 Data requirement model (Reproduced with permission from Hwang et al. 2007 Copyright © Elsevier)

- Define the time period required for maintenance work, including steps and time.
- Define various parameters required for the maintenance period, such as action and failure code, materials and its cost, tools, information system required, occurrence of failure or fault, start and end date of work etc.

3. Information on supporting system

- Defining user friendly images and index.
- Collect data from the asset management database to identify the different index demands and the information required for each of these indexes.

15.4.4 FUNCTION REQUIREMENTS

This section gives details on the relationship between data enlisted and their corresponding function requirements (Table 15.2).

TABLE 15.2
Function Requirements for this Study

1. Equipment reliability	Mean time between failure (in unit of days)
1.1 MTBF Formula	MTBF is calculated at tag/equipment. Tab and levels above tag level will be rolled-up. The number of the operable days were calculated from the hobs done on that tab with failure or success status = 'F' or 'S'
Additional	Sum (No. of operable days of all tags in the selected time period)/sum (no. of failures at all tags in the period selected)
	Ranked by MTBF of Top Ten of equipment
1.2 MTA Formula	Mean time availability (in percentage %)
	MTA is a percentage calculated at the Tag/equipment level. Levels above tag level will be rolled-up
	No. of available days in selected time period * 100/no. of days in selected time period
1.3 MTTR Formula	Mean time to repair (in unit of days)
	MTTR is calculated at work order. Tab and levels above tag level will be rolled-up
	Sum (TRUNC(End_date) − TRUNC(Start_date) + 1) for all jobs/no. of jobs
1.4 MTTT Formula	Mean time to turnaround (in unit of days)
	MTTT is calculated at work order. Tab and levels above tag level will be rolled-up
	Sum (TRUNC(End_date) − TRUNC(Occur_date) + 1) for all jobs/no. of jobs
2. Job counts	Total of jobs raised, completed
2.1 Job counts	Unit in charge Fault codes; diagnostic counts; work type
Dimensions	Equipment type, equipment class
Date range	Year to date, from date to date
Filters	Work status = completion
2.2 Job started on-time	Jobs started on-time (as planned) (in percentage %)
	Percentage of jobs started on/before the planned date
2.3 Job completed on-time	Jobs completed on-time (as planned) (in %)
	Percentage of jobs completed on/before the planned date

2.4 No. of work orders	Work orders that has been approved but not progressed and all the delayed work
on holding	Sum (delayed work orders)
Formula	Persons/unit in charge
Dimensions	Contract number
Date range	Year to date, from date to date
Filters	Work type = work order and work status
Additional	Ranked by work unit of Top Ten
3.Equipment failure	Failure counts by tag
analysis	Sum (work orders)
3.1 Failure time	Failure codes
Formula	Equipment/tag classes
Dimensions	Year to date, from date to date
Date range	Ranked by failure time of Top Ten of all tags
Additional	
3.2 Failure time by	Failure counts by component
component	Sum (work orders)
Formula	Failure codes
Dimensions	Equipment classes
Date range	Year to date, from date to date
Additional	Ranked by failure time of component of Top Ten
4. Cost	Persons/unit in-charge; by work type; by tag; by tag classes; by equipment classes
4.1 Maintenance cost	YTD, from date to date
Dimensions	Internal labour, parts, contract, tools, total actual work order
Date range	Ratio between labour and contract labour
Measure	Work status = completion
Filters	
4.2 Contract cost	Contract number; persons/unit in-charge; by contract type
Dimensions	Year to date, from award date to date
Date range	Purchase request amount by work order
Measurements	Ranked by contract cost of Top Ten of all contacts
Additional	
4.3 Materials cost	Work units, materials classes, equipment classes
Dimensions	Year to date, from award date to date
Date range	Materials issues-materials returns
Measurements	

(Reproduced with permission from Hwang et al. 2007 Copyright © Elsevier)

15.5 SUMMARY

It has been seen that the vertical integration of all the divisions in a petrochemical plant needs to be strongly emphasized. For any failure or error in any strata of the integration, the effect could be seen all throughout the vertical area. Any occurrence of economic failure, manufacturing laybacks, calamities, fatalities would impact the petrochemical plant in a grave manner. Thus, it became of utmost requirement to focus on the maintenance aspect of a plant so as to avoid any risk of plant accidents or plant failure. This could be done primarily by improving the knowledge of the professionals on the work undergoing in a plant and the techniques associated with a petrochemical plant. Keeping under consideration all the tools and provision of proper training to the skilled professionals the losses can be significantly minimized.

A large number of the present manufacturing system uses a very high-end production technique which is highly dependent on a computerized system and other modern maintenance facilities. In order to achieve a full-scale reliability and functioning of the plant a highly skilled official such as a manager with well-trained staff and a properly designed strategy for maintenance must be available.

With almost full automation of the industry becoming more and more prevalent, it becomes important to maintain a good performance of all the facilities by routine checking and maintenance. Herein, EIS is of greater importance as it helps in collecting all the necessary data for monitoring the maintenance work in a petrochemical plant by providing adequate knowledge on various issues and proper ways of handling special cases if need be rising.

Thereby, an efficient software was needed for maintaining the data and also be capable of sustaining the quick and efficient management response at urgent times. The availability of an efficient metric on techniques to assist an enterprise and help in perceiving new approaches is a fundamental condition for efficient management. This would give an added advantage from the past, where the maximum reliance was placed on limited information and experience on the personal level.

However, with increasingly complex operations, more compatible matrices would be required to analyze the condition. Thus, with the progress of time some random data variable in the current situation could be utilized as a viable resource for a future management system.

REFERENCES

Chang, M.K., Wu, R.C., Lee, C.P. and Hsu, C.M., 2000, Inspection methodology of petrochemical plant pipe decomposition and preventative technology review, *Inst. Occ. Saf. Heal. J.*, 8(3): 329–343.

Chien, J.F., Liu, C.M., Wen, Y.P. and Hsue, C.H., 2000, Research of cable car maintenance performance measurement index establishment for metro rapid mechanical plant, Rapid-Transit *Technology*, 22: 71–78.

Errington, J. and Bullemer, P.T., 1998, Designing for abnormal situation management, Proceedings of the 1998 AIChE Conference on Process Plant Safety, Houston, TX, USA.

Huang, H.H. and Chou, Y.W., 2002, The Application of integrated analysis mode on EIS structure, Electronic Business Management Conference, National Chiao Tung University, Taipei, Taiwan, ROC.

Hwang, W. T., Tien, S. W., Shu, C.M., 2007, Building an Executive Information System for Maintenance Efficiency in Petrochemical Plants—An Evaluation, *Proc. Safety Environ. Protec.* 85(2), 139-146.

Levitt, J., 2003, *Complete Guide to Preventive and Predictive Maintenance*, 49–52 (Industrial Press, New York, NY, USA).

Lorenzo, D.K., 1990, A manager's guide to reducing human error, Chemical Manufacturers' Association, Washington, DC, USA, 1–1.

Mather, D., 2002, *CMMS: A Timesaving Implementation Process*, 86–106 (CRC Press, Boca Raton, FL, USA).

Pintelon, L.M. and Wassenfove, L.V., 1990, A maintenance management tool, *OMEGA Int. J. Manag. Sci.*, 18(1): 59–70.

Waeyenbergh, G. and Pintelon, L., 2002, A framework for maintenance concept development, *Int. J. Prod. Econ.*, 77: 299–313.

Wireman, T., 1999, *Developing Performance Indicators for Managing Maintenance*, 167–180 (Industrial Press, New York, NY, USA).

16 A Study on the Challenges in Emerging Economies to Industry 4.0 Initiatives for Supply Chain Sustainability

16.1 INTRODUCTION

With the modernization of industries, it has become an utmost necessity to introduce changes in the manufacturing, supply chains and business policies of existing industries to be on a par with the newly developed modern industries. To do so, the concept of Industry 4.0 can be taken into consideration (Hermann et al., 2016; Mangla et al., 2015; Govindan et al., 2016; Luthra et al., 2017). In efforts to make industry more sustainable it is very important to adapt to changing business environment and customer preferences. Different technologies have emerged in the recent times, such as 3D printing, data analytics, and so on, which could profoundly affect supply chain management (Almada-Lobo, 2016). The concept of Industry 4.0 is very intriguing considering the fact that it is an amalgamation of efficient resource use, employee welfare, operation safety, the protection of the environment, and the adoption of smart processes for the overall aim of industrialization as a whole (Hermann et al., 2016; Liao et al., 2017). Although several negative implications may also be seen for industrialization such as the impact on the environment, along with high energy consumption and health risks to employees. Industry 4.0 tries to take into account all such risks associated with the advent of industrialization, focusing on the smart production system with advanced cyber networks and interactions. Thus, there exists much more flexibility and economic benefits combined with better interaction of connected elements by cyber interaction. With the advantages in Industry 4.0, there also exist several challenges in its implementation. Hence, detailed research needs to be carried out, keeping in mind the industrial value chain system to identify the existing challenges. The concept in itself is new to a developing country such as India and needs proper practice and comprehension (Hofmann and Rüsch, 2017, Forbes, 2016). At present, manufacturing industry in India accounts for more than 16 % of the gross domestic product and employs nearly 13% of the employed workforce (IBEF, 2016). In recent years, several steps have been taken by the Government of India for the implementation of Industry 4.0. These include programs such as "Make in India", "Digital India" and "Smart city" (Abhishek, 2017). It has been found that India has become more accommodating towards the adoption of modern technologies as per the NRI (Network readiness index) value of 91 out of 139 nations (Grant Thornton Report, 2017). Regarding knowledge at the managerial level, the adoption of automation in the manufacturing industries of India is still in the very earliest stages and a wide range of opportunities exists (BRICS Business Council, 2017). According to a report by Grant Thornton, the Indian government is constantly looking out for ways to implement Industry 4.0 and the modern technologies that come along with it (Grant Thornton Report, 2017). It has become a necessity for the Indian economy to identify the various challenges that come along with the adoption of Industry 4.0 for a sustainable supply chain management. The coupling of "Make in India" and "Industry 4.0" could go a long way to creating opportunities for the local mass with the added benefits of ecological and social development. Thus, the main objective of the current work is to identify the key challenges associated with Industry 4.0 followed by analyzing the challenges with a goal to prioritize these challenges in order to achieve a successful implementation of Industry 4.0. A survey was conducted for the same, which included distributing a set of questionnaires which helped in obtaining relevant information regarding the identified challenges (Field, 2009). Tools such as Explanatory Factor Analysis (EFA) and Analytical Hierarchy Process (AHP) were used for the survey. The EFA helped in subdividing the challenges into different dimensions for maintaining the sustainable supply chain. The identified challenges were then mapped out and organized.

16.2 UNDERSTANDING INDUSTRY 4.0

This section basically covers the previous literature and reports with regard to the Industry 4.0 initiative and maintaining a good supply chain sustenance.

16.2.1 INDUSTRY 4.0

Industrial revolution has played a major role in the advancement of current industries. The fourth industrial revolution is said to have brought considerable attention for scholars all across the world (Liao et al., 2017) The three previous industrializations mainly aided in the provision of groundwork for industrialization, power and computers. The fourth revolution thus provided the groundwork to Industry 4.0 with the application of technology, automation, intelligent production, remote operation, 3D printing, etc. (Basl, 2017; Khan et al., 2017; Duarte and Cruz-Machado, 2017). The initiative "Industry 4.0" can be defined as a concept which can impact the entire business in terms of its products and deliverables (Hofmann and Rüsch, 2017). The concept mainly helps in having a grip over the production system of an industry compared to the older version of the same, focusing primarily on the transformations that may take place with the advent of modernization of manufacturing industry. Nonetheless, the concept needs assured understanding for reaping the benefits for real practice in the industry (Hofmann and Rüsch, 2017).

It is of great necessity to develop modern technologies in order to maintain a sustainable supply chain in an industry. Not only modernization, but green manufacturing coupled together can lead to a healthy set-up as per the Industry 4.0 initiative. Thus, new business trends would become more common in developing countries like India (Duarte and Cruz-Machado, 2017). Flexibility of supply chains is also a significant need for sustainability (Bechtsis et al., 2017; de Sousa Jabbour et al., 2018c). A global cyber network also needs to be developed for an enhanced data transfer and control which would be highly flexible; this would result in a smart value chain, thereby enhancing the overall performance of the business and the associated activities (Gilchrist, 2016; Lin et al., 2017). The initiative would thus result in a highly coordinated connection between equipment, materials and goods (Branke et al., 2016) and would help control and monitor the different production variables (de Sousa Jabbour et al., 2018b). Thus, a sustainable culture and superior production methods have become the necessary outcomes for an Industry 4.0 initiative, helping business organizations to advance towards sustainable development (de Sousa Jabbour et al., 2018a). Smart factories can reconsider the sustainable parameters for the optimal usage of technology and resources. In addition to the various advantages associated with the applicability of Industry 4.0, it is also of greater importance to make the managers understand the concept of Industry 4.0 to reap its benefit (Brettel et al., 2014). Thus, an ordered classification is necessary to focus specifically on the definition of Industry 4.0 and to overcome the challenges associated with it for sustainable supply chain management.

16.2.2 CHALLENGES TO INDUSTRY 4.0 INITIATIVES FOR SUSTAINABILITY IN SUPPLY CHAINS

In an attempt to assist the development of a sustainable supply chain management, intensive literature review has been done and expert opinion has also been taken into consideration. As per the researches and opinions 18 challenges were identified and have been explained further (Table 16.1).

16.3 METHODOLOGY

The method followed for the present work has been presented in Figure 16.1. A detailed literature review was performed in this chapter, focusing mainly on certain keywords relevant to Industry 4.0 (Biel and Glock, 2016; Liao et al., 2017). Google Scholar, Scopus and a number of databases of Google were used extensively to carry out the identification of research work related to the current chapter, focusing mainly on supply chain management along with the sustainability and challenges in an industry. The accumulated articles were then analyzed for the quality of their English, and articles published in peer-reviewed high-impact factor journals (Mangla et al., 2018). Challenges that may come into the picture have been considered and the analysis for studying such challenges were carried out using a questionnaire meant for random people using EPA. Different tools and methodologies have been followed in addition to the usage of the AHP method.

TABLE 16.1

Research Methodology for this Study

Challenges	Concepts	Author/Source/Year
Low understanding on Industry 4.0 implications	There is a very low understanding on Industry 4.0 implications among both the researchers and practitioners. Literature clearly demands highly organized and focal research for a specific definition of the Industry 4.0. Industrial and practicing managers undoubtedly understood the importance of Industry 4.0 adoption in manufacturing environment; however, they are still unsure on its exact consequences/implications on accomplishing sustainability objectives in supply chains	Almada-Lobo (2016); Hofmann and Rüsch (2017)
Poor research & development on Industry 4.0 adoption	Industry 4.0 has been inferred by different practicing managers in their own way. Mostly business organizations are facing different problematic issues in effective adoption of Industry 4.0 so as to lacking in accurate decision strategies during this business transformation. The prime reason behind this is a lack of focused research on addressing the various aspects of Industry 4.0 adoption. The scientific focused research would provide necessary theoretical foundations to Industry 4.0 driven sustainability in supply chains.	Schmidt et al. (2015); Hermann et al. (2016)
Legal issues	Industry 4.0 tends to develop a cyber-physical network where various machines, sensors, facilities, and humans are interlinked to the internet and exchanges data with each other. This cyber physical network may emerge with several complex legal issues. To help industries, legal issues should be taken into account while adopting modern technological procedures and ideas. Data privacy and security issues needs to be considered in developing data driven sustainable business models of Industry 4.0	Schröder (2016); Möller et al. (2017a)
Poor company's digital operations vision and strategy	Industry 4.0 describes an innovative approach to business operations and especially the manufacturing organizations by the digital transformation, which requires a clear digital operations vision and mission. During this transformation, organizations fail to apparently illustrate its Industry 4.0 vision and strategy. So far, organizations seem to struggle when transforming the visionary ideas of Industry 4.0 to a missionary level of developing the sustainability of supply chains.	Erol et al. (2016)
Low management support and dedication	In order to develop an effective Industry 4.0 concept, management support and dedication to accept the changes is very crucial. Industry 4.0 calls for a revolutionary transformation in business processes and supply chain activities, thus, most relevant management practices should be established Organizations should focus on improving their capabilities in terms of employee training and development, knowledge management programs, for Industry 4.0 driven sustainable business development. This could not possible without management support and dedication.	Gökalp et al. (2017); Savtschenko et al. (2017); Shamim et al. (2017)
Profiling and complexity issues	In recent years, supply chains are becoming global and characterized by highly complex structures. Therefore, workforce should be trained to know the essential processes, their dependencies, and data interpretation to accept digitization in the manufacturing environment. Business professionals generally lacks competencies on managing the complexity issues related to data analysis, space or time, usage of particular instructions, in effective Industry 4.0 adoption. This lack of roadmaps and guides supporting its implementation, as well as its high complexity makes "Industry 4.0" too uncertain for achieving sustainability in supply chains.	Erol et al. (2016); Ras et al. (2017);

Challenges	Concepts	Author/Source/Year
Lack of digital culture	Digitization is the foremost requirement for initiating Industry 4.0 in business environment. Further, Industry 4.0 generally of interdisciplinary in nature which requires digitization to connect different elements of a network.	Ras et al. (2017); Schuh et al. (2017)
Reluctant behavior towards Industry 4.0	Most of industries are still unfamiliar and unsure with the topic of Industry 4.0. Due to the ignorance of possible benefits, majority of industries are reluctant to adopt Industry 4.0 based technologies.	Müller et al. (2017b); Theorin et al. (2017); Perales et al. (2018)
Unclear economic benefit of digital investments	In Industry 4.0, prime emphasis is given on its technical competence and knowhow, whereas the economic discussion is still in its infancy. The lack of clearly defined return on investment could be seen as a one of major challenge to Industry 4.0 initiatives for accomplishing sustainability in the supply chain.	Kiel et al. (2017); Marques et al. (2017)
Lack of global standards and data sharing protocols	In Industry 4.0 initiatives, systems generally coupled to an intelligence mechanism to communicate freely. To achieve success in this, industries need to follow global standards and data sharing protocols. It has been notices that industries lacks in standards and protocols in data transfers in adopting sustainability oriented modern information interface technologies in business networks.	Branke et al. (2016)
Lack of infrastructure and internet based networks	High infrastructure, information technology based facilities and technologies are crucial in effective adoption of Industry 4.0 concepts. Poor internet connectivity is an imperative barrier to Industry 4.0 initiatives. Further, in Indian context, internet based technology are not to be recognized equally in urban and rural areas which can impede the sustainable business growth.	Leitão et al. (2016); Bedekar (2017); Pfohl et al. (2017)
Lack of competency in adopting/ applying new business models	Current industrial system needs highly customized and flexible environment to compete globally. In this sense, industries need to adopt new business models. Integration of multiple systems pushed the data to the big data due to the deluge of data generation in the manufacturing processes. Industrial big data analytics increase the productivity of enterprises. Prediction of new events from big data provides a concrete foundation for planning new projects. As it is not necessary that all the new insights will be workable and only some events are interesting out of million events, so revealing these insights are a challenge for data scientists to write suitable algorithms in adopting/applying new business models.	Khan et al. (2017); Saucedo-Martinez et al. (2017)
Poor existing data quality	Data quality is one of foremost requirement in making decisions in successful Industry 4.0 adoption. In Industry 4.0, several machines, sensors, manufacturing systems and facilities are interconnected so as generates big data. The available big data may help managers to practice Industry 4.0 innovations for a sustainable future. This could not possible without higher data quality.	Santos et al. (2017)

(Continued)

Table 16.1: *(Continued)*
Research Methodology for this Study

Challenges	Concepts	Author/Source/Year
Lack of integration of technology platforms	The integration of technology is very essential in effective communication and higher productivity. Industries are facing difficulties in designing a flexible interface to integrate various heterogeneous components. Cyber physical networks many different components, which needs to be integrated and supported for an effective data exchange and analysis in manufacturing environment. Thus, it is significant to design and develop a platform to integrate technology for developing an effective Industry 4.0 driven sustainable supply chain.	Zhou et al. (2015)
Problem of coordination and collaboration	Collaboration and transparency among members is important in understanding the organizational policies in adopting concepts of Industry 4.0 and improving supply chain sustainability. The coordination and collaboration with suppliers is necessary for better communication mechanisms, with high compatibility issues of hardware and software which should require standardized interfaces, and synchronization of data to get better synchronization with manufacturers.	Lee et al. (2014); Duarte and Cruz-Machado (2017); Pfohl et al. (2017)
Security issues	One of the Industry 4.0 features is the ability to connect across organizational environments, which has the potential to make the supply chain more efficient. However, the supply chain systems have inherent security vulnerabilities, which are exploited by attackers. One of the security vulnerabilities starts with the supplier, which is vulnerable to phishing attacks and the stolen of privileged credentials, resulting in mass data exposure. The major vulnerability is in the top of the supply chain, reaching the rest of the organizational processes through its dependent actors. Security is the prime requirement to transform a factory into smarter factor and a supply chain into smarter value chains.	Sommer (2015); Wang et al. (2016); Pereira et al. (2017)
Lack of governmental support and polices	Government policies and directions are crucial in developing supply chain sustainability through Industry 4.0. Clearly, there is a lack of definite government guidelines and directions on Industry 4.0 in most of the economies including India. In addition to this, governments are also unsure on probable consequences of Industry 4.0. As a resultant, policy analysts and government bodies have not revealed the roadmap for transforming the traditional business functions into smarter and sustainable processes.	BRICS Business Council (2017)
Financial constraints	In Industry 4.0, financial constraints are considered to be a very important challenge among business organizations for developing their capabilities in terms of advanced equipment and machines, facilities and sustainable process innovations	Dawson (2014); Theorin et al. (2017); Nicoletti, (2018)

(Reproduced with permission from Luthra et al. 2018, Copyright © Elsevier)

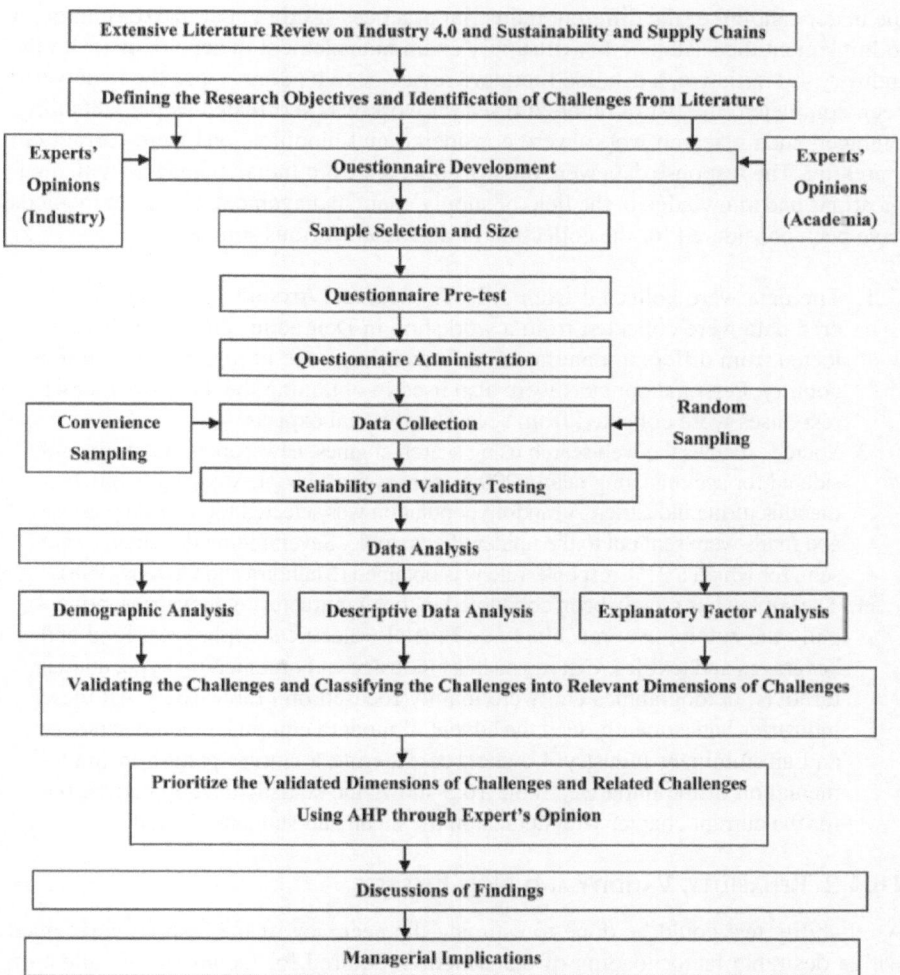

FIGURE 16.1   Research methodology for this study (Reproduced with permission from Luthra et al. 2018, Copyright © Elsevier).

16.4 DATA COLLECTION AND RESULTS

The present chapter focuses mainly on the Indian manufacturing sector with nearly 97 responses, which was considered sufficient enough to represent a population as huge as India and thereby give a brief description about its time, cost and resources. The data have been analyzed based on various tools and the corresponding results have been presented after the detailed analysis.

16.4.1 INSTRUMENT DEVELOPMENT AND DATA COLLECTION

A detailed pool of data could be collected based on the response obtained on the different developments of an instrument, thereby providing a broader knowledge of the understanding of the different industrial practices (Luthra et al., 2016a). In order to implement the sustainability of supply chain management in accordance with the Industry 4.0 initiative, a detailed literature review is to be done. Once the review had been completed, the list of detailed research work was shortlisted as per suitability. Eighteen such research works were considered and modifications were carried out thereafter. The response data were mainly collected from Indian managers with qualifications and knowledge in the field of supply chain management. Different methods have been considered for the collection of data from various sources:

1. The data were collected from different sources. Around 13 of the considered data were collected from a workshop in Dehradun. The rest were collected from different manufacturing industries in the northern region of the country. Personal contacts were also used in obtaining the data. Another 41 responses were collected from several industrial experts.
2. Sources such as the web search using search engines, newspapers etc. were considered for accumulating nearly 200 responses from people working in different manufacturing industries. A random population was selected for the said purpose and mails were sent out to the unidentified people. Several reminder emails were sent, for which a 21% response rate was obtained (Malhotra and Grover, 1998).
3. The survey has mostly been conducted in the manufacturing industries, principally the automotive; machinery and metals, electrical appliances, food and beverages and textiles. Criteria such as industry and organization type, annual turnover, demographics etc. were mainly focused on (Table 16.2). All these industries had evidently seen the advent of modern equipment and machinery and an imminent Industry 4.0 start, considering a similar pattern in implementation of the initiative. Table 16.3 shows the statistical study carried out for the current chapter with details on the mean and standard deviation.

16.4.2 RELIABILITY, VALIDITY AND NON-BIASNESS

A reliability test could be done to estimate the accuracy of the responses received with a desirable factor loading of 0.5 which is required for a convergent validation (Hair et al., 2006, Field, 2009). The current work had a value of more than the required number. As per Cronbach's alpha value, which basically determines the internal reliability for determining the consistency of the survey, a value not less than 0.5 shall be preferred. After checking the reliability, the next step was to check the discrimination validity followed by the inspection of biasness. The biasness test was done considering 96 responses, of which 56% were early responses corresponding to the convenience sampling and 44% were late responses corresponding to random sampling. No major changes have been observed as per the t-test with a P value of greater 0.05. The random responses were subdivided again into early and late responses, of which 11 were early and 31 were late. A Chi-square test suggested that there was no significant difference between the two subdivisions of responses.

16.4.3 EXPLANATORY FACTOR ANALYSIS (EFA)

For the analysis of data and the reduction of the same, an explanatory factor analysis was taken into consideration (Hair et al., 2006). The aforementioned test, along with

TABLE 16.2

Summary on Demographics of Indian Organizations

Organization	Criteria	Respondents	Percentage
Industry type	Automotive	33	34.38
	Metals and machinery	26	27.08
	Electrical equipment and appliances	11	11.46
	Food and beverage	9	9.37
	Textile	5	5.21
	Others	12	12.50
Organization type	Private sector	54	56.25
	Public sector	5	5.21
	Multinational corporation	37	38.54
	Others	0	0
Annual turnover (In million rupees)	Less than or equal to 500	18	18.75
	501-1000	31	32.29
	1001-5000	36	37.50
	5001-10,000	8	8.33
	>10,000	3	3.13
Kind of business	O.E.M.	38	39.58
	Supplier	58	60.42
		29	30.21
Supplier information	Less than or equal to 50	33	34.37
	51-100	19	19.79
	101-200	15	15.63
	>200		
Automation level		74	77.08
	Yes	18	18.75
	No	4	4.17
	In progress		

(Reproduced with permission from Luthra et al. 2018, Copyright © Elsevier)

the reliability test, were done with the help of a statistical software such as SPSS version 20.0. The KMO value of 0.76 was obtained after a detailed analysis, which was significantly lower than the minimum set value of 0.6 (Kaiser, 1974; Hair et al., 2006). The current case has an eigenvalue of discontinuity greater than 1, a factor loading above 0.5 and the alpha value of 0.7 (Hu and Hsu, 2010; Luthra et al., 2016a). Table 16.4 gives the detailed analysis of all the results obtained. Few challenges that account for nearly 74% variance in the result have been listed out as issues related to organizational, strategic, legal and ethics and technological. The results obtained after EFA analysis have been subdivided into different dimensions and are discussed as below:

Organizational challenges (OR): The first dimension mainly includes challenges that account for 24% of variance which are mainly hurdles at the organizational stage

TABLE 16.3

Challenges to Industry 4.0 for Sustainability Orientation in Supply Chain

Challenges to Industry 4.0 Initiatives for Sustainability Orientation in Supply Chains	Mean	Standard Deviation
Low understanding of industry 4.0	4.208	0.579
Poor research and development of industry 4.0	4	0.681
Legal issues	4.427	0.660
Poor vision on digital operation and strategy	4.219	0.668
Low management support and dedication	4.260	0.669
Profiling and complexity issues	4.437	0.577
Lack of digital culture	4.031	0.623
Reluctant behaviour	4.26	0.700
Unclear economic benefit for digital investment	4.021	0.768
Lack of global standards and data sharing protocols	3.833	0.627
Lack of infrastructure and internet based networks	3.865	0.450
Lack of competency in adopting/applying new business models	4.208	0.614
Poor data quality	3.854	0.562
Lack of integrating technology	3.844	0.604
Coordination and collaboration problems	4.427	0.661
Security issues	4.458	0.614
Lack of governmental support and policies	3.990	0.733
Financial constraints	4.281	0.644

(Reproduced with permission from Luthra et al. 2018, Copyright © Elsevier)

such as social and economic point of views for the sustainable supply chain management as per the Industry 4.0 initiative.

Legal and ethical issues (LE): The second dimension accounted for 20% of variance with four elaborated hurdles that are primarily focused on the ethical and legal matters related to the sustenance of Industry 4.0.

Strategic challenges (ST): The third dimension mainly dealt with Industry 4.0 diffusion with a concern on the strategic front consisting of 15% of variance.

Technological challenge (TE): The fourth dimension had a variance of 14.4% with mainly four underlying challenges. Theoretical analysis was laid down considering the factor analysis and later validated to know their priority for a favorable Industry 4.0 outcome with sustainable supply chain through the analytical hierarchy process.

16.4.4 ANALYTICAL HIERARCHY PROCESS (AHP)

AHP is a tool suggested by Professor Thomas L Saaty which aids in converting complex issues into an ordered structure of varying magnitudes (Dey and Cheffi, 2013). Due to its simplicity in usage, it is considered to be a better analytical tool than ANP, although it might have certain inconsistencies due to human error (Mangla et al., 2015; Luthra et al., 2017, Gandhi et al., 2016). For the sustenance of Industry 4.0,

TABLE 16.4
Factor Analysis Results for Challenges to Industry 4.0 Initiatives for Sustainability Orientation in Supply Chains

Dimension	Challenges	Item Loading	Eigen Value	Cumulative Percentage
Organizational (OR)	Financial constraints (OR1)	0.933	4,403	24,461
	Low management support and dedication (OR2)	0.908		
	Reluctant behavior towards Industry 4.0 (OR3)	0.884		
	Poor company digital operations vision and mission (OR4)	0.872		
	Lack of competency in adopting/ applying new business models (OR5)	0.848		
	Low understanding on Industry 4.0 implications (OR6)	0.848		
Legal and ethical issues (LE)	Legal issues (LE1)	0.915	3,628	44,617
	Problem of coordination and collaborations (LE2)	0.898		
	Security issues (LE3)	0.888		
	Profiling and complexity issues (LE4)	0.787		
Strategic (ST)	Lack of governmental support and polices (ST1)	0.834	2,763	59,969
	Poor research & development on Industry 4.0 adoption (ST2)	0.816		
	Unclear economic benefit of digital investments (ST3)	0.771		
	Lack of digital culture (ST4)	0.735		
Technological (TE)	Lack of global standards and data sharing protocols (TE1)	0.816	2,594	74,379
	Poor existing data quality (TE2)	0.794		
	Lack of integration of technology platforms (TE3)	0.759		
	Lack of infrastructure and internet based networks (TE4)	0.637		

(Reproduced with permission from Luthra et al. 2018, Copyright © Elsevier)

this analysis helps in identifying the underlying challenges (Luthra et al., 2016b). Numerous steps are associated with the AHP tool, which can be outlined as below:

1. Objective of the study:
   The AHP tool helps in identifying and prioritizing challenges, taking into consideration expert opinions. In accordance with the discussions, a classification of challenges was identified. The experts were mainly a combination of

academicians and industry professionals from the areas of operations, information safety, engineering, production and supply chain selected based on their expertise in the relevant areas. Moreover, each expert had a work experience of more than ten years. The classifications done by the experts is shown in Figure 16.2 which has mainly three subdivisions as follows

Level 1 for resourcefulness of orientation in supply chains

Level 2 for the challenges in the four dimensions

Level 3 for the 18 identified challenges

2. Comparison between identified challenges
A comparative study was carried out in the second step wherein the pair-wise comparison was made for the challenges as per the dimensions and those as per the Saaty scale.

3. Estimation of consistency ratio and the relative weights
Table 16.5 shows the relative weights considering the comparisons made in the previous steps. The pattern of the challenges has been reported as OR (0.30) > TE (0.28) > ST (0.24) > LE (0.18). Table 16.6 shows the priority weights with all the results within acceptable permissible limits. AHP results showed a pattern of OR (0.30) and TE (0.28) as the primary dimensional challenges.

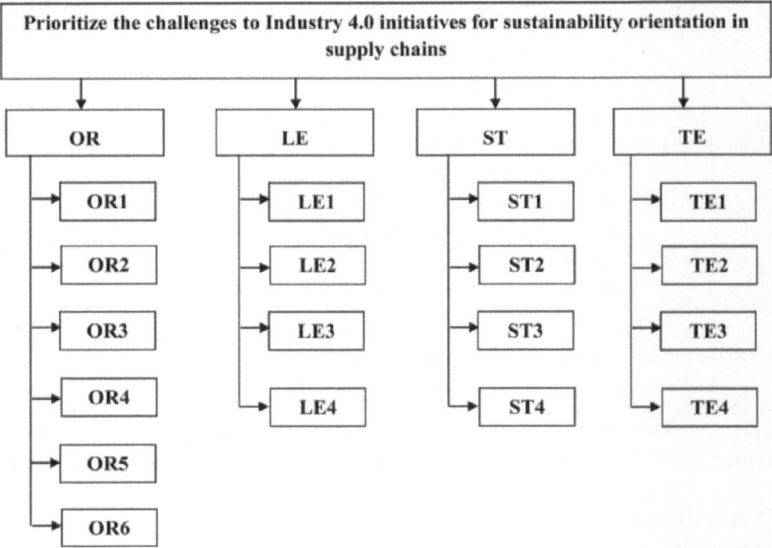

FIGURE 16.2 The developed decision hierarchy of challenges (Reproduced with permission from Luthra et al. 2018, Copyright © Elsevier).

TABLE 16.5

Pair Wise Comparison Matrix for four Dimensions of Challenges and their Computed Priority Weights

Major Dimensions of Challenges	OR	LE	ST	TE	Priority Weight	Rank
Organizational (OR)	1	1	2	1	0.2976	1st
Legal and ethical	1	1	0.5	0.5	0.1760	4th
issues (LE)	0.5	2	1	1	0.2454	3rd
Strategic (ST)	1	2	1	1	0.2810	2nd
Technological (TE)						

(Reproduced with permission from Luthra et al. 2018, Copyright © Elsevier)

As per the global ranking of challenges identified, 4 out of 5 belong to OR and TE. The 5 globally recognized challenges are as follows:

Non-availability of standards as per global levels belonging to TE1 challenge

Insufficient policies and back-up from the government belonging to ST1 challenge

Financial setbacks as per OR1 challenge

Low-grade infrastructure and IT networks as per TE4 challenge

Meagre support from the management and dedication as per OR2 challenge

16.5 DISCUSSION

The main aim of the present work is to identify all the challenges using the EFA analysis and grouping them into 18 main challenges which are then further subdivided into 4 dimensional challenges: Organization (OR), Technological (TE), Strategic (ST) and Legal and ethical (LE). The identified challenges are then arranged in a sequential manner to identify the most critical challenge as OR > TE > ST > LE. Organizational challenge thus proves to be the most critical challenge and plays a very important role in the supply chain management in the manufacturing sector of a developing country. In order to inculcate Industry 4.0, a practice on the development of the business culture, work specialization, work policies, good leadership must be given primary importance. Hence, it is the duty of the people in the top management to bring about all these changes dynamically to make the adoption of Industry 4.0 more efficient. Financial constraints (OR1) plays a major role in case of small enterprises due to the difficulties faced during the adoption of Industry 4.0, mainly because of various financial setbacks. Followed by OR1 Challenges is the OR2 challenge, which is mainly related to the low grades of support from the management and work dedication. The OR2 challenge plays a major role in diffusing Industry 4.0 for a better production environment. This is then followed by OR4 low-grade digital

TABLE 16.6

Ranking of Challenges to Industry 4.0 Initiatives for Sustainability Orientation in Supply Chains

Main Dimensions	Relative weights	Sub Challenges	Relative Weights	Relative Rankings	Global Weights	Global Ranking
Organizational (OR)	0.2976	Financial constraints (OR1)	0.2818	1st	0.0839	3rd
		Low management support and dedication (OR2)	0.2175	2nd	0.0647	15th
		Reluctant behavior towards Industry 4.0 (OR3)	0.1087	5th	0.0324	14th
		Poor company digital operations vision and mission (OR4)	0.1754	3rd	0.0522	10th
		Lack of competency in adopting/applying new business models (OR5)	0.1010	6th	0.0301	15th
		Low understanding on Industry 4.0 implications (OR6)	0.1154	4th	0.0343	13th
Legal and ethical issues (LE)	0.1760	Legal issues (LE1)	0.1480	4th	0.0260	17th
		Problem of coordination and collaborations (LE2)	0.1630	3rd	0.0287	16th
		Security issues (LE3)	0.3629	1st	0.0639	6th
		Profiling and complexity issues (LE4)	0.3261	2nd	0.0574	9th
Strategic (ST)	0.2454	Lack of governmental support and polices (ST1)	0.3465	1st	0.0850	2nd
		Poor research & development on Industry 4.0 adoption (ST2)	0.2036	3rd	0.0499	11th
		Unclear economic benefit of digital investments (ST3)	0.2463	2nd	0.0604	7th
		Lack of digital culture (ST4)	0.2036	3rd	0.0499	11th
Technological (TE)		Lack of global standards and data sharing protocols (TE1)	0.3301	1st	0.0927	1st
		Poor existing data quality (TE2)	0.1748	4th	0.0491	12th
		Lack of integration of technology platforms (TE3)	0.2069	3rd	0.0581	8th
		Lack of infrastructure and internet based networks (TE4)	0.2883	2nd	0.0810	4th

advancement in the industry which would, in the long run, cause a failure in the established roadmaps and visions for the successful implementation of Industry 4.0. OR6, which corresponds to the modest knowledge of the Industry 4.0, followed by OR3, which further corresponds to the unenthusiastic behavior for the Industry 4.0. Lastly, the incompetence of adopting new business models as OR5 also comes into consideration. As far as maintaining the Industry initiative 4.0 is concerned, it is said to be heavily dependent on man-made skills and performances. Skills such as leadership, management, manufacturing diligence are all governed by human efficiency in maintaining a eco-friendly operation of supply chains as per the Industry 4.0 initiative (De Sousa Jabbour et al., 2018a).

The next important dimension is called the Technological challenge. It is again subdivided into four main headings. The absence of standards as per global demands (TE1) is said to occupy the highest priority. The non-availability of appropriate infrastructure (TE3) also plays a key role and data sharing networks. It has been suggested that a common platform should be established for storing data and developing new production systems. A good connectivity is a requisite need for maintaining the up-to-date data and devices in rural as well as urban areas. Following TE3 is TE4, which accounts for the absence of well-placed platforms for the available technologies necessary for the diffusion of multiple cyber components. The last challenge would be the TE2 challenge which corresponds to the bad management of existing data which is of very important concern in having a successful Industry 4.0 implementation.

It has been strongly suggested by several researchers that the proposed concept of Industry 4.0 is novel and has found very limited use in the present industrial scenario (De Sousa Jabbour et al., 2018b).

The next important dimension is defined as the strategic challenge (ST), which is again further subdivided into various groups. Several suggestions have been made regarding the importance of support from the government side which could well benefit a traditional industry by converting it into a modern factory of the future and could also remove the barricades on the way of reaching Industry 4.0 (Kagermann, 2015; Muller et al., 2018). In a country such as India several initiatives has been taken by the government to aid the growth of Industry 4.0, including the "Make in India", "GST", the creation of green corridors for creating trade and commerce routes yet a great deal of understanding is required on the part of the industry to inculcate the changes suggested by the government for the implementation of Industry 4.0. There also exists the blurred picture of benefits associated with digital investments (ST3) challenge, which mainly corresponds to the half-hearted knowledge on the economic benefits of implementing Industry 4.0 (Hofmann and Rüsch, 2017). A lack of efficient research in the industry leads to the existence of challenges such as inadequate R&D for Industry 4.0 adoption (ST2) and the absence of digital culture (ST4).

The last dimension of this entire discussion includes the issues related to legal rights and ethics (LE). The LE3 challenge, which is a security-based issue under this dimension, deals mainly with the implementation of appropriate steps required for solving the various technological challenges. This is followed by the LE4 challenge, which correlates to the profiling and complexity issues, and the LE2 challenge, which is based on problems associated with coordination and collaborations, followed by the last challenge, which is associated with legal issues (LE1). It is of high relevance that the legal concerns are given importance as during the advancement of an industry towards technology 4.0 several personal details and sensitive data are shared whose safety is to be taken care of by the legal team with a proper ethical approach. There also exists a need to set down steps to counterfeit discrepancies that might arise in the long run, such as cybercriminal acts owing to the misuse of company data in the long run (Müller et al., 2018).

16.6 SUMMARY

The main aim of this work was to identify the underlying challenges that would arise when an industry would take the initiative of implementing Industry 4.0 for a sustainable supply chain management. The initiative not only promotes sustainability concepts for the managerial levels but also focuses widely on the control measures, environmental protection and process safety for a community welfare in the entire supply chain.

16.6.1 THEORETICAL BENEFACTION

It has been suggested that the initiative Industry 4.0 is a new concept in developing nations like India and needs a clear understanding for implementation as per business practice (de Sousa Jabbour et al., 2018a). Eighteen key challenges have already been identified, as already mentioned previously through extensive literature studies, and these were scrutinized based on nearly 96 responses obtained unanimously by the distribution of a questionnaire-based survey to different manufacturing sectors in India. The EFA tool was used to lay the foundation of the selected challenges. Different dimensions have been identified along with their respective challenges. The AHP tool has designated different ranks to the identified challenges which suggests the following pattern based on importance as Organizational challenge> Legal and ethical> Strategic challenge> Technological challenge.

The Industry 4.0 initiatives, with their greener products, processes and operations, has the scope of a smart manufacturing system and a value added supply chain management. In developing economies, Industry 4.0 can act as an initiation of diminishing probable hurdles that may arise in modification of manufacturing systems and supply chain managements. At the same time it would also create a healthy environment for the overall welfare of the staff and the industry with considerable economic gains.

16.6.2 MANAGERIAL BENEFACTION

The Industry 4.0 initiative focuses primarily on a sustainable supply chain for the industries. This provides an opportunity for the engineers and the managers to consider matters related to the different challenges faced during the maintenance of a viable supply chain. Such in-depth knowledge on the challenges can help the people in charge to make suitable designs, optimize processes and control systems etc.

Thereby, in the long run this identification of various challenges would help in eliminating hurdles and developing modern technologies like the IoT, robotics design and product manufacturing in a sustainable environment. Managers could in turn help in employing superior control devices such as robots, sensors etc. for improved productivity and creating a safe working environment. Thus, the implementation of Industry 4.0 is also largely dependent on cyber networks. It is also a major duty of the manager to take care of the different aspects of the new technologies taken into consideration being sustainable to the proper functioning of the industry and thereby sustaining the ultimate goal of a suitable supply chain. This would, in turn, not only impose increased productivity but also result in flexibility of production and operations without disrupting the environment through the minimal consumption of energy and resources. In case of higher requirements of resources or energy, sustainable aspects of manufacturing must be brought under consideration which would require the assistance of engineers, managers and the stakeholders. Industry 4.0 can thus act as an enormous opportunity for employment generation along with sustainable business providing better job satisfaction, maintaining the resources as per needs and supplying the desired energy for maintaining the supply chain.

16.6.3 SHORTCOMINGS AND FUTURE PROPOSALS

The present chapter suggests 18 challenges, keeping in mind a developing nation like India. However, several other challenges may play an important role considering other nations in the picture. There also exist several scopes of extending the present research work such as bringing about a comparative study between the data of different developing countries. In addition to this a comparative study on the same nation considering a broader geographic region could also be carried out. Sensitivity analysis can also be carried out to validate the results. The opinions suggested by the experts might be subjective to biases. Hence, a system should be established in the future wherein such discrepancies could be minimized. SEM or Structural equation modelling may be carried out in the future studies for expanding the test framework.

Notably, all listed challenges obtained mean value greater than 3. This suggests that all identified challenges are significant.

REFERENCES

Abhishek, R., 2017. *Towards Smart Manufacturing – Industry 4.0 and India*, Online available at: http://www.makeinindia.com/article/-/v/towards-smart-manufacturing-industry-4-0-and-india (last Accessed 08 December 2017).

Almada-Lobo, F., 2016. The Industry 4.0 revolution and the future of manufacturing execution systems (MES). *J. Innov. Manage.* 3 (4), 16–21.

Basl, J., 2017. Pilot study of readiness of Czech companies to implement the principles of Industry 4.0. *Manage. Prod. Eng. Rev.* 8 (2), 3–8.

Bechtsis, D., Tsolakis, N., Vlachos, D., Iakovou, E., 2017. Sustainable supply chain management in the digitalization era: the impact of Automated Guided Vehicles. *J. Clean. Prod.* 142, 3970–3984.

Biel, K., Glock, C.H., 2016. Systematic literature review of decision support models for energy-efficient production planning. *Comput. Ind. Eng.* 101, 243–259.

Branke, J., Farid, S.S., Shah, N., 2016. Industry 4.0: a vision for personalized medicine supply chains? *Cell Gene Ther. Insights* 2 (2), 263–270.

Brettel, M., Friederichsen, N., Keller, M., Rosenberg, M., 2014. How virtualization, decentralization and network building change the manufacturing landscape: an industry 4.0 perspective. *Int. J. Mech. Ind. Sci. Eng.* 8 (1), 37–44.

BRICS Business Council, 2017. Skill development for industry 4.0. In: A White Paper by BRICS Skill Development Working Group. BRICS Business Council, India Group, Online available at: http://www.globalskillsummit.com/Whitepaper-Summary.pdf (last Accessed 22 October 2017).

de Sousa Jabbour, A.B.L., Chiappetta Jabbour, C.J., Sarkis, J., Gunasekaran, A., Furlan Matos Alves, M.W., Ribeiro, D.A., 2018a. Decarbonisation of operations management-looking back, moving forward: a review and implications for the production research community. *Int. J. Prod. Res.*, 1–23.

de Sousa Jabbour, A.B.L., Jabbour, C.J.C., Foropon, C., Godinho Filho, M., 2018b.Whentitans meet-can industry 4.0 revolutionize the environmentally-sustainable manufacturing wave? The role of critical success factors. *Technol. Forecast. Soc. Change* 132, 18–25, doi:10.1016/j.techfore.2018.01.017.

de Sousa Jabbour, A.B.L., Jabbour, C.J.C., Godinho Filho, M., Roubaud, D., 2018c. Industry 4.0 and the circular economy: a proposed research agenda and original roadmap for sustainable operations. *Ann. Oper. Res.*, 1–14, doi:10.1007/s10479-018-2772-8.

Dey, P.K., Cheffi, W., 2013. Green supply chain performance measurement using the analytic hierarchy process: a comparative analysis of manufacturing organizations. *Prod. Plann. Control* 24 (8–9), 702–720.

Duarte, S., Cruz-Machado, V., 2017. Exploring linkages between lean and green sup-ply chain and the industry 4.0. In: *International Conference on Management Science and Engineering Management*, Springer, Cham, July, pp. 1242–1252.

Field, A., 2009. *Discovering Statistics Using SPSS*, 5th edition. Sage publications, Thousand Oaks, CA.

Forbes, 2016. India to Be World's Fastest Growing Economy: Keeping It Going Will Be the Difficult Trick. A Report by Forbes, Online available at: http://www.forbes.com/sites/timworstall/2016/02/08/india-to-be-worlds-fastest-growing-economy-keeping-it-going-will-be-the-difficult-trick/ (Last Accessed 04 March 2017).

Gandhi, S., Mangla, S.K., Kumar, P., Kumar, D., 2016. A combined approach using AHP and DEMATEL for evaluating success factors in implementation of green supply chain management in Indian manufacturing industries. *Int. J. Logist. Res. Appl.* 19 (6), 537–561.

Gilchrist, A., 2016. Introducing industry 4.0. In: *Industry 4.0*. A press, Berkley, CA, pp. 195–215.

Govindan, K., Seuring, S., Zhu, Q., Azevedo, S.G., 2016. Accelerating the transition towards sustainability dynamics into supply chain relationship management and governance structures. *J. Clean. Prod.* 112, 1813–1823.

Grant Thornton Report, 2017. *India's Readiness for Industry 4.0- A Focus on Auto-motive Sector*, Online available at: http://www.grantthornton.in/globalassets/1.-memberfirms/india/assets/pdfs/indiasreadinessforindustry4afocusonautomotive sector.pdf (last Accessed 28 August 2017).

Hair Jr., J.F., Black, W.C., Babin, B.J., Anderson, R.E., Tatham, R.L., 2006. *Multivariate Data Analysis: A Global Perspective*, 7th edition. Pearson publications, Upper Saddle River, Boston.

Hermann, M., Pentek, T., Otto, B., 2016. Design principles for Industry 4.0 scenarios. In: *2016 49th Hawaii International Conference on System Sciences (HICSS)*, January, IEEE, pp. 3928-3937.

Hofmann, E., Rüsch, M., 2017. Industry 4.0 and the current status as well as future prospects on logistics. *Comput. Ind.* 89, 23–34.

Hu, A.H., Hsu, C.W., 2010. Critical factors for implementing green supply chain management practice: an empirical study of electrical and electronics industries in Taiwan. *Manag. Res. Rev.* 33 (6), 586–608.

IBEF, 2016. *India Brand Equity Foundation- A Report on Indian Manufacturing Sector*, Online available at: http://www.ibef.org/industry/manufacturing-sector-india.aspx (Last Accessed 24 February 2017).

Kagermann, H., 2015. Change through digitization-Value creation in the age of Industry 4.0. In: *Management of Permanent Change*. Springer Fachmedien, Wiesbaden, Germany. pp. 23–45.

Kaiser, H.F., 1974. An index of factorial simplicity. *Psychometrika* 39 (1), 31-36.

Khan, M., Wu, X., Xu, X., Dou, W., 2017. Big data challenges and opportunities in the hype of Industry 4.0. In: 2017 IEEE International Conference on Communications (ICC). May, IEEE, pp. 1-6.

Liao, Y., Deschamps, F., Loures, E.D.F.R., Ramos, L.F.P., 2017. Past, present and future of Industry 4.0-A systematic literature review and research agenda proposal. *Int. J. Prod. Res.* 55 (12), 3609–3629.

Lin, K.C., Shyu, J.Z., Ding, K., 2017. A cross-strait comparison of innovation policy under Industry 4.0 and sustainability Development Transition. *Sustainability* 9(5), 786.

Luthra, S., Garg, D., Haleem, A., 2016b. The impacts of critical success factors for implementing green supply chain management towards sustainability: an empirical investigation of Indian automobile industry. *J. Clean. Prod.* 121, 142–158.

Luthra, S., Govindan, K., Kannan, D., Mangla, S.K., Garg, C.P., 2017. An integrated framework for sustainable supplier selection and evaluation in supply chains. *J. Clean. Prod.* 140, 1686–1698.

Luthra, S., Mangla, S.K., Xu, L., Diabat, A., 2016a. Using AHP to evaluate barriers in adopting sustainable consumption and production initiatives in a supply chain. *Int. J. Prod. Econ.* 181, 342–349.

Malhotra, M.K., Grover, V., 1998. An assessment of survey research in POM: from constructs to theory. *J. Oper. Manage.* 16 (4), 407–425.

Mangla, S.K., Kumar, P., Barua, M.K., 2015. Risk analysis in green supply chain using fuzzy AHP approach: a case study. *Resour. Conserv. Recycl.* 104, 375–390.

Mangla, S.K., Luthra, S., Jakhar, S., Tyagi, M., Narkhede, B., 2018. Benchmarking the logistics management implementation using Delphi and fuzzy DEMATEL. *Benchmarking: Int. J.* 25, 6.

Müller, J., Dotzauer, V., Voigt, K.I., 2017a. Industry 4.0 and its impact on reshoring decisions of German manufacturing enterprises. In: *Supply Management Research*. Springer Gabler, Wiesbaden, pp. 165–179.

Müller, J.M., Maier, L., Veile, J., Voigt, K.I., et al., 2017b. Cooperation strategies among SMEs for implementing industry 4.0. In: Kersten, W. (Ed.), Proceedings of the Hamburg International Conference of Logistics (HICL) ?23. Digitalization in Supply Chain Management and Logistics, October 2017, Epubli, pp. 301–318 (ISBN: 9783745043280).

17 A Detailed Study on the Spatial Characteristics of Heavy Metal Pollution and Ecological Risk of Mining Area

17.1 INTRODUCTION

Heavy metals from the mining industry will enter the soil mainly as a result of human activities such as mining. Over time, such heavy metal accumulation in the soil would inhibit plant growth and have a negative impact on the environment. At the same time, these heavy metals are highly injurious to human health (Li et al., 2014; Zhao et al., 2012; Wuana and Okieimen, 2011). Severe effects have been found on the liver, kidneys and central nervous system due to the heavy metals copper and zinc, which are mainly emitted as the result of industrial activities (Sani et al., 2017). Severe neurological impacts were also observed, along with skin disorders and other cancers in the body, by the presence of arsenic (Adio et al., 2017). Hence, it has become extremely important to analyze the effects of soil pollution due to heavy metals.

17.2 HEAVY METAL POLLUTION

A large number of methods have been suggested for the analysis of heavy metals, including the single factor evaluation method, the Nemero index, the Hakanson method, the pollution load index, the geographical accumulation index, the fuzzy coefficient method, and the gray clustering method (Hakanson, 1980; Müller, 1969; Fan et al., 2012). The type of method to be used for any particular analysis depends upon several factors, such as the content of the heavy metals, toxicity, migration and transformation behaviors (Hakanson, 1980). This particular procedure for evaluation has been applied in several areas of China and other countries with a very high heavy metal pollution. The Simple risk index (E_i) and Comprehensive risk index (RI_i) needs to be evaluated and adjusted based on the type of heavy metal present in the soil and their corresponding toxicity. As per the Hakanson method, a few standard values are prescribed for the toxicity coefficient

($T^{PCB} = T^{Hg} = 40$) and the total amount of toxicity coefficients among eight pollutants ($133 >$ the maximum toxicity coefficient, $T^{Hg} = 40$). For a variation of content and type of heavy metals, there arises different assessment standards. In order to have accurate results, it is necessary to modify the evaluation zone (Peng et al., 2007; Fu et al., 2009; Jian-Hua et al., 2011). The results obtained as per the latest modification in the evaluation process gives more accurate results than the old ones. The present chapter also investigates the accurate location of high-risk level areas, identified by GIS technology. A visual interpretation of the risk dispersal by determining the value of unspecified points helps in conducting an evaluation of soil samples by the spatial interpolation method (Chabukdhara and Nema, 2013; Sun et al., 2010). All the previous data on the detection of heavy metals was found to be focused mainly on large areas and very little research has been carried out on the quantitative research. This chapter focuses on the extensive research on the townships with the risks being divided based on the area and area ratio. The results thus obtained from the case study in this chapter are more precise and relatable to the existing scenarios compared to the older researches; thus, the outcomes can provide new dimensions for gaining more details on the soil pollution by heavy metals. A large polymetallic mining area in Suxian, China was selected for the case study in this chapter, focusing mainly on Zn, As, Cu, Pb and Hg as the main contaminants. This chapter considers the effect of these metals in soil along with their distribution and environmental threats to the nearby villages and towns.

17.3 SAMPLE COLLECTION AND ANALYSIS

17.3.1 LOCATION

The Suxian district in China was selected as the location for the carrying out of the impact of heavy metal pollution in soil. Located in the southern hemisphere of Hunan province, the district is characterized by monsoon humidity with heavy rainfalls. The average rainfall is around 1470 mm and the annual average temperature remains somewhere between 17-18 °C.

The district is also characterized by the large-scale availability of minerals such as the ShiZhuyuan metal, Qiao Kou Pb-Zn ore, Ma Nao Mn ore, and Xu Jiadong, Jie Dong and Qi Fengdu coal mines. The region also boasts of around 50–51 different varieties of non-ferrous metals, mainly Lead (Pb), Zinc (Zn), Tin (Sn), Bismuth (Bi) and Molybdenum (Mo). It also has around 18 deposits, of which Tang Xi, Da Kuishang, Ao Shang and Bai Lutang are the largest. The Shizuyuan mine is said to be the largest polymetallic mine, with around 143 different minerals.

17.3.2 SAMPLE COLLECTION AND ANALYSIS

The samples were collected in 2015 for a time period from July to September. The obtained soil samples were then analyzed and tested as per the following procedure described by Bao.

An intermediary area was established based on the location of the mining area and soil samples were collected from the surface layer (0–20 cm) from different points.

For each 9 m² area around 5–6 soil samples were collected. 100 g of the soil samples was taken out and mixed properly before being packed and labelled with date, type of soil, location etc.

The present case study focused on accumulating around 167 soil samples with different contents of Zn, Pb, As, Cu and Hg and testing them as per the GB 15618–1995 method.

0.1–0.2 g of dried and powdered soil samples were taken in a digestion tank and cooked with $HCl-HNO_3-HClO_4-HF$.

As, Hg analysis was carried out using atomic fluorescence spectroscopy (AFS), Cu, Pb, Zn were analyzed using inductively coupled plasma atomic emission spectroscopy (ICP-OES). pH detection was done using a pH meter.

17.3.3 DATA SOURCE AND PROCESSING

Geographic details of the township must be collected, including maps of the township, village map of Suxian district, mining areas of Suxian etc. All these data were collected from Resources and Environmental Science Data Center of the Chinese Academy of Sciences. The exact location of the sampling points was confirmed by the use of GPS technology. Tools such as Microsoft Excel 2010, IBM SPSS Statistics 19, and ArcGIS10.1 (ESRI Inc.) were used to process the samples and geographic data.

17.4 RESEARCH SURVEY

17.4.1 DATA PROCESSING

The mining area, river locations, and soil sampling areas were determined by a digital map acquired from the China resource satellite application center along with the aid of GPS.

17.4.2 POTENTIAL ECOLOGICAL RISK INDEX METHOD

The quantitative and qualitative determination of heavy metal pollution in the mining areas of Suxian district in China was carried out using the Hakanson analysis method. The method took into consideration the importance of toxicity to explore the potential risk posed by heavy metal pollution in the Suxian region (Hakanson, 1980),

$$E_q^p = T^p \times C_q^p = T^p \times \frac{C_M}{C_R} \tag{17.1}$$

The comprehensive potential ecological risk index RI_q is calculated as follows:

$$RI_q = \sum_{p=1}^{n} E_q^p \tag{17.2}$$

where E_q^p is the index of a single potential ecological risk of heavy metal p at sampling point q; RI_p is the index of comprehensive potential ecological risk at sampling point q; T_p is the toxic response coefficient of heavy metal p (As = 10 Cu = 5Hg = 40 Pb = 5 Zn = 1) (Suresh et al., 2012; Feng et al., 2017). C_q^p is the pollution coefficient of heavy metal p at sampling point q; C_M is the measured value of heavy metal p at sampling point q; and C_R is the reference content of heavy metal p.

The Hakanson method mainly considers the coefficient of toxicity being determined by eight main types of pollutants/heavy metals. The method of calculation for the toxic coefficient is described in detail by Hakanson, Zheng and Shi and Fernandez and Carballeira. For the present case study, five pollutants were considered; hence the results obtained by the Hakanson method might show a slight deviation from the normal (Hakanson, 1980; Fernández and Carballeira, 2001). Most research published to date does not take into account the changes that have to be incorporated if the number of pollutants is not as per the prescribed method (Wei, 2010; Hu et al., 2013). The changes and the adjustments made during such analysis still remain unclear and unverified (Qing and Shu, 2008).

Keeping all these bottlenecks in mind, the present paper tries to address all the possible shortfalls existing with different evaluation methods for determining heavy metal pollution. Evaluation methods have been modified based on the type of heavy metal present (Li, 2016; Jian-Hua et al., 2011; Fernández and Carballeira, 2001). The terms E_p and RI_p in the Hakanson method is mainly calculated considering the presence of eight pollutants, this adjustment on E_p is done by introducing a first boundary value

Where E_i = non-production index (C = 1) × the maximum toxicity coefficient considering the number of involved heavy metals

The calculations result in grade values at least twice larger than the last. The value of RI_p is adjusted as follows:

RI = (Hakanson 1st grade boundary value/total toxic coefficient of 8 pollutants) = 150/133 = 1.13

Considering the case of the most toxic metal in this case study, mercury gives a toxic coefficient of 40. The total toxic contents for Hg, As, Zn, Pb and Cu are found to be 61. The RI_p value is (61×1.13) ≈ 70. A detailed comparison of E and RI (Hankanson and current research) is shown in Table 17.1.

TABLE 17.1

Comparison of Hakanson Classification Standard in this Study in E and RI Grading Standards

E_i	Hakanson method	<40	40-80	80-160	160-320	≥320
	Present research	<40	40-80	80-160	160-320	≥320
RI_i	Hakanson method	<150	150-300	300-600	≥600	-
	Present research	<70	70-140	140-280	280-560	-≥560
Ecological risk level		Slight	Moderate	Strong	Quite strong	Extremely strong

(Data obtained with permission from Chen et al. Copyright © 2018, Elsevier)

17.4.3 IDW Interpolation of Heavy Metals in Soil

The Inverse distance weighted (IDW) is an interpolation method which is based on a concept that objects in close proximity to one another have similar properties with regard to details such as sampling points, characteristics of data, and so on (Spokas et al., 2003; Gong et al., 2014; Ferguson, 1996). The aim is to consider the distance between the sample point and the interpolation point as the weighted value which is then averaged. Closer points (sample and interpolation) would give larger weighted value compared to points situated far from each other. The formula for IDW method considering a series of points represented by X_i, Y_i, Z_i ($i = 1, 2, \ldots\ldots, n$) distributed on the same plane, then, acquiring Z value through weight value.

$$Z(m) = \sum_{p=1}^{n} w_p z_p \bigg/ \sum_{p=1}^{n} w_p \qquad (17.3)$$

$$w_p = d_p^{-u} \qquad (17.4)$$

where $Z(m)$ is an interpolated predicted value, Z_p is the available point, n is the total number of available known points used in interpolation, d_p is the distance between predicted value and the point p and w_p is the assigned weightage to point p. With an increase in distance between predicted values and known value, the weightage decreases (Shepard, 1968), and u is the weighing power showing that decrease.

17.5 DETAILED ANALYSIS OF RESULTS

This section deals with the in-depth analysis of the soil such as their heavy metal characteristic value, the heavy metal soil pollution, and their potential ecological risk assessment.

17.5.1 Soil Heavy Metal Pollution

17.5.1.1 Characteristic Value Analysis of Heavy Metals in Soil

Table 17.2 shows the statistical data of the five heavy metals focused in the current research responsible for soil pollution in the Suxian region mainly by the determination of certain indices and parameters like the Nemero index, the single factor index, the coefficient of variation, and the average metal content. Once these parameters have been determined, the next step is to determine the soil pollution limiting values as shown in Table 17.2. The standard limits for As, Zn and Hg are 40 mg/g, 200 mg/g and 0.3 mg/g, respectively. Analysis studies suggested that As and Zn had already exceeded the prescribed limit with values as high as 78 and 203 mg/g, respectively whereas the mercury content more or less remained the same, suggesting that high pollution risk may exist due to the presence of As and Zn in more than prescribed safety limits. As mainly stays in soil in inorganic form, Pb and Zn are found in non-residual states, whereas Cu is found as a sulfophilic element (Huang et al., 2012; Fang, 2016). The

TABLE 17.2
Pollution of Soil with Heavy Metals

	Mean Value (mg/Kg)	Standard Deviation	Scale (mg/kg)	Skewness	Kurtosis	Co-efficient of Variation	Background Value (mg/kg)	Two Levels of Environment Quality Standard (mg/kg)	Index of Single Factor	The Nemerow Index
As	78.70	98.70	1.31-547.17	2.61	6.92	125.00	14.00	40.00	5.62	11.60
Cu	37.96	54.36	3.01-611.46	7.72	78.11	143.00	26.00	150.00	10.41	
Hg	0.31	0.24	0.02-1.28	0.71	0.70	77.00	0.096	0.30	1.46	
Pb	160.19	265.33	0.29-2566.78	5.42	42.71	166.00	27.00	250.00	5.93	
Zn	202.79	201.56	22.02-1686.8	3.63	20.22	99.00	94.40	200.00	2.14	

(Data obtained with permission from Chen et al. Copyright © 2018, Elsevier)

single factor index method, which basically shows the pattern by which heavy metals get deposited in the soil, can be seen in detail, as described by Guo. The pattern for the considered heavy metals is as follows Hg> Pb> As> Zn> Cu, all having a value greater than one. Hg shows the highest accumulation in soil with a single factor index value of 10.41. The Nemero index, on the other hand, focuses on the effect of the heavy metals with very high concentrations impacting the quality of the soil (Kowalska et al., 2016; Yang et al., 2014). As per the Nemero index, a value less than one is considered to be non-polluting in nature, metals having values between 1 and 2 pose slight caution with regard to heavy metal polluting the soil and for metals having values greater than 3 are considered to be high-risk metals causing soil pollution. An analysis of the Nemero index in the present case study suggested a value of 12.1 which definitely posed a threat in terms of polluting the soil. Coefficient of variation is also calculated which helps in determining the variation of metals at each sampling point (Table 17.1) (Manta et al., 2002). A value greater than 1 suggests the high influence of human activity on the metal pollution of soil. The metals considered in the present case study follow the pattern as described Pb> Cu> As> Zn> Hg. The Suxian district, being highly rich in minerals, is widely exploited by human activities such as mining, which brings about huge imbalances in the environment (Saleh, 2015b). Results suggest that a very high Single index factor and the Nemero index factor are evidence for a strong soil pollution scenario and that the cause is extensive human activity.

17.5.1.2 Heavy Metal Pollution in the Case Study Area

The heavy metal content in the soil is basically analyzed using the IDW interpolation in ArcGIS (Figure 17.1). The area in which mining activities are conducted is marked as red. Mercury has been found to have a very high accumulation rate compared to the other four metals. This accumulation is basically found at the point of intersection of three towns: Tangxi, Bailu and Da Kuishan. Smaller regions had a high distribution of arsenic, lead and zinc, whereas the mercury content could be found mainly in the regions of Qi Fengdu, Wu Lipai, Ma Touling, Qiaokou, Aoshang and Liangtian with extremely high values in the Shi Zhuyuan–Manaoshan area. Smaller areas such as Qi Fengdu may also be influenced by the presence of mercury, but relatively less in comparison to the above-mentioned areas. Heavy metal accumulation is as a result of various human activities, including the processing of minerals and transportation, especially in mining areas. The Suxian mining industry saw a wide-scale development during the early 1980s; however, these developments were mainly substandard. As a result, large numbers of bottlenecks were observed, which proved extremely harmful for the environment at large. The riverbeds were polluted due to the wastewater from the mining industry finding its way to various rivers; moreover, during monsoons the rain would also wash away the accumulated heavy metals into farmlands, thereby destroying crops in the neighboring area. The wastewater and the waste ores were found to be the main source of all these causes. Thus the junction points of various towns in the township were mainly found to be the primary location for heavy metal soil pollution.

17.5.2 Ecological Risk Assessment of Heavy Metals

17.5.2.1 Single Factor Ecological Risk Assessment of Heavy Metals in Soil

The various ecological risks and impacts of the heavy metals considered for the case study was carried out using the interpolation method known as the ArcGIS

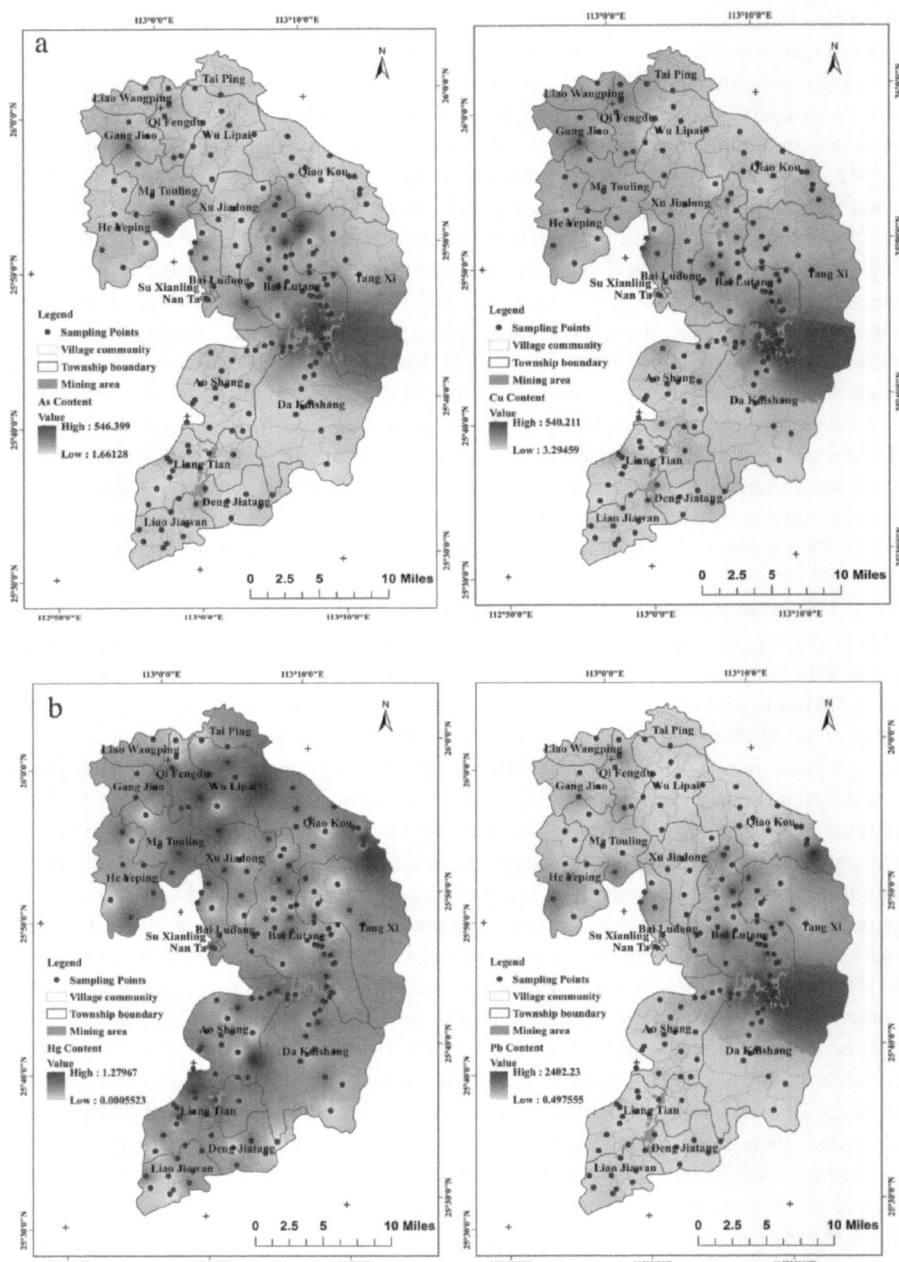

FIGURE 17.1 (a) The distribution of As (left) and Cu (right) contents in soil. (b) The distribution of Hg (left) and Pb (right) contents in soil. (c) The distribution of Zn contents in soil(Reproduced with permission from Chen et al. Copyright © 2018, Elsevier).

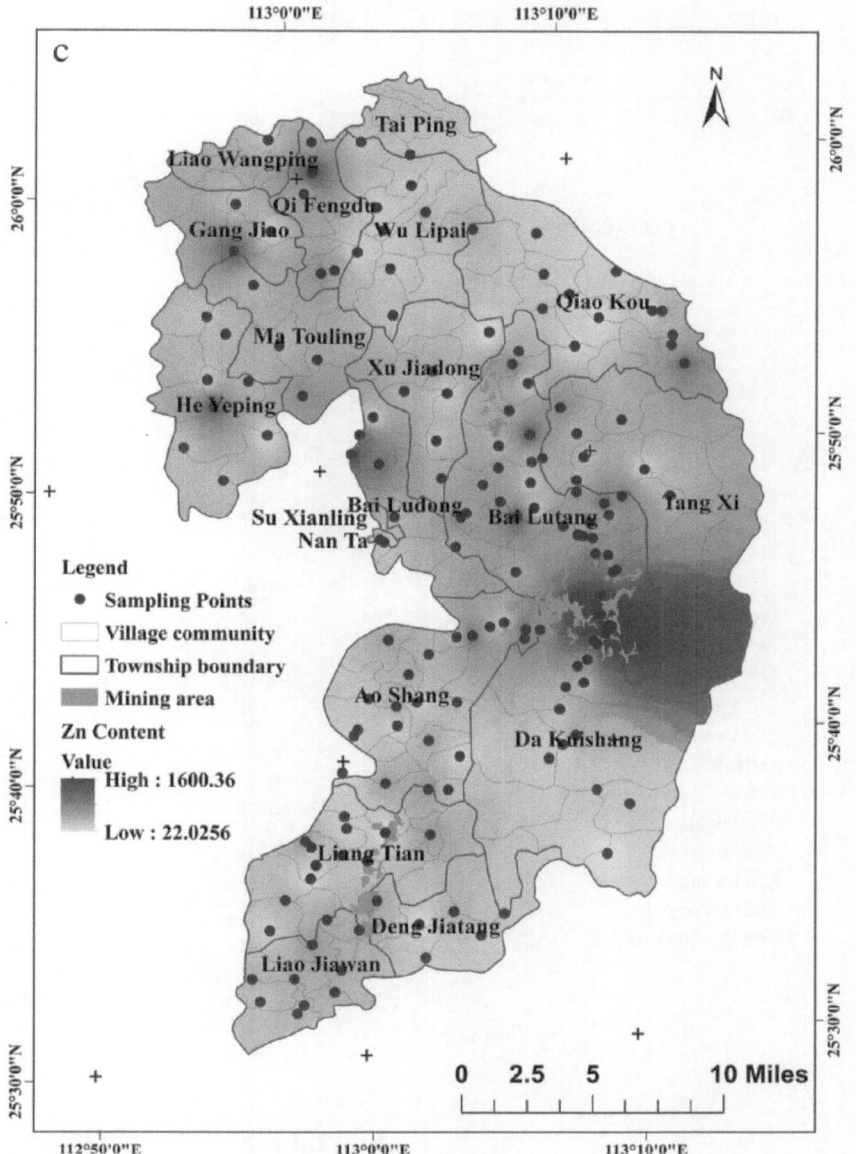

FIGURE 17.1 (*Continued*). (a) The distribution of As (left) and Cu (right) contents in soil. (b) The distribution of Hg (left) and Pb (right) contents in soil. (c) The distribution of Zn contents in soil(Reproduced with permission from Chen et al. Copyright © 2018, Elsevier).

technique. Figure 17.2 well presents the different results of the risk assessment for the study. Mercury was found to pose a profound risk on the townships of the Suxian district, whereas metals such as zinc and copper posed medium-level threats. High-risk level areas which show the Hg pollution is represented by a deep colour. A pollution risk of level 3 for As and Pb has been observed in townships

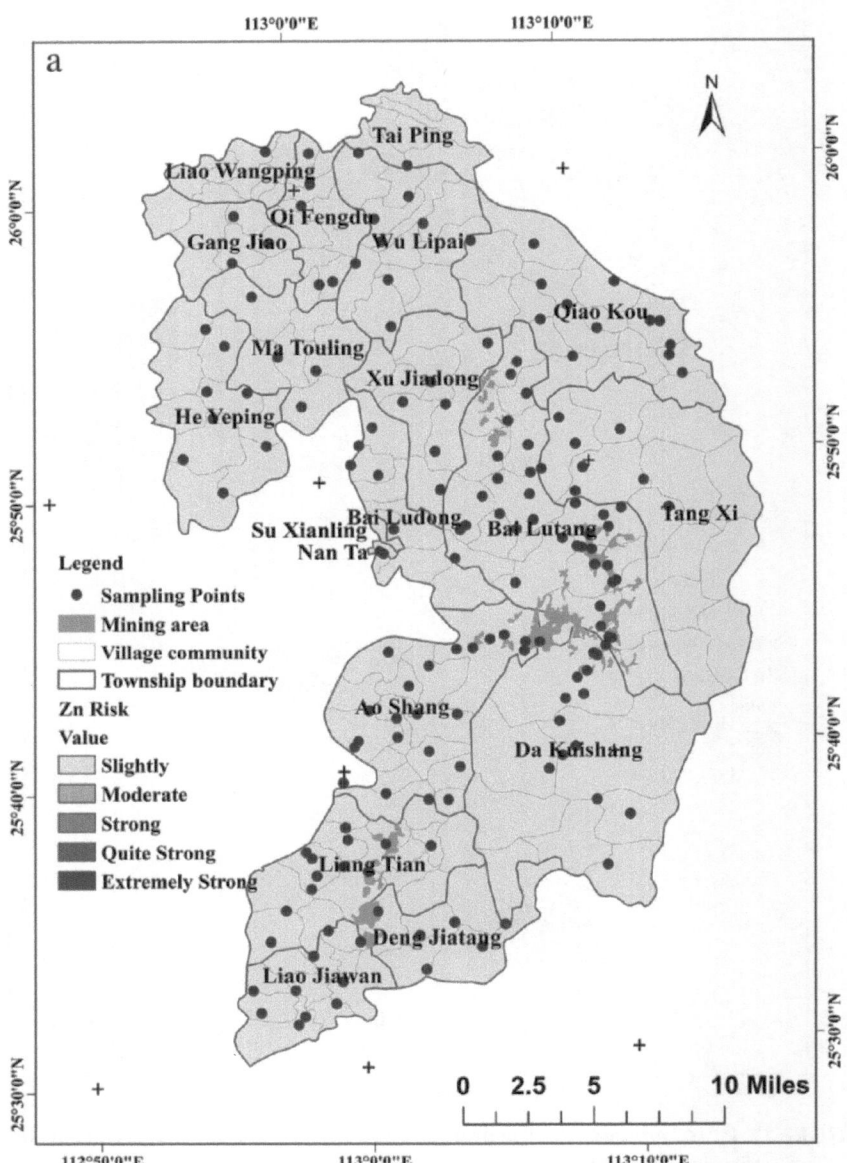

FIGURE 17.2   (a) Single-factor ecological risk index of Zn in soil. (b) Single-factor ecological risk index of As (left) and Cu (right) in soil. (c) Single-factor ecological risk index of Hg (left) and Pb (right) in soil(Reproduced with permission from Chen et al. Copyright © 2018, Elsevier).

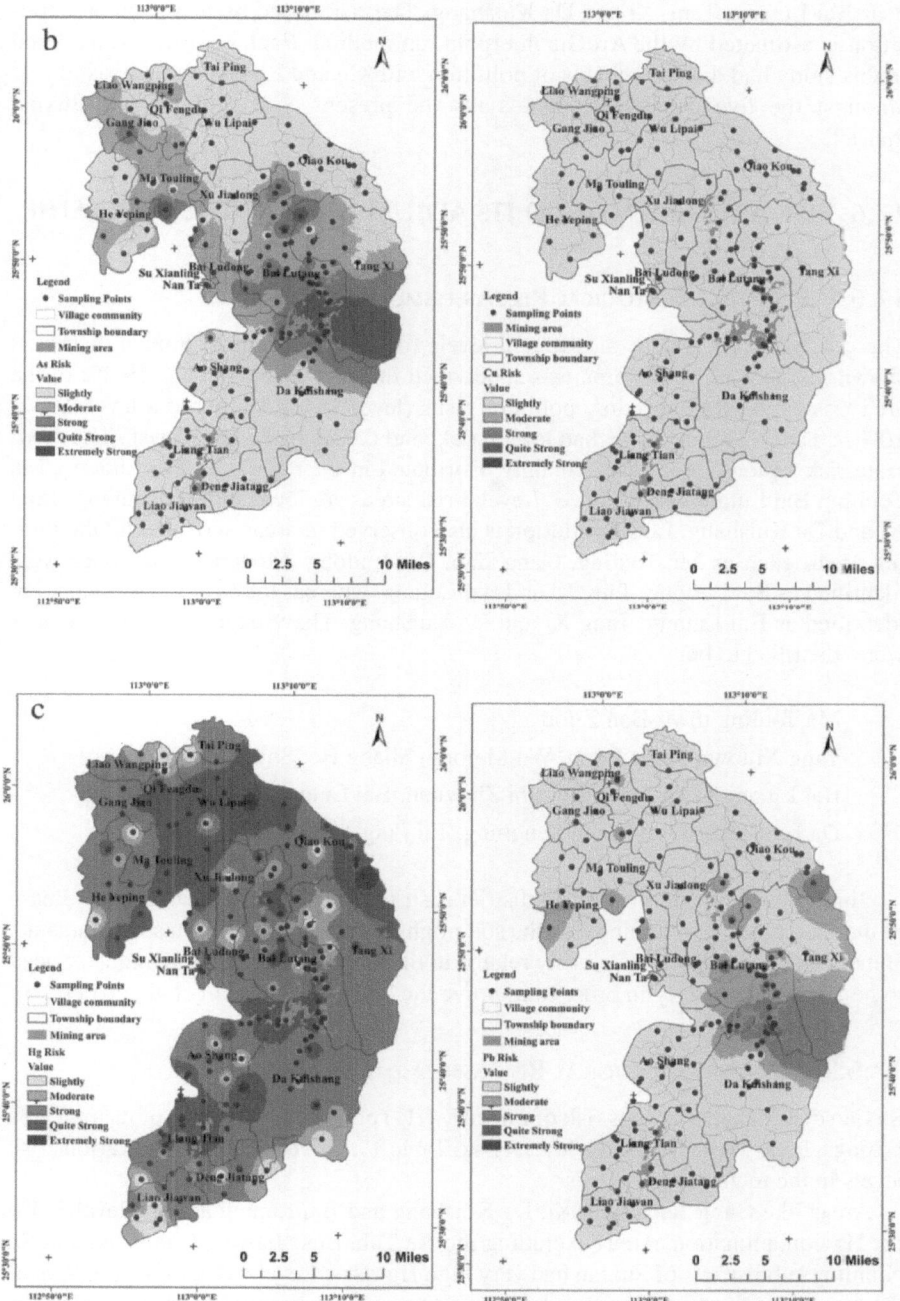

FIGURE 17.2 (*Continued*). (a) Single-factor ecological risk index of Zn in soil. (b) Single-factor ecological risk index of As (left) and Cu (right) in soil. (c) Single-factor ecological risk index of Hg (left) and Pb (right) in soil(Reproduced with permission from Chen et al. Copyright © 2018, Elsevier).

near Bai Lutang, Tang Xi and Da Kuishang. The risk posed by heavy metal deposition is estimated by the ArcGis interpolation method. Each township considered in this study had different levels of pollution with Cu and Zn posing the least threat amongst the five considered metals for the present case study in the Suxian district.

17.6 RISK ASSESSMENT AND ITS ADJUSTMENT AT THE TOWNSHIP SCALE

17.6.1 POTENTIAL ECOLOGICAL RISK ASSESSMENT OF ARSENIC

The risk posed by arsenic at different levels for the townships in Suxian district is shown in Table 17.3. The area was subdivided into five levels of risk: 59.4% of the area was exposed to very low pollution risks (level 1); 24.5% posed a level 2 risk; 10% had a level 3 risk; 6.1% had level 4 risk; and 0.02% had a level 5 risk. The moderate-risk regions (level 2) are mainly distributed in the regions of Bai Ludong, Ma Touling, Bai Lutang and Tang Xi. Level 3 risk areas are located in Bai Lutang, Tang Xi and Da Kuishang. Local pollution is also observed in areas surrounding the mining areas, such as Ma Touling, Gang Jiao, Bai Ludong. Moderate-risk zones were identified as Bai Ludong, Tang Xi and Bai Lutang, whereas the strong risk areas were identified as Bai Lutang, Tang Xi and Da Kuishang. The villages with level 4 risks were identified to be:

  Ma Touling town-Ban Zilou

  Tang Xi town-Heng Long, Wu Malong, Shang He, Shi Hu

  Bai Lutang town-Dong Bo, Shi Zhuyuan, Bai Lutang, Xiang Shan-ping

  Da Kuishang town-Liang Sangoing, Tai Pingtou.

Since a large area in the Suxian district is surrounded by mining areas, the release of the minerals due to mining into the soil might result in pollution caused by arsenic. Hence, locations which are closely related to the areas near the mining industry need to be analyzed properly in order to improve the soil quality (Saleh et al., 2011).

17.6.2 POTENTIAL ECOLOGICAL RISK ASSESSMENT OF MERCURY

Suxian district had a serious risk of mercury (Hg) pollution, with 56% of its total area posing a level 3 risk and 26% a level 4 risk. Table 17.4 gives details of the various risk levels in the townships.

Areas like Gang Jiao, Tang Xi, Da Kuishang and Bai Lutang are at a level 3 risk for Hg contamination. Areas excluding the Na Tub, Liao Jiawan, Gang Jiao and Su Xianling subdistricts of Suxian had very high Hg risks. Level 4 risk exists for 40% of the total areas for Ao Shang, Wu Lipai and Qiao Kou, whereas level 5 risk exists for nearly 85% of the total areas for Wu Lipai, Liang Tian and Aiao Kou. Hg leakage has been one of the major causes for such a high rise in risk level of different areas in Suxian (Saleh, 2015a). Nearly 50% of its area is at either moderate or very high risk

TABLE 17.3

Potential Ecological Risk Status of As in Each Village/town

Township	Level 1 (Slight)		Level 2 (Moderate)		Level 3 (Strong)		Level 4 (Quite Strong)		Level 5 (Extreme Strong)	
	Area, m²	%	Area, m²	%	Area, m²	%	Area, m²	%	Area, m²	%
Liao Wangping	1612.30	2.28	582.81	2	346.13	2.89				
Gang Jiao	996.98	1.41	2348.97	8.06						
Qi Fengdu	2002.84	2.83	1242.53	4.27						
Tai Ping	2588.60	3.66	2.96	0.01						
Wu Lipai	6822.07	9.65	106.50	0.37						
Ma Touling	550.26	0.78	3579.66	12.29	562.10	4.70	349.09	4.8		
He Yeping	4819.23	6.82	2218.80	7.62	159.76	1.33				
Xu Jiadong	4413.93	6.24	1136.03	3.90	5.92	0.05				
Qiao Kou	8987.62	12.71	745.52	2.56	174.55	1.46				
Bai Ludong	251.46	0.36	3121.11	10.71	739.60	6.18				
Bai Lutang	402.34	0.57	4819.23	16.54	4848.82	40.50	3121.1	42	29.5	1
Tang Xi	2597.48	3.67	5262.99	18.06	2479.14	20.70	3242.4	44		
Su Xianling	165.67	0.23								
Nan Ta	109.46	0.15								
Ao Shang	7567.59	10.70	1251.40	4.29	325.42	2.72	130.17	1.7		
Liang Tian	10700.5	15.13	2387.43	8.19	2322.34	19.39	417.13	5.7		
Da Kuishang	7303.25	10.33	112.42	0.39						
Deng Jiatang	4523.39	6.40	82.84	0.28						
Liao Jiawan	4271.93	6.04	124.25	0.43						
Total	70690.9	59.37	29125.4	24.46	11963.7	10.05	7259.9	6.1	29.5	0.02

Note: % (except the "Total" column) represents the proportion of the area for a certain risk level to the total area of the same risk level; the % in the "Total" column represents the proportion of area for a certain risk level to the total area in research zone.
(Data obtained with permission from Chen et al. Copyright © 2018, Elsevier)

TABLE 17.4
Potential Ecological Risk Status of Hg in Each Village/town

Township	Level 1 (Slight) Area, m²	%	Level 2 (Moderate) Area, m²	%	Level 3 (Strong) Area, m²	%	Level 4 (Quite Strong) Area, m²	%	Level 5 (Extreme Strong) Area, m²	%
Liao Wangping	38.46	0.75	641.98	4.42	1514.7	2.26				
Gang Jiao	38.46	0.75	3121.11	21.4	532.51	0.80				
Qi Fengdu	411.22	8.01	295.84	2.04	502.93	0.75	2032.4	6.4	2.9	0.28
Tai Ping	88.75	1.73	147.92	1.02	1849	2.76	505.89	1.6		
Wu Lipai	201.77	3.92	322.47	2.22	1118.28	1.67	5135.7	16	150	14.06
Ma Touling	153.84	3	269.21	1.85	2141.88	3.20	2476.1	7.8		
He Yeping	207.09	4.03	491.09	3.38	4848.82	7.25	1650.7	5.2		
Xu Jiadong	224.82	4.38	597.60	4.11	3159.56	4.72	1573.8	5		
Qiao Kou	124.25	2.42	301.76	2.08	4132.88	6.18	4644.6	14	704	65.38
Bai Ludong	106.50	2.07	307.67	2.12	3124.07	4.67	573.9	1.8		
Bai Lutang	372.76	7.25	1529.49	10.5	8946.20	13.37	2372.6	7.5		
Tang Xi	461.51	8.98	1470.32	10.1	9694.68	14.49	1955.5	6.2		
Su Xianling	165.67	3.22								
Nan Ta	2.96	0.06	2.96	0.02	53.25	0.08	50.29	0.1		
Ao Shang	227.80	4.43	541.39	3.72	4946.44	7.39	3396.2	10	162	15.02
Liang Tian	204.13	3.97	431.93	2.97	4067.80	6.08	2665.5	8.4	50.2	4.64
Da Kuishang	606.47	11.79	1922.96	13.2	11026	16.48	2272	7.2		
Deng Jiatang	458.55	8.91	1067.98	7.35	2973	4.44	106.5	0.3		
Liao Jiawan	1035.44	20.11	1059.11	7.28	2245.43	3.36	56.21	0.1		
Total	5129.86	4.31	14522.7	12.2	66877.6	56.17	31468	26	1070	0.9

Note: % (except the "Total" column) represents the proportion of the area for a certain risk level to the total area of the same risk level; the % in the "Total" column represents the proportion of area for a certain risk level to the total area in research zone.
(Data obtained with permission from Chen et al. Copyright © 2018, Elsevier)

levels of heavy metal pollution of soil and thus extensive testing must be carried out to curb the occurrence of pollution by Hg and similar heavy metals and to regulate the transportation and exploitation activities near the metal mining areas.

Studies were also performed in order have secure an insight into the pollution scenario caused by the lead (Pb). The risk level for Pb pollution was identical to that of arsenic, i.e. level 3. The majority of the area had lower Pb pollution risk levels. The level 3 risk of pollution was profound in Tang Xi-Wu, mainly Ba Lutang, Tang Xi and Da Kuishang. By comparison, level 4 and level 5 risks were found in Bai Lutang and Da Kuishang, respectively, which are relatively closer to the main mining areas and other activities such as smelting, ore dressing, etc. (Table 17.5). After the analysis of the township areas, the analysis was further continued into villages to identify the potential risks of Pb pollution. Areas like Heng Long, Wu Malong, Shang He, Dong Bo, Shi Zhuyuan, Bai Lutang, Xiang Shanping, Liang San-ping, Tai Pingtou, Bao Anling and BaiXi were identified as areas with considerable risk of such pollution.

17.6.3 COMPREHENSIVE ECOLOGICAL RISK EVALUATION OF HEAVY METALS IN SOIL

Figure 17.3 gives detail of the risk levels of the considered metals in Suxian district with majority of the areas having a level 3 or above risk. Areas near the mining site of Shi Zhuyuan, such as Tang Xi, Da Kuishang and Bai Lutang, pose a level 4 threat. The meeting junction of all these three townships had a level 5 risk. Table 17.6 shows the area of each township that is threatened by heavy metal pollution. Strongly polluted areas are mainly observed for Ma Touling, Bai Lutang, Da Kuishing, Qiao Kou, Tang Xi and Ao Shang, all of which have a threat of level 4 or more. The main reason for such high-risk pollution scenarios in these areas is because of the fact that large activities occur in the mining area, such as wastewater discharge, leaching, the tailing of heavy metals, other domestic and agricultural activities (Gupta et al., 2012; Saleh, 2015c). It has thus been suggested that the control of such damaging activities should be given immediate attention and control measures should be implemented, mainly in the Bai Lutang and the Tang Xi areas.

17.7 RESULTS AND DISCUSSION

17.7.1 HEAVY METALS IN SOIL BASED ON TOWNSHIP SCALE

The soil around the mining areas is prone to large-scale heavy metal pollution risk. Detailed study of the pollution causing metals still remains a constraint due to shortfalls in the analysis methods and availability of different standards for establishing the specific pollution areas. Specializing in the recognition of small-scale regions for risks in the heavy metal pollution area. To determine the different levels of pollution in the Suxian district, one needs to consider the district into divisions of townships. Such division would help determine the entire condition of soil pollution by heavy metal pollution near the mining areas.

TABLE 17.5
Potential Ecological Risk Status of Hg in Each Village/town

Township	Level 1 (Slight) Area, m²	%	Level 2 (Moderate) Area, m²	%	Level 3 (Strong) Area, m²	%	Level 4 (Quite Strong) Area, m²	%	Level 5 (Extreme Strong) Area, m²	%
LiaoWangping	2195.1	2.30								
Gang Jiao	3635.87	3.81	56.21	0.39						
Qi Fengdu	2890.36	3.03	355.01	2.49						
Tai Ping	2591.56	2.71								
Wu Lipai	6928.57	7.25								
Ma Touling	4526.35	4.74	514.76	3.61						
He Yeping	5943.43	6.22	1254.36	8.79						
Xu Jiadong	5555.88	5.82								
Qiao Kou	8600.07	9.00	639.02	4.48	656.77	7.25	11.83	5.9		
Bai Ludong	3609.25	3.78	502.93	3.52						
Bai Lutang	5783.67	6.05	4337.13	30.4	3008.6	33.20	82.84	41	8.88	33.33
Tang Xi	6431.56	6.73	4115.13	28.8	3035.3	33.49				
Su Xianling	165.67	0.17								
Nan Ta	109.46	0.11								
Ao Shang	8517.23	8.92	541.39	3.79	215.96	2.38				
Liang Tian	7419.67	7.77								
Da Kuishang	11617.6	12.16	1943.67	13.6	2144.8	23.65	103.5	50	17.7	66.67
Deng Jiatang	4606.23	4.82								
Liao Jiawan	4396.18	4.60								
Total	95523.7	80.23	14259.4	11.9	9061.5	7.61	198.21	0.1	26.6	0.02

Note: % (except the "Total" column) represents the proportion of the area for a certain risk level to the total area of the same risk level; the % in the "Total" column represents the proportion of area for a certain risk level to the total area in research zone.

(Data obtained with permission from Chen et al. Copyright © 2018, Elsevier)

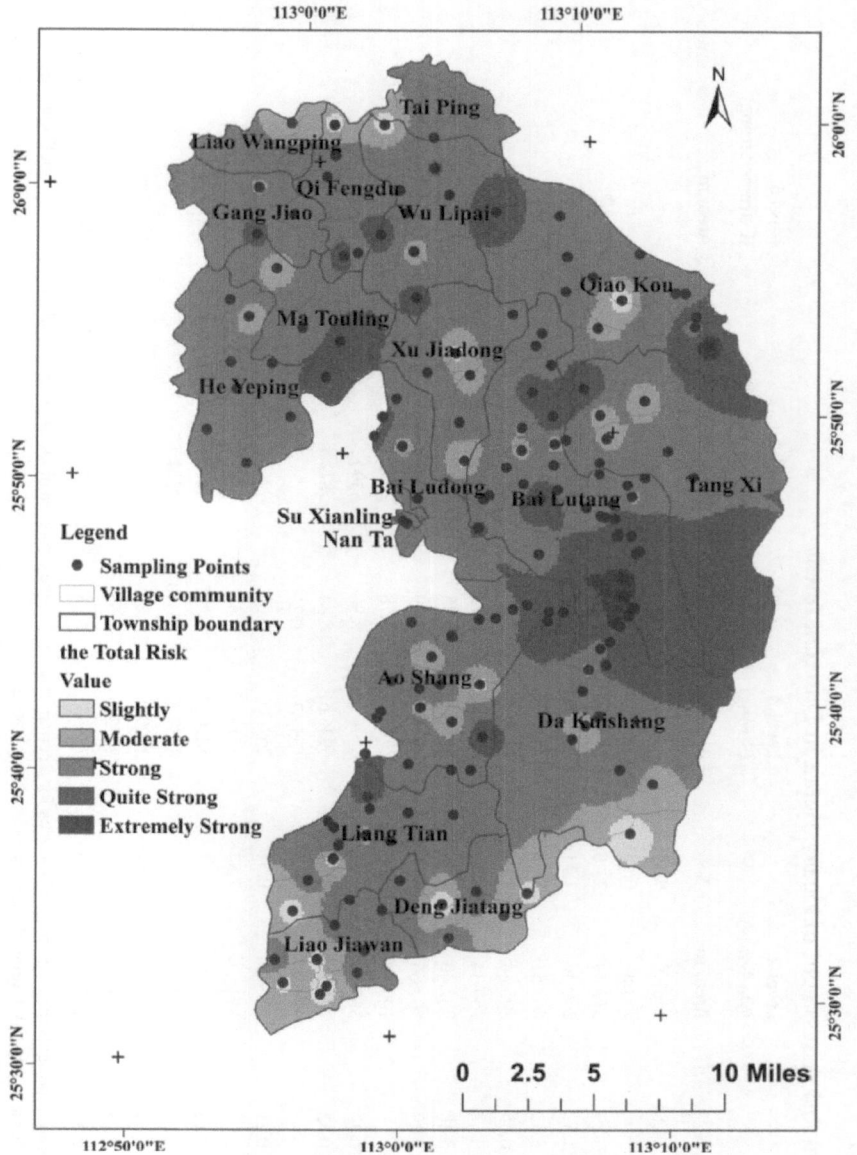

FIGURE 17.3 Total potential ecological risk index of the Suxian District (revision evaluation criteria) (Reproduced with permission from Chen et al. Copyright © 2018, Elsevier).

The case study in this chapter would provide a detailed assessment of the statistics at different levels of the township, thereby aiding the proper efficient management of the soil pollution scenario. In order to give an example of the results that might be useful in the control of soil pollution, arsenic is considered as the metal of importance and in order to determine the various ecological risks associated with it the

TABLE 17.6
Total Potential Ecological Risk Assessment of Heavy Metals in Different Towns

Township	Level 1 (Slight)		Level 2 (Moderate)		Level 3 (Strong)		Level 4 (Quite Strong)		Level 5 (Extreme Strong)	
	Area, m²	%	Area, m²	%	Area, m²	%	Area, m²	%	Area, m²	%
LiaoWangping	692.27	24.84	1502.8	9.89	3464.29	4.42	147.92	0.6		
Gang Jiao	91.71	3.29	79.88	0.53	2493.93	3.18	346.13	1.5		
Qi Fengdu	44.38	1.59	313.59	2.06	2286.84	2.92				
Tai Ping	106.50	3.82	260.34	1.71	5322.16	6.78	1038.4	4.7		
Wu Lipai	23.67	0.85	461.51	3.03	3106.95	3.96	1615.2	7.4		
Ma Touling			295.84	1.95	6830.95	8.72	106.5	0.5		
He Yeping	91.71	3.28	260.34	1.71	4372.52	5.58	124.25	0.6		
Xu Jiadong	147.92	5.29	967.40	6.36	6964.07	8.89	1928.8	8.8		
Qiao Kou			707.06	4.65	3748.29	4.78	263.3	1.2	159	17
Bai Ludong	11.83	0.42	100.59	0.66	6896.03	8.80	5390.2	24		
Bai Lutang	8.88	0.32	387.55	2.55	6925.61	8.84	5872.4	26	535	59
Tang Xi			775.10	5.09	165.67	0.21				
Su Xianling					26.63	0.03				
Nan Ta	5.92	0.21	76.92	0.51	6718.53	8.57				
Ao Shang	82.84	2.96	114	7.54	5810.30	7.41	1304.6	5.9		
Liang Tian	201.17	7.18	1020	6.71	8937.33	11.40	387.5	1.8	20.7	2.3
Da Kuishang	633.10	22.59	2772	18.21	2209.92	2.82	3304.5	12	180	19
Deng Jiatang	236.67	8.44	2159	14.19	2073.84	2.65				
Liao Jiawan	408.26	14.56	1914	12.58						
Total	2786.81	2.34	15203	12.77	78353	65.8	21830	18.33	896.39	0.75

Note: % (except the "Total" column) represents the proportion of the area for a certain risk level to the total area of the same risk level; the % in the "Total" column represents the proportion of area for a certain risk level to the total area in research zone.

(Data obtained with permission from Chen et al. Copyright © 2018, Elsevier)

single factor ecological risk level and the arsenic ecological risk condition for each subdivided township has been considered as shown in Figure 17.2 and Table 17.3. Figure 17.2 shows a visual interpretation of the various risk-level areas in the townships with detailed positions and area. A township with very high-risk level consisting of various subdivision of villages is further distributed into

Level 1 village and Level 2 village

Townships with less or moderate risk levels of heavy metal pollution

Townships and villages are classified into different levels for carrying out the precise treatment based on the extent of pollution.

Level 1: Villages with very high risk level of pollution

Level 2: Villages considerably high risk level of pollution and townships with moderate risk levels

Utilizing a smaller scale for identifying the pollution risk level helps in better management of soil pollution problems and also solves high metal pollution risks in small regions.

17.7.2 The Adjustment of the Potential Ecological Risk Assessment Domain

The determination of an index such as Risk index (RI) helps in determining the risk levels of pollutants in various levels in a district or a village. One such method is the Hakanson method, which mainly requires the presence of at least eight heavy metals to carry out the risk index analysis. Adjustments were made in the evaluation method and the evaluation standards to have a more accessible approach and obtain results with good practical viability (Figure 17.3 and Figure 17.4). Analysis suggested that identification of risk areas almost remained the same as the original method, but showing a slight lower risk level when standards were not adjusted. Without adjustments in standard adjustment level 3 was also obtained. Table 17.6 shows the comparative results of the risk levels before and after adjustment of assessment domain. As already mentioned, adjustments made in the standards causes a decrease in the level 1 and level 2 areas by almost 15% and 50%, respectively. However, without adjustments the analyzed risk levels were not at par with the real-life scenario. With adjustments, level 3 was the main risk level for 66% of the area of Suxian district with some regions also predicting level 5 risk hazard which could not be found in the original research analysis. Hence, adjustments must be made in the domains to portray the real-life scenario.

17.7.3 Selection of Spatial Interpolation Methods for Heavy Metals in Soil

A large variety of interpolation methods are available for carrying out the analysis of heavy metal soil pollution such as IDW, Spline and Kriging. There exists no definite conclusion on determining the best method from the three; however, literature research suggest that IDW might be one of the best methods (Ferguson, 1996; Weber and Englund, 1992; Laslett et al., 1987). In addition, a definite standard does not exist and a variation in primary research objectives might not produce accurate results

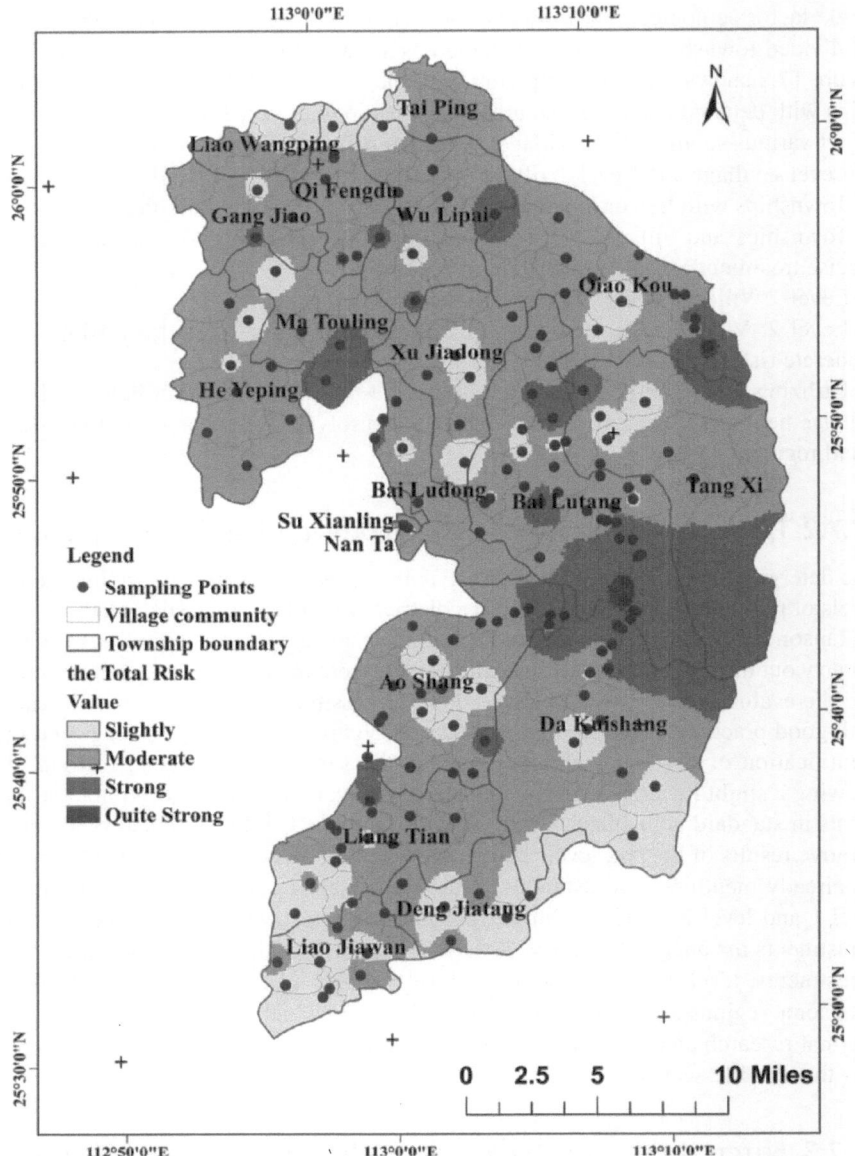

FIGURE 17.4 The Hakanson total potential ecological risk index of Suxian District (Reproduced with permission from Chen et al. Copyright © 2018, Elsevier).

when it comes to the scenario of analysis of risk levels for heavy metal pollution (Mcshane et al., 1997; Dlamini and Chaplot, 2012).

It has also been observed that IDW is mainly suitable for non-normal distributions (Xia, 1968). Table 17.7 shows the K-S results to determine if the heavy metals considered in this chapter follow a normal distribution. Results suggest that bilateral

TABLE 17.7
The Significant Remediation Village and Level for as High Pollution Risk

Pollution Remediation Level	Township	Village Name
Level 2 (quite strong ecological risk)	Tang Xi	Heng Long, Ma Wulong, Shang He, Shi Hu
	Bai Lutang	Dong Po, shi Zhuyuan, Bai Lutang, Xiang
Level 2 (Strong ecological risk)	Ma Touling	Shanping, Jin Tiancun
	Da Kuishang	Ban Zilou
	Tang Xi	Tai Pingtou, Liang Sanping
	Bai Lutang	Xiao Xi, Zhu Dui, Guan Shan
	Bai Ludong	Guan Shandong, Xia Baishui, Ping Tian, Yang Xi
	Gang Jiao	Suo Shiqiao, Long Menchi
		Yun Feng

(Data obtained with permission from Chen et al. Copyright © 2018, Elsevier)

significant values were < 0.1, thereby not meeting the requirements for the tests clearly suggesting a normal distribution.

The interpolation method may also be used with smoothing effect, which is comparably less in case of IDW interpolation. A high degree of smoothing may result in deviation of accuracy of the results. A comparison with other interpolation methods suggested that the IDW method gave much accurate and relevant pollution index, thereby making it more prevalent in heavy metal pollution research studies. Another method, known as the Kriging interpolation method, is also said to give a very high accuracy results, which is accompanied with algorithms. The use of continuous variables would give results closer to true values (Kravchenko and Bullock, 1999; Weber and Englund, 1992). It has also been suggested that a combination of different interpolation methods will improve the precision of results (Dlamini and Chaplot, 2012). The interpolation method has an added advantage in comparison to other techniques of analysis. The old interpolation method, along with the high accuracy surface modelling method, would provide a sufficient advantage for maintaining the accuracy related to the interpolation data. At the same time, it should also be kept in mind that the methods described above are still in the nascent stages and cannot guarantee accurate results for all of the studies. The IDW method thus can be used as a good analysis technique for precise results.

17.8 SUMMARY

A combination of the Hakanson method for risk analysis and ArcGIS technology helps to explain the heavy metal content at different levels of risk. High-risk metal pollution levels were further detected in the township level and the risk levels for different areas and the area ratio for each township were also analyzed. The results and analysis obtained from this chapter is as below:

1. Based on the type of heavy metals present in the soil an extensive study was done using the Hakanson assessment method and the IDW interpolation method. Pollution analysis suggested that areas which showed a higher

content of heavy metals posed higher threats and the data obtained were much more consistent compared to the actual conditions.

2. A detailed study was carried out to identify the areas which poses risk of metal pollution in the township and then locate specifically the areas which have extremely greater risk levels.

3. Mercury was found to show a very high accumulation threats compared to the other four metals. A high concentration of heavy metals was found to exist at the junction of Bai Lutang and Da Kuishang, in Bai Lutang, Tang Xi and to the south of Tang Xi. In addition, a high content of mercury was observed in Ao Shang, Liang Tiang, Ma Touling, Qi Fengdu and Wu Lipai. The risk posed by each metal was basically determined by analyzing the Nemero index and the single factor.

4. Areas of Bai Lutang and Da Kuishang had a high risk of being polluted by arsenic and lead. But the major pollutant in several areas of the Suxain district was found to be mercury, which posed a level 4 threat. In areas of Liang Tian and Qiao Kou it was found to pose a level 5 risk. Suxian region also had other major pollutants like zinc and copper, but these metals posed relatively lower level of threats when compared to mercury. The entire Suxain district as a whole had 83% of its total area under the high-risk region of level 3 and above. Thus, the mining activities posed a very high threat of soil pollution throughout this area.

Note: % (except the "Total" column) represents the proportion of the area for a certain risk level to the total area of the same risk level; the % in the "Total" column represents the proportion of area for a certain risk level to the total area in research zone.

Note: % (except the "Total" column) represents the proportion of the area for a certain risk level to the total area of the same risk level; the % in the "Total" column represents the proportion of area for a certain risk level to the total area in research zone.

REFERENCES

Adio, S.O., Omar, M.H., Asif, M., et al., 2017. Arsenic and selenium removal from water using biosynthesized nanoscale zerovalent iron: a factorial design analysis. *Process Saf. Environ. Prot.*, 107, 518–527

Chabukdhara, M., Nema, A.K., 2013. Heavy metals assessment in urban soil around industrial clusters in Ghaziabad, India: probabilistic health risk approach. *Ecotoxicol. Environ. Saf.* 87(1), 57.

Chen, Y., Jiang, X., Wang, Y., Zhuang, D., 2018. Spatial characteristics of heavy metal pollution and the potential ecological risk of a typical mining area: A case study in China. *Proc. Safety Environ. Prot.* 113, 204–219.

Dlamini, P., Chaplot, V., 2012. On the interpolation of volumetric water content in research catchments. *Phys. Chem. Earth A/B/C* 50–52 (2), 165–174.

Fan, L.Q., Chen, F.H., Fan, Y.L., 2012. Comprehensive assessment of soil environmental quality with improved grey clustering method: a case study of soil heavy metals pollution. *J. Agric. Sci. Appl.* 1 (3), 67–73.

Fang, Z.Q., 2016. *Pollution Characteristics of Heavy Metal in Soil from Lead and Zinc Mine and Its Stabilization Study.* China University of Mining & Technology, Beijing.

Feng, S., Renmei, L., Amjad, Ali, et al., 2017. Spatial distribution and risk assessment of heavy metals in soil near a Pb/Zn smelter in Feng County, China. *Ecotoxicol. Environ. Saf.* 139, 254–262.

Ferguson, R.B., 1996. Comparison of kriging and inverse-distance methods for mapping soil parameters. *Soil Sci. Soc. Am. J.* 60(4), 1237–1247.

Fernández, J.A., Carballeira, A., 2001. Evaluation of contamination, by different elements, in terrestrial mosses. *Arch. Environ. Contam. Toxicol.* 40 (4), 461–468.

Fu, C.A., Guo, J.S., Pan, J., et al., 2009. Potential ecological risk assessment of heavy metal pollution in sediments of the Yangtze River within the Wanzhou section, China. *Biol. Trace Elem. Res.* 129 (1–3), 270–277.

Gong, G., Mattevada, S., O'Bryant, S.E., 2014. Comparison of the accuracy of kriging and IDW interpolations in estimating groundwater arsenic concentrations in Texas. *Environ. Res.* 130 (24), 59–69.

Gupta, V.K., Ali, I., Saleh, T.A., et al., 2012. Chemical treatment technologies for waste-water recycling—an overview. *RSC Adv.* 2 (16), 6380–6388.

Hakanson, L., 1980. An ecological risk index for aquatic pollution control. A sedimentological approach. *Water Res.* 14 (8), 975–1001.

Hu, Y., Liu, X., Bai, J., et al., 2013. Assessing heavy metal pollution in the surface soils of a region that had undergone three decades of intense industrialization and urbanization. *Environ. Sci. Pollut. Res.* 20 (9), 6150.

Huang, X.X., Zhu, X.F., Tang, L., 2012. Pollution characteristics and their comparative study of heavy metals in the gold and iron mine soil of the upstream area of Miyun Reservoir, Beijing, *Acta Sci. Circumst.* 32 (6), 1520–1528.

Jian-Hua, M.A., Wang, X.Y., Hou, Q., et al., 2011. Pollution and potential ecological risk of heavy metals in surface dust on urban kindergartens. *Geogr. Res.* 30 (3), 486–495.

Kowalska, J., Mazurek, R., Gasiorek, M., et al., 2016. Soil pollution indices conditioned by medieval metallurgical activity—a case study from Krakow (Poland). *Environ. Pollut.* 218, 1023–1036.

Kravchenko, A., Bullock, D.G.A., 1999. Comparative study of interpolation methods for mapping soil properties. *Agron. J.* 91 (3), 393–400.

Laslett, G.M., Mcbratney, A.B., Pahl, P.J., et al., 1987. Comparison of several spatial prediction methods for soil pH. *Eur. J. Soil Sci.* 38 (2), 325–341.

Li, L., Cui, J., Liu, J., et al., 2016. Extensive study of potential harmful elements (Ag, As, Hg, Sb, and Se) in surface sediments of the Bohai Sea, China: sources and environmental risks. *Environ. Pollut.* 219, 432.

Li, K., Gu, Y., Li, M., et al., 2017. Spatial analysis, source identification and risk assessment of heavy metals in a coalmining area in Henan, Central China. Int. Biodeterior. Biodegradation, Available online 18 April 2017.

Li, Z., Ma, Z., van der Kuijp, T.J., et al., 2014. A review of soil heavy metal pollution from mines in China: pollution and health risk assessment. *Sci. Total Environ.* 468–469, 843.

Manta, D.S., Angelone, M., Bellanca, A., et al., 2002. Heavy metals in urban soils: a case study from the city of Palermo (Sicily), Italy. *Sci. Total Environ.* 300 (1–3), 229–243.

Mcshane, L.M., Meier, K.L., Wassermann, E.M., 1997. A comparison of spatial prediction techniques for an exploratory analysis of human cortical motor representations. *Stat. Med.* 16 (12), 1337–1355.

Müller, G., 1969. Index of geo accumulation in sediments of the Rhine River. *Geo J.* 2 (3), 109–118.

Peng, H., Liu, Y., Li, J., et al., 2007. An ecological risk assessment for heavy metals of the lead-zinc ore tailings. *Chin. Geog. Sci.* 3 (2), 217–224.

Qing, X., Shu, H.L., 2008. Heavy metal pollution of surface soil and its evaluation of potential ecological risk: a case study of different functional areas in Baotou City. *J. Nat. Disasters* 17(6), 6–12.

Saleh, T.A., 2015a. Mercury sorption by silica/carbon nanotubes and silica/activated carbon: a comparison study. *Aqua* 64 (8),892–903.

Saleh, T.A., 2015b. Isotherm, kinetic, and thermodynamic studies on Hg(II) adsorption from aqueous solution by silica-multiwall carbon nanotubes. *Environ. Sci. Pollut. Res.* 22 (21), 16721–16731.

Saleh, T.A., 2015c. Applying Nanotechnology to the Desulfurization Process in Petroleum Engineering. IGI, ISBN-13: 978-1466695450.

Saleh, T.A., Agarwal, S., Gupta, V.K., 2011. Synthesis of MWCNT/MnO$_2$, and their application for simultaneous oxidation of arsenite and sorption of arsenate. *Appl. Catal. B Environ.* 106 (1–2), 46–53.

Sani, H.A., Ahmad, M.B., Hussein, M.Z., et al., 2017.Nanocomposite of ZnO with montmorillonite for removal of lead and copper ions from aqueous solutions. *Process Saf. Environ. Prot.* 109, 97–105.

Shepard, D., 1968. A two-dimensional interpolation function for irregularly-spaced data. In: ACM National Conference, ACM, pp. 517–524.

Spokas, K., Graff, C., Morcet, M., et al., 2003. Implications of the spatial variability of landfill emission rates on geospatial analyses. *Waste Manage.* 23 (7), 599–607.

Sun, C., Bi, C., Chen, Z., et al., 2010. Assessment on environmental quality of heavy metals in agricultural soils of Chongming Island, Shanghai City. *J. Geogr. Sci.* 20 (1), 135–147.

Suresh, G., Sutharsan, P., Ramasamy, V., et al., 2012. Assessment of spatial distribution and potential ecological risk of the heavy metals in relation to granulometric contents of Veeranam lake sediments, India. *Ecotoxicol. Environ. Saf.* 84 (10), 117–124.

Weber, D.D., Englund, E.J., 1992. Evaluation and comparison of spatial interpolators II. *Math. Geol.* 24 (4), 381–391.

Wuana, R.A., Okieimen, F.E., 2011. Heavy metals in contaminated soils: a review of sources, chemistry, risks and best available strategies for remediation. *ISRN Ecol.* 2011, 2090–4614.

Yang, G., Shao, C., Ju, M., 2014. Heavy metal contamination assessment and partition for industrial and mining gathering areas. *Int. J. Environ. Res. Public Health* 11 (7), 7286–7303.

Zang, F., Wang, S., Nan, Z., et al., 2017. Accumulation, spatio-temporal distribution, and risk assessment of heavy metals in the soil-corn system around a polymetallic mining area from the Loess Plateau, northwest China. *Geoderma* 305, 188–196.

Zhao, H., Xia, B., Fan, C., et al., 2012. Human health risk from soil heavy metal contamination under different land uses near Dabaoshan Mine, Southern China. Sci. *Total Environ.* 417–418 (7385), 45–54.

18 Evaluation of Human Factors Risk and Management in Process Safety in Engineering

18.1 INTRODUCTION

With the growth of the economy and the development of science and technology, there has actually been a significant decrease in the number of deaths and accidents. However, the safety situation is still grim. In an analysis of accident-induced factors, human factors led to the vast majority of accidents. This chapter through investigation found that some 70%–80% of aviation accidents, 60% of petrochemical accidents, 90% of iron and steel metallurgy accidents, and 90% of road traffic accidents are caused by human factors (Cai et al., 2008). Therefore, human factors are the main factors leading to accidents. Many scholars have applied a wide variety of research methods to analyse and study human factors resulting in accidents. Early human factors are confined to the individual's perspective. Farmer and Chamber introduced the concept of the accident-prone tendency, suggesting that some people have personal traits such as clumsiness or carelessness that easily lead to accidents (Tarlor et al., 2001). In 1931, Heinrich's book *Industrial Accident Prevention* he expounded his accident causation theory. Heinrich suggested that human ancestry and the social environment caused people to make errors. A mistake by a person can lead to an unsafe action, which leads to the occurrence of an accident (Heinrich et al., 1980). In 1972, Wigglesworth proposed a new chain of accident causation. He believed that, because of a lack of knowledge and education, people will make mistakes and cause accidents. Subsequently, Bode and Roose-fort introduced the concept of "management" into the cause of an accident (Fu, 2013). Stewart (2002, 2011) divided "safety management" into two levels, which are more specific given the cause and root causes of an accident. From the development course of an accident cause chain, the human factor has evolved from an individual level to an organizational level. At the same time, scholars have also studied a range of individual human factors. Research by Kumar (2016) shows that if there are no emergency rescue measures, emergencies will be difficult to deal with. Harvey (2016) believes that if companies do not develop safety rules and regulations, then employee awareness of safety will be weak, thereby increasing the probability of an accident. Konstantin (2010) believes that a good safety culture can improve employee safety awareness and safety skills, thereby reducing the incidence of accidents. Nie et al. (2016) claims that appropriate psychological counselling and a reasonable amount of rest time can effectively improve the working status of employees. Although the previous research has been of great help in improving the risk of accidents involving people, it is still confined to the study of individual factors.

18.2 ASSESSMENT OF HUMAN FACTORS

With the increasing amount of research on human factors, a series of human factors analysis models have been developed. Among them, the Human Factor Analysis and Classification System (HFACS) is the most widely used and accepted method. The HFACS was developed by Shappell and Wiegmann based on Reason's "Swiss-Cheese" model and was applied in aviation accident analysis (Shappell and Wiegmann, 2001, 2003, 2004; Wiegmann and Shappell, 1997, 2001a,b,c, 2003). The framework is valuable in the analysis and classification of the human factors involved in accidents. Currently, HFACS is applied in different fields and has been applied effectively. For example, Celik and Cebi (2009) investigated the human factors in ship accidents using the HFACS. Similarly, Daramola (2014) used the framework to investigate air accidents in Nigeria during the period from 1985 to 2010. Australian scholars, Patterson and Shappell (2010), conducted an analysis of 508 human-induced mine accidents in the Queensland area using the HFACS. Michael used the HFACS to analyse 263 coal mine accidents in Australia between 2007 and 2008 (Lenne et al., 2012). Ruth Madigan applied the HFACS to line incidents in the UK rail system (Madigan et al., 2016) and Christine Chauvin analysed collisions at sea (Chauvin et al., 2013). Soner et al. (2015) used the HFACS-FCM in fire prevention modelling on board ships. Chen et al. (2013) used the HFACS-MA framework for maritime accident investigation and analysis. Zhan et al. (2017) used the Human Factors Analysis and Classification System-Railway Accident (HFACS-RAs) framework to identify and classify human and organizational factors involved in railway accidents. It is clear, therefore, that the HFACS has made an important contribution in the analysis of accidents and has provided a range of important research results. However, from the application point of view, most of the studies involve a qualitative analysis and occur after the accident. The safety management of human factors is an important means to prevent accidents. Methods for realizing risk assessment and management of human factors in daily production has still not been addressed. In view of this, a new method for human risk assessment and management was proposed in this chapter. Based on the HFACS framework, a human factors risk assessment model was established. Using the assessment model and the set pair analysis method, this case study calculates the connection number to get the safety score, risk level and risk development interval of each factor. The risk development trend of each factor is studied by using the partial connection number. Using the established SPA–Markov chain risk prediction model, the future development of risk is predicted. The ABC analysis and "S-O-R" model were used for human risk management to address the safety management link. The application results show that this method can effectively improve the safety of human factors. Finally, the common unsafe human factors and accident paths are summarized, and effective safety stimulus measures for different types of human factors were introduced.

18.3 HUMAN FACTORS RISK ASSESSMENT MODEL

The HFACS framework is an important factor analysis method in accident investigation and is now used in aviation, power, mining and other areas of accident investigations. It was developed from Reason's "Swiss-Chess" model of human error (Reason, 1990) and provides an organizational framework for accident analysis. Therefore, the HFACS addressed human errors and divided them into four different levels (Shappell and Wiegmann, 2003). Level 1 (Organizational Influences) includes three factors: resource management, organizational climate and organizational process. Level 2 (Unsafe Supervision) includes four factors: inadequate supervision, planned inappropriate operations, failed to correct a known problem and supervisory violations. Level 3 (Preconditions for Unsafe Acts) includes seven factors: physical environment, technological environment, physical/mental limitations, adverse mental states, adverse physiological states, crew resource management and personal readiness. Level 4 (Unsafe Supervision) includes four factors: decision errors, skill-based errors, perceptual errors and violations. In modern society, most enterprises place greater emphasis on safe production, and on taking many safety measures. Therefore, in such contexts unsafe acts may not lead to accidents. For example, if the unsafe operation of power systems leads to leakage, the relay protection system will be quickly shut down to protect the whole power system. In the chemical industry, when a tank leaks, there are often emergency measures to prevent accidents. However, when the final defense system fails (emergency failure), this usually results in an accident. Based on this case, the revised accident model was developed. The model is described diagrammatically in Figure 18.1. According to the revised accident model and the HFACS framework, the human factors risk assessment model was established. This model includes 5 levels (organizational influences, unsafe supervision, preconditions for unsafe acts, unsafe acts and emergency failure) and 25 human factors, as shown in Figure 18.2.

18.4 APPLIED METHODOLOGY

18.4.1 SET PAIR ANALYSIS (SPA)

The set pair analysis method is a system research method proposed by Zhao (1989). In this method, the certainty and uncertainty between systems are studied from three angles of "identity-discrepant-contrast". It has been applied to several fields. Chong et al. (2017) applied it to coal mine hazards. Su et al. (2009) used the SPA method to assess urban ecosystem health. Li et al. (2016) conducted a risk assessment of water pollution based on integrated k-means clustering and the setpair analysis method. Wei et al. (2016) used it to predict the analysis model of integrated carrying capacity. Tao et al. (2014) applied it to perform the multifunctional assessment and for the

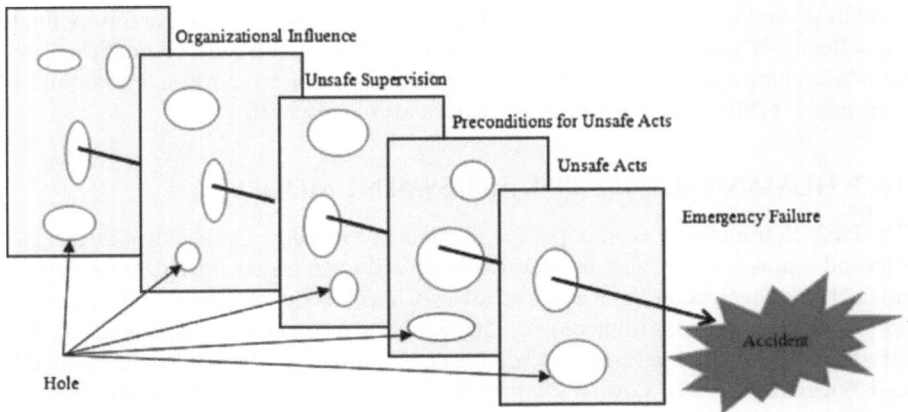

FIGURE 18.1 Accident causation model (Reproduced with permission from Xuecai & Deyong 2018 Copyright © Elsevier)

zoning of a crop production system. In a system, two sets, *A* and *B*, which have a certain relationship, form a set pair H (*A*, *B*). Assuming the set has *N* characteristics, where *S* is the total number of identical characteristics, *P* is the total number of contrary characteristics and *F = N-S-P* is the total number of discrepancy characteristics. The connection degree "μ" was given by Zhao (2000a,b) as follows:

$$\mu = \frac{S}{N} + \frac{F}{N}i + \frac{P}{N}j = a + bi + cj \tag{18.1}$$

Definition 1.

A = S/N, b = F/N, c = P/N represent the degree of "identity", "discrepancy" and "contrast", respectively, and *a + b + c* = 1. In this connection degree, *I* is the uncertainty coefficient, *I* ∈ [-1,1], and j is a contrast coefficient that usually equals -1 (Zhao, 2000 a,b).

Definition 2.

When considering the weight, the contact degree is as follows:

$$\mu = a + bi + cj = \sum_{k=1}^{S} w_k + \sum_{k=s+1}^{S+F} w_k i + \sum_{k=S+F=1}^{N} w_k$$

$$j = WRE = \left(w_1, w_2, \ldots, w_n\right) \begin{pmatrix} a_1 & b_1 & c_1 \\ a_2 & b_2 & c_2 \\ \vdots & \vdots & \vdots \\ a_n & b_n & c_n \end{pmatrix} \begin{pmatrix} 1 \\ i \\ j \end{pmatrix} \tag{18.2}$$

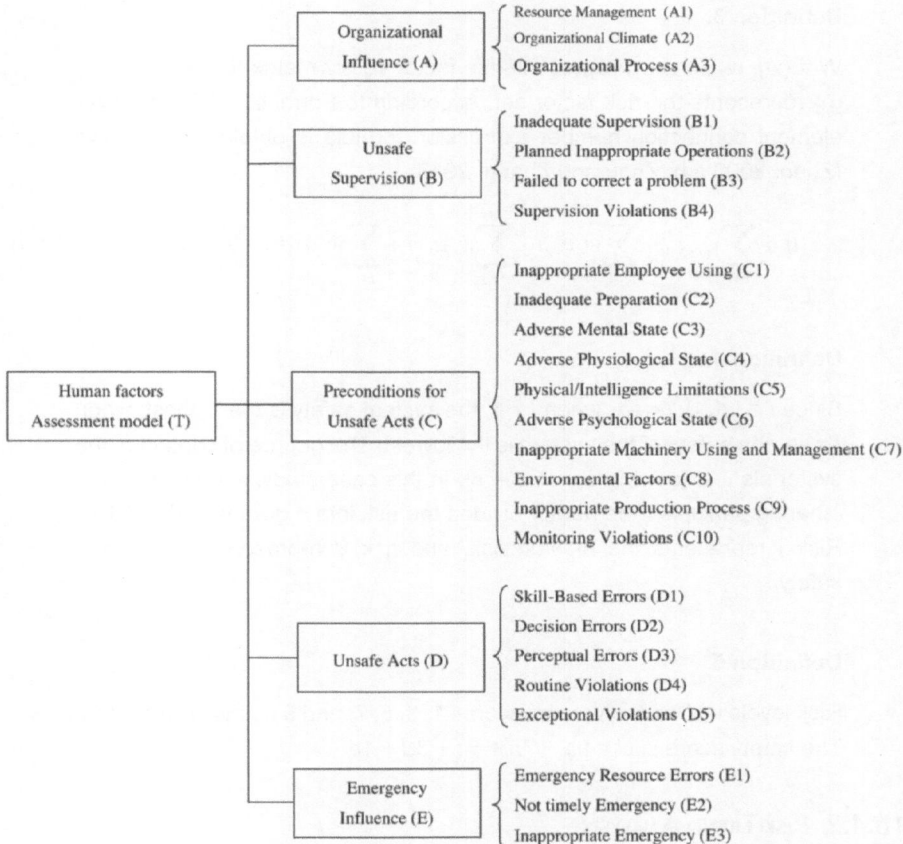

FIGURE 18.2 The human factors risk assessment model (Reproduced with permission from Xuecai & Deyong 2018 Copyright © Elsevier)

where W is the weight coefficient vector, R is the "identity-discrepant-contrast" assessment matrix, and E is the connection degree matrix (Zhao, 2000 a,b; Chong et al., 2017).

If the b_i terms expand, then the multiple contact number is

$$\mu = a + b_1 i_1 + b_2 i_2 + \ldots + b_n i_n + c_j \tag{18.3}$$

When n = 3, the five-element connection number is

$$\mu = a + b_1 i_1 + b_2 i_2 + b_3 i_3 + c_j = a + bi + cj + dk + el \tag{18.4}$$

In the application process, the five-element connection number can be associated with the level of risk; a represents the ratio of risk V, b represents the ratio of risk IV, c represents the ratio of risk III, d represents the ratio of risk II, and e represents the ratio of risk I.

Definition 3.

$W = (w_1, w_2, \ldots, w_n)$ represents the factor vector matrix. $U = \{\mu_1, \mu_2, \ldots, \mu_n\}$ represents the risk factor set. According to Formula (18.2), the five-element connection number expression formula is obtained as follows (Zhao, 2000 a,b; Zhou and Zhang, 2013)

$$\mu = \sum_{r=1}^{n} w_r \mu_{r1} + \sum_{r=1}^{n} w_r \mu_{r2} i + \sum_{r=1}^{n} w_r \mu_{r3} j + \sum_{r=1}^{n} w_r \mu_{r4} k + \sum_{r=1}^{n} w_r \mu_{r5} l \qquad (18.5)$$

Definition 4.

Since $l \in [-1,1]$, $j = -1$, when $l = 1$, the system safety is the highest; when $l = -1$, the safety of the system is the lowest. The degree of contact of the system is $\mu \in [a - b - c, a + b - c]$. In this case study, according to the "sharing principle", the author divided the risk into 5 grades (Table 18.1). Risk I represents the highest risk, and risk V represents the highest safety.

Definition 5.

Risk levels I–V were given the scores 1, 3, 5, 7, and 9 points (Table 18.1). The safety score is $U = 9a + 7b + 5c + 3d + 1e$.

18.4.2 RISK TREND ANALYSIS

In the SPA method, the partial connection number is an adjoint function of the connection number. Using the partial connection number, the development trend of the system can be predicted (Zhao, 2005; Wu, 2009).

TABLE 18.1
Risk Classification Table

Risk level	I	II	III	IV	V
Risk change					
Contact degree	$-1 \leq \mu < -0.6$	$-0.6 \leq \mu < -0.2$	$-0.2 \leq \mu < 0.2$	$0.2 \leq \mu < 0.6$	$0.6 \leq \mu \leq 1$
Score	1	3	5	7	9
Scoring interval	[1.0, 2.6]	[2.6, 4.2]	[4.2, 5.8]	[5.8, 7.4]	[7.4, 9.0]

(Adapted with permission from Xuecai et al. 2018 Copyright © Elsevier)

Definition 6.

For multiple connection numbers $\mu = a + b_1 i_1 + b_2 i_2 + \ldots + b_n i_n + cj$, its 1st order partial connection number is as follows:

$$\partial \mu = \partial a + i_1 \partial b_1 + i_2 \partial b_2 + \ldots + i_{n-2} \partial_{n-2} \quad (18.6)$$

where $\partial_a = \dfrac{a}{a + b_1}, \partial_{b_1} = \dfrac{b_1}{b_1 + b_2}, \partial_{b_2} = \dfrac{b_2}{b_2 + b_3}, \ldots, \partial b_{n-2} = \dfrac{b_{n-2}}{b_{n-2} + c}$

Definition 7.

For multiple connection numbers $\mu = a + b_1 i_1 + b_2 i_2 + \ldots + b_n i_n + cj$, its 2nd order partial connection number is as follows:

$$\partial^2 \mu = \partial(\partial \mu) = \partial^2 a + i_1 \partial^2 b_1 + i_2 \partial^2 b_2 + \ldots + i_{n-3} \partial^2 b_{n-3} \quad (18.7)$$

where $\partial^2 \mu = \dfrac{\partial a}{\partial a + \partial b_1}, \partial^2 b_1 = \dfrac{\partial b_1}{\partial b_1 + \partial b_2}, \ldots, \partial^2 b_{n-3} = \dfrac{\partial b_{n-3}}{\partial b_{n-3} + \partial b_{n-2}}$

Definition 8.

For multiple connection numbers $\mu = a + b_1 i_1 + b_2 i_2 + \ldots + b_n i_n + cj$, its $n-1$-order partial connection number is as follows:

$$\partial^{n-1} \mu = \partial^{n-2}(\partial \mu) = \partial^{n-1} a \quad (18.8)$$

where $\partial^{n-1} a = \dfrac{\partial^{n-2} a}{\partial^{n-2} a + \partial^{n-2} b_1}$

18.4.3 SPA–Markov Risk Prediction Method

In the production process, human factors are uncertain and change dynamically. Therefore, to predict the human factor risk, a prediction model must be chosen that can handle the characteristics of humans. The Markov chain has the discrepancy of time and state. This characteristic is consistent with the changing law of human factors and can be used to predict the risk of human factors. In this case study, a human factor evaluation model based on the SPA method and the Markov chain was proposed and applied.

Definition 9.

Assume at time T that the contact pojytfjhgdegree is

$$\mu(t) = a(t) + bi(t) + cj(t) + dk(t) + el(t) =$$

$$\sum_{k=1}^{A(t)} \omega_k(t) + \sum_{k=A(t)+1}^{A(t)+B(t)} \omega_k(t)i + \sum_{k=A(t)+B(t)+1}^{A(t)+B(t)+C(t)} \omega_k(t)j$$

$$+ \sum_{k=A(t)+B(t)+C(t)+1}^{A(t)+B(t)+C(t)+D(t)} \omega_k(t)k + \sum_{k=A(t)+B(t)+C(t)+D(t)+1}^{A(t)+B(t)+C(t)+D(t)+E(t)} \omega_k(t)l \qquad (18.9)$$

Where

$$\sum_{k=1}^{A(t)} \omega_k(t) + \sum_{k=A(t)+1}^{A(t)+B(t)} \omega_k(t)i + \sum_{k=A(t)+B(t)+1}^{A(t)+B(t)+C(t)} \omega_k(t)j +$$

$$\sum_{k=A(t)+B(t)+C(t)+1}^{A(t)+B(t)+C(t)+D(t)} \omega_k(t)k + \sum_{k=A(t)+B(t)+C(t)+D(t)+1}^{A(t)+B(t)+C(t)+D(t)+E(t)} \omega_k(t)l = 1$$

After taking safety measures, the safety status change during time [t, t+T], where T is a safety management cycle. In this period, $A(t_1)$ holds the a level, $A(t_2)$ becomes the b level, $A(t_3)$ becomes the c level, $A(t_4)$ becomes the d level, and $A(t_5)$ becomes the e level.

Then, the transfer matrix vector in time [t, t+1] is:

$$\vec{P}_A = \begin{pmatrix} P_{11} & P_{12} & P_{13} & P_{14} & P_{15} \end{pmatrix}$$

$$where \; \alpha(t) = \sum_{k=1}^{A(t)} \omega_k(t); \qquad p_{11} = \frac{\sum_{k=1}^{A(t)} \omega_k(t)}{\alpha(t)}; \qquad p_{12} = \frac{\sum_{k=A(t_1)+1}^{A(t_1)+B(t_2)} \omega_k(t)}{\alpha(t)};$$

$$p_{13} = \frac{\sum_{k=A(t_1)+B(t_2)+1}^{A(t_1)+B(t_2)+C(t_3)} \omega_k(t)}{\alpha(t)}; \qquad p_{14} = \frac{\sum_{k=A(t_1)+B(t_2)+C(t_3)+1}^{A(t_1)+B(t_2)+C(t_3)+D(t_4)} \omega_k(t)}{\alpha(t)};$$

$$p_{15} = \frac{\sum_{k=A(t_1)+B(t_2)+C(t_3)+D(t_4)+1}^{A(t_1)+B(t_2)+C(t_3)+D(t_4)+E(t_5)} \omega_k(t)}{\alpha(t)}$$

Similarly, the transfer vector matrix of $b(t)$, $c(t)$, $d(t)$, and $e(t)$ can be obtained. Thus, the transfer matrix of the system during the time [t, t + T] is

$$P = \begin{bmatrix} P_{11} & P_{12} & P_{13} & P_{14} & P_{15} \\ P_{21} & P_{22} & P_{23} & P_{24} & P_{25} \\ P_{31} & P_{32} & P_{33} & P_{34} & P_{35} \\ P_{41} & P_{42} & P_{43} & P_{44} & P_{45} \\ P_{51} & P_{52} & P_{53} & P_{54} & P_{55} \end{bmatrix} \qquad (18.10)$$

Based on the state transition probability matrix determined above, the evaluation model of human factors is established by the degree of association. That is, at $t + T$, the safety factor for human factors is

$$\mu_{A \sim B}(t+T) = a(t+T) + b(t+T)i + c(t+T)j$$
$$+ d(t+T)k + e(t+T)l = \left[a(t), b(t), c(t), d(t), e(t)\right] P(1, i, j, k, l)^T \quad (18.11)$$

Assuming that the state transition probability matrix of the safety evaluation index is the same between each change period T, that is, the state transition probability matrix P is a constant matrix, the safety assessment value of the human factor after n change periods is

$$\mu_{A \sim B}(t+nT) = a(t+nT) + b(t+nT)i + c(t+nT)j$$
$$+ d(t+nT)k + e(t+nT)l = \left[a(t), b(t), c(t), d(t), e(t)\right] P^{nT} (1, i, j, k, l)^T \quad (18.12)$$

The formula satisfies the *C–K* equation (*C–K, Chapman–Kolmogorov*). According to the trail of the Markov chain, P^{nT} will stabilize as the management cycle increases. Therefore, the steady-state values of human factor risk can be obtained from the following equations:

$$(abcde).(I - P) = 0 \quad (18.13)$$

18.5 ASSESSMENT AND MANAGEMENT PROCEDURE

18.5.1 ASSESSMENT PROCEDURE

Step 1: Establishment of an assessment team

The establishment of an assessment team is the first step in carrying out an assessment, primarily to facilitate the management, monitoring, statistics and analysis of the data. Evaluation team members can be selected by employees.

Step 2: Determine the weight of coefficient vector (W)

The determination of the human factor weight is an important part of the assessment work. There are many methods to determine the weight of the current stage, which are divided into two main categories: the subjective weighting method and the objective weighting method. To determine the weight more accurately, the combination of the expert scoring method and the entropy weight method were chosen to determine the index weight.

(1) Subjective weights (W_1)

The assessment team selects several experts to score the factors and then takes the mean as the subjective weight of the item. The team of experts can be composed of chemical plant staff, scholars, safety management personnel, etc. Moreover, the greater the number of experts, the better. The subjective weights are denoted by W_1.

(2) Objective weight (W_2)

The concept of entropy is derived from thermodynamics and is a measurement of the uncertainty of the system. The entropy method determines the weights according to the amount of information reflected by the variability degree of the assessment index values. The entropy weight method is defined as follows (Dai, 2000):

Assuming there are m experts and n assessment indicators that constitute a raw data matrix, $X = (x_{ij})_{m \times n}$. Where $i = 1, 2, \ldots, m$; $j = 1, 2, \ldots, n$; and X_{ij} denotes the weight value given by the i-th expert to the j-th index.

Because each assessment index usually has different dimensions, they cannot be directly compared, so the index value matrix must be standardized. For the human factor assessment model, the more important the factor, the greater the weight coefficient. According to the following formula, the normalized matrix $Z = z(x_{ij})_{m \times n}$ is obtained.

$$Z_{xy} = \frac{X_{xy} - \min\limits_{j} X_{ij}}{\max\limits_{j} X_{ij} - \min\limits_{j} X_{ij}} \tag{18.14}$$

According to the concept of information entropy, the entropy value of the j-th assessment index is defined as e_j.

$$e_j = -\frac{1}{\ln^m} \sum_{i=1}^{m} f_{ij} \ln^{f_{ij}} \tag{18.15}$$

Where $f_{ij} = Z_{ij} / \sum\limits_{i=1}^{m} Z_{ij}$. If $f_{ij} = 0, f_{ij} \ln^{f_{ij}} = 0$.

Then, the entropy weight of the j-th assessment index W_2 is defined as:

$$W_2 = (1 - e_j) / \left(n - \sum_{j=1}^{n} e_j \right) \tag{18.16}$$

(3) Combined weight (W_3)

To reduce the deviation of weights in the decision-making process, the most suitable weighting method is to use half the subjective weight plus half the objective weight. The combined weight W is calculated as follows:

$$W_3 = \frac{W_1 + W_2}{2} \tag{18.17}$$

(4) True weight (W)

For each level, the combined weight is the actual weight of the occupied system. For each factor, the combined weight is relative to the weight of the layer to which it belongs. Therefore, the actual weight of each element is equal to the combined weight of the element multiplied by the combined weight of its layer relative to the system.

Step 3: Construct the "identity-discrepant-contrast" assessment matrix (R)

As mentioned in Definition 2, the five-element connection number can be associated with the level of risk. In this chapter, the "identity-discrepant-contrast" assessment matrix was obtained by means of questionnaires. In the questionnaire, each factor has five options: "V", "IV", "III", "II" and "I", corresponding to a–f. Respondents fill out the questionnaire according to their own views. The assessment team analysed the questionnaires and calculated the statistics on the choice of each factor. Then, the assessment matrix was constructed.

Step 4: Calculate the connection number and partial connection number

After obtaining W and R, using Formula (18.5), the connection number is calculated. Next, using Formulas (18.6)–(18.8), the partial connection number is obtained.

Step 5: Risk analysis and risk trend analysis

After calculation, each factor, each level and whole system will have a connection number. Then, the assessment team needs to calculate the safety score, the connection degree interval, the risk development interval and the trend of risk changes, thereby determining the risk level of the human factors.

Step 6: Risk prediction

After several rounds of evaluation, the SPA–Markov chain risk prediction method established in this case study can be used to predict the development of human risk.

18.5.2 MANAGEMENT PROCEDURE

The ABC analysis (Antecedent–Behaviour–Consequences) method is a new safety management method based on the BBS (Behaviour Based Safety) method. The S-O-R (Stimulate–Organism–Reaction) mode was proposed by Mehrabian and Russell in 1974. The two models reveal the characteristics of human behaviour and have been applied in many areas. This chapter deals with the ABC analysis and S-O-R model for human risk management, which are illustrated in Figure 18.3.

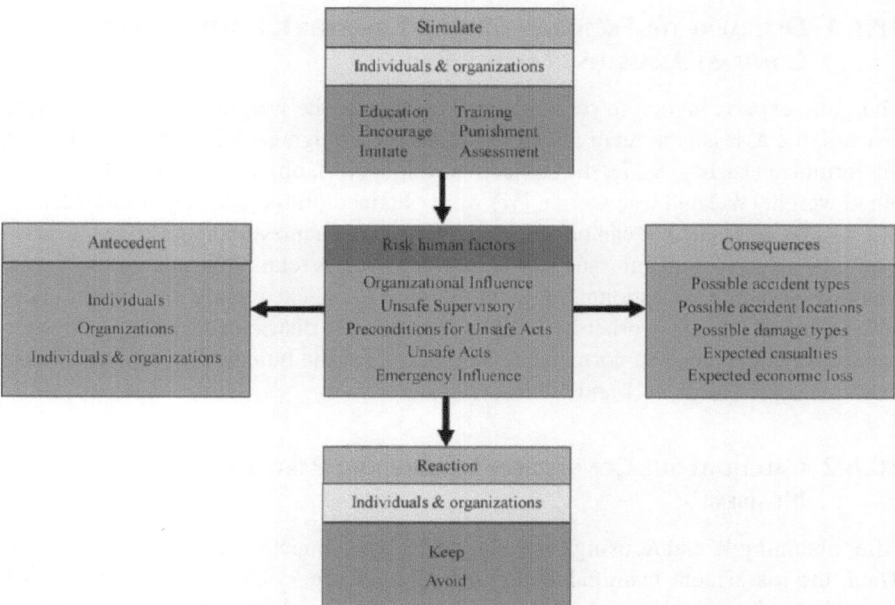

FIGURE 18.3 ABC analysis and "S-O-R" model in human factor risk reduction (Reproduced with permission from Xuecai & Deyong 2018 Copyright © Elsevier)

Step 1: Antecedent–Behaviour–Consequences analysis

After the risk assessment, various risk human factors were found. The ABC analysis method was used to determine the prerequisite for the occurrence of the risk factors. Simultaneously, a brief prediction of possible accidents needs to be performed. In the "antecedent" analysis, the assessment team should distinguish between individual or organizational reasons, or the result of the interaction between the two. At the same time, the assessment team also analyses the types of risks that may arise, the location of the risk, the degree of risk, the expected damage, and the economic loss.

Step 2: Stimulate–Organism–Reaction analysis

Normally, we need to take measures to improve unsafe factors. When using the "S-O-R" model in risk reduction, different "Stimulates" (safety measures) are enacted on the risk factors. Managers need to choose an effective "Stimulus" approach based on the different "Reactions" to get the most targeted safety measures. The entire risk assessment and risk management process is shown in Figure 18.4.

18.6 APPLICATION

In Shaanxi province, a chemical plant was selected as a case study to demonstrate the method. The assessment and management of human factors is a complex process that encompasses multiple levels and factors. Through multiple stages of risk assessment and risk management, human factor safety levels will be greatly improved.

18.6.1 DETERMINE THE FACTOR WEIGHT (W) AND THE IDENTITY-DISCREPANT-CONTRAST ASSESSMENT MATRIX (R)

The more experts invited to score, the more accurate the weight. In this assessment process, the assessment team invited 10 experts to weigh each factor. According to the formulae (18.14–18.17), the subjective weight (W_1), objective weight (W_2), combined weight (W_3) and true weight (W) were obtained (Table 18.1 and Table 18A.1). The assessment matrix R can be obtained by means of a questionnaire. In this assessment, the assessment team issued 200 questionnaires. A total of 18 categories of staff was involved in the questionnaire, including managers, excavators, equipment operators, and bolt support workers. All 200 questionnaires distributed by the assessment team were collected. The normalized data matrix for the human factor is the assessment matrix R (Table 18.1 and Table 18A.1).

18.6.2 CALCULATE THE CONNECTION NUMBER AND PARTIAL CONNECTION NUMBER

After obtaining W and R, using Formula (18.5), the connection number is calculated. Then, the assessment team calculated the safety score, connection degree interval and risk development interval. At the same time, the partial connection number of the factors and their trends were calculated and shown in Table 18.2.

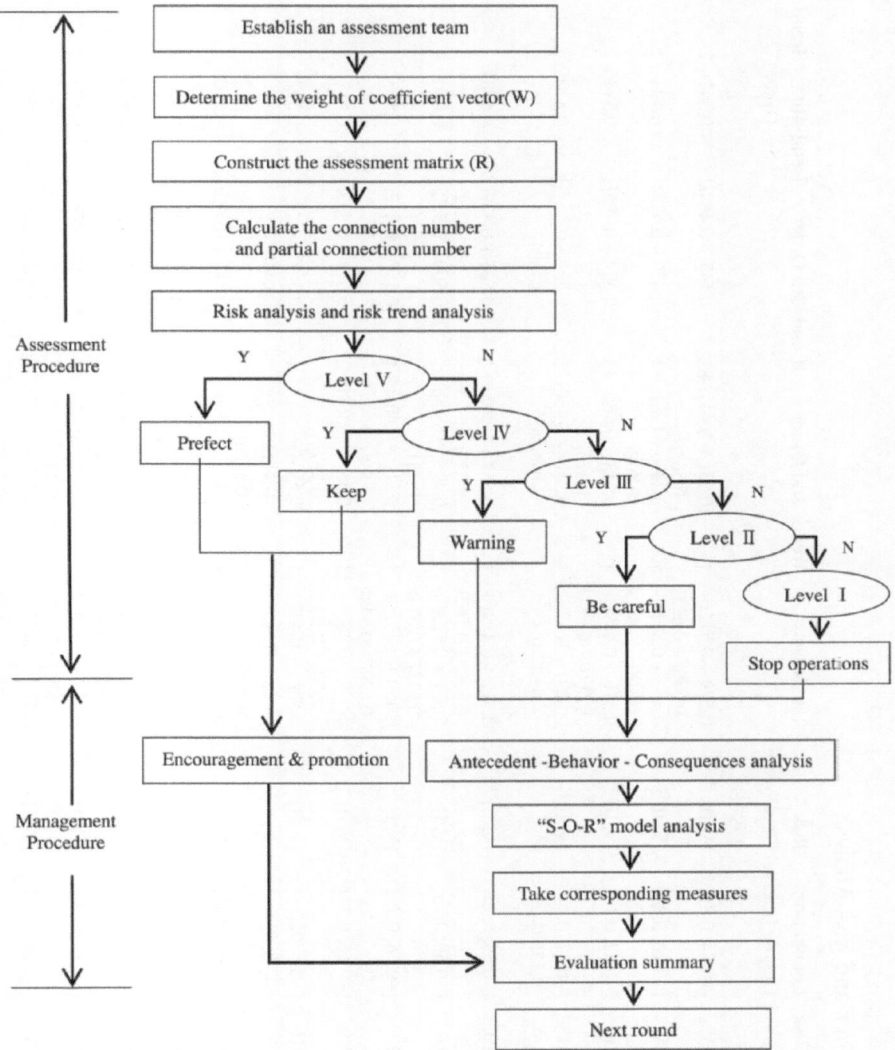

FIGURE 18.4 Assessment and management procedure (Reproduced with permission from Xuecai & Deyong 2018 Copyright © Elsevier)

18.6.3 RISK ANALYSIS

From Table 18.2, the connection number of 5 levels and the entire systems were determined. To better compare the safety state, this case study compares the data of each level and the entire system shown in Figure 18.5 (a) and (b), and compares the data of each human factor as shown in Figure 18.5 (c) and (d). To describe the range of risk changes intuitively, the assessment team also calculated the connection degree

TABLE 18.2

Human Factor's Connection Number and Risk Analysis

Factor	The Five-Element Connection Number	Safe Score	Risk Level	Connection Degree Interval	Risk Development Interval	1st Order	Trend	2nd Order	Trend	3rd Order	Trend	4th Order	Trend
A1	$0.330 + 0.240i + 0.230j + 0.125k + 0.075l$	6.25	IV	[−0.34, 0.85]	[V, II]	$0.579 + 0.511i + 0.648j + 0.625k$	↑	$0.531 + 0.441i + 0.509j$	↓	$0.547 + 0.464i$	↓	0.541	→
A2	$0.250 + 0.300i + 0.225j + 0.140k + 0.085l$	5.98	IV	[−0.50, 0.83]	[V, II]	$0.455 + 0.571i + 0.616j + 0.622k$	↑	$0.443 + 0.481i + 0.498j$	↑	$0.479 + 0.492i$	↑	0.494	↑
A3	$0.240 + 0.225i + 0.235j + 0.185k + 0.115l$	5.58	III	[−0.52, 0.77]	[V, II]	$0.516 + 0.489i + 0.560j + 0.617k$	↑	$0.513 + 0.466i + 0.476j$	↓	$0.524 + 0.495i$	↓	0.514	→
A	$0.2865 + 0.2550i + 0.2296j + 0.1423k + 0.0866l$	6.0250	IV	[−0.4270, 0.8268]	[V, II]	$0.5291 + 0.5262i + 0.6174j + 0.6217k$	↑	$0.5014 + 0.4601i + 0.4986j$	↓	$0.5214 + 0.4801i$	→	0.5206	→
B1	$0.210 + 0.200i + 0.290j + 0.170k + 0.130l$	5.38	III	[−0.58, 0.74]	[V, II]	$0.512 + 0.408i + 0.630j + 0.567k$	↑	$0.557 + 0.393i + 0.517j$	↓	$0.586 + 0.427i$	↓	0.578	→
B2	$0.250 + 0.230i + 0.210j + 0.190k + 0.120l$	5.60	III	[−0.50, 0.76]	[V, II]	$0.521 + 0.523i + 0.525j + 0.613k$	↑	$0.499 + 0.499i + 0.461j$	↓	$0.500 + 0.520i$	↑	0.490	↑
B3	$0.180 + 0.220i + 0.290j + 0.170k + 0.140l$	5.26	III	[−0.64, 0.72]	[V, I]	$0.450 + 0.431i + 0.630j + 0.548k$	↑	$0.511 + 0.406i + 0.535j$	↑	$0.557 + 0.432i$	↓	0.563	→
B4	$0.220 + 0.190i + 0.310j + 0.150k + 0.130l$	5.44	III	[−0.56, 0.74]	[V, II]	$0.537 + 0.380i + 0.674j + 0.536k$	→	$0.585 + 0.361i + 0.557j$	↓	$0.619 + 0.393i$	↓	0.612	→
B	$0.2149 + 0.2097i + 0.2756j + 0.1697k + 0.1301l$	5.4192	III	[−0.5702, 0.7398]	[V, II]	$0.5061 + 0.4321i + 0.6189j + 0.5660k$	↑	$0.5394 + 0.4111i + 0.5223j$	→	$0.5675 + 0.4404i$	→	0.5630	→
C1	$0.225 + 0.255i + 0.285j + 0.145k + 0.090l$	5.76	III	[−0.55, 0.82]	[V, II]	$0.469 + 0.472i + 0.663j + 0.617k$	↑	$0.498 + 0.416i + 0.518j$	↑	$0.545 + 0.445i$	↓	0.550	→
C2	$0.210 + 0.220i + 0.235j + 0.185k + 0.150l$	5.31	III	[−0.58, 0.70]	[V, II]	$0.488 + 0.484i + 0.560j + 0.552k$	↑	$0.502 + 0.464i + 0.503j$	↑	$0.520 + 0.479i$	↓	0.520	→
C3	$0.180 + 0.235i + 0.255j + 0.180k + 0.150l$	5.23	III	[−0.64, 0.70]	[V, I]	$0.434 + 0.480i + 0.586j + 0.545k$	↑	$0.475 + 0.450i + 0.518j$	↑	$0.513 + 0.465i$	↓	0.525	→
C4	$0.210 + 0.250i + 0.240j + 0.180k + 0.120l$	5.50	III	[−0.58, 0.76]	[V, II]	$0.457 + 0.510i + 0.571j + 0.600k$	↑	$0.472 + 0.472i + 0.488j$	↑	$0.500 + 0.492i$	↓	0.504	→
C5	$0.260 + 0.290i + 0.320j + 0.080k + 0.050l$	6.26	IV	[−0.48, 0.90]	[V, II]	$0.473 + 0.475i + 0.800j + 0.615k$	↑	$0.499 + 0.373i + 0.565j$	↑	$0.572 + 0.397i$	↓	0.590	→

Factor	The Five-Element Connection Number	Safe Score	Risk Level	Connection Degree Interval	Risk Development Interval	1st Order	2nd Order	Trend 3rd Order	Trend 4th Order	Trend
C6	$0.195 + 0.205i + 0.250j + 0.200k + 0.150l$	5.19	III	$[-0.61, 0.70]$	[V, I]	$0.488 + 0.451i + 0.556j + 0.571k$ ↑	$0.520 + 0.448i + 0.493j$ +↓	$0.537 + 0.476i$↓	0.530	→
C7	$0.225 + 0.235i + 0.240j + 0.220k + 0.080l$	5.61	III	$[-0.55, 0.84]$	[V, II]	$0.489 + 0.495i + 0.522j + 0.733k$ ↑	$0.497 + 0.487i + 0.416j$ +↓	$0.505 + 0.539i$↑	0.484	←
C8	$0.250 + 0.260i + 0.270j + 0.140k + 0.080l$	5.92	IV	$[-0.50, 0.84]$	[V, II]	$0.490 + 0.491i + 0.659j + 0.636k$ ↑	$0.500 + 0.427i + 0.509j$ +↓	$0.539 + 0.456i$↓	0.542	→
C9	$0.220 + 0.240i + 0.280j + 0.160k + 0.100l$	5.64	III	$[-0.56, 0.80]$	[V, II]	$0.478 + 0.462i + 0.636j + 0.615k$ ↑	$0.509 + 0.420i + 0.508j$ +↓	$0.548 + 0.453i$↓	0.547	→
C10	$0.195 + 0.240i + 0.265j + 0.190k + 0.110l$	5.44	III	$[-0.61, 0.78]$	[V, I]	$0.448 + 0.475i + 0.582j + 0.633k$ ↑	$0.485 + 0.449i + 0.479j$ +↓	$0.519 + 0.484i$↓	0.518	→
C	$0.2147 + 0.2408i + 0.2617j + 0.1725k + 0.1103l$	5.5542	III	$[-0.5706, 0.7794]$	[V, II]	$0.4717 + 0.4792i + 0.6027j + 0.6100k$ ↑	$0.4959 + 0.4429i + 0.4970j$ ↑	$0.5282 + 0.4712i$ →	0.5285	→
D1	$0.250 + 0.260i + 0.270j + 0.120k + 0.100l$	5.88	IV	$[-0.50, 0.80]$	[V, II]	$0.490 + 0.491i + 0.692j + 0.545k$ ↑	$0.500 + 0.415i + 0.559j$ +↑	$0.547 + 0.426i$↓	0.562	→
D2	$0.270 + 0.280i + 0.300j + 0.090k + 0.060l$	6.22	IV	$[-0.46, 0.88]$	[V, II]	$0.491 + 0.483i + 0.769j + 0.600k$ ↑	$0.504 + 0.386i + 0.562j$ +↑	$0.567 + 0.407i$↓	0.582	→
D3	$0.300 + 0.280i + 0.310j + 0.060k + 0.050l$	6.44	IV	$[-0.40, 0.90]$	[V, II]	$0.517 + 0.475i + 0.838j + 0.545k$ ↑	$0.522 + 0.362i + 0.606j$ +↑	$0.591 + 0.374i$↓	0.612	→
D4	$0.160 + 0.225i + 0.245j + 0.230k + 0.140l$	5.07	III	$[-0.70, 0.70]$	[V, II]	$0.416 + 0.479i + 0.516j + 0.622k$ ↑	$0.465 + 0.481i + 0.453j$ +↓	$0.491 + 0.515i$↑	0.488	←
D5	$0.310 + 0.300i + 0.270j + 0.080k + 0.040l$	6.52	IV	$[-0.38, 0.92]$	[V, II]	$0.508 + 0.526i + 0.771j + 0.667k$ ↑	$0.491 + 0.406i + 0.536j$ +↑	$0.548 + 0.431i$↓	0.560	→
D4	$0.160 + 0.225i + 0.245j + 0.230k + 0.140l$	5.07	III	$[-0.70, 0.70]$	[V, II]	$0.416 + 0.479i + 0.516j + 0.622k$ ↑	$0.465 + 0.481i + 0.453j$ +↓	$0.491 + 0.515i$↑	0.488	←
D5	$0.310 + 0.300i + 0.270j + 0.080k + 0.040l$	6.52	IV	$[-0.38, 0.92]$	[V, II]	$0.508 + 0.526i + 0.771j + 0.667k$ ↑	$0.491 + 0.406i + 0.536j$ +↑	$0.548 + 0.431i$↓	0.560	→

Factor	The Five-Element Connection Number	Safe Score	Risk Level	Connection Degree Interval	Risk Development Interval	1st Order	Trend	2nd Order	Trend	3rd Order	Trend	4th Order	Trend
D	$0.2502 + 0.2649i + 0.2774j + 0.1210k + 0.0865l$	5.9426	IV	[−0.4996, 0.8270]	[V, II]	$0.4857 + 0.4885i + 0.6963j + 0.5831k$	↑	$0.4986 + 0.4123i + 0.5442j$	↑	$0.5474 + 0.4310i$	→	0.5594	↓
E1	$0.270 + 0.260i + 0.250j + 0.120k + 0.100l$	5.96	IV	[−0.46, 0.80]	[V, II]	$0.509 + 0.510i + 0.676j + 0.545k$	↑	$0.500 + 0.430i + 0.553j$	↑	$0.538 + 0.437i$	↓	0.551	→
E2	$0.190 + 0.240i + 0.220j + 0.190k + 0.160l$	5.22	III	[−0.62, 0.68]	[V, I]	$0.442 + 0.522i + 0.537j + 0.543k$	↑	$0.459 + 0.493i + 0.497j$	↑	$0.482 + 0.498i$	↑	0.492	←
E3	$0.180 + 0.220i + 0.230j + 0.200k + 0.170l$	5.08	III	[−0.64, 0.66]	[V, I]	$0.450 + 0.489i + 0.535j + 0.541k$	↑	$0.479 + 0.478i + 0.497j$	↑	$0.501 + 0.490i$	↓	0.506	→
E	$0.2192 + 0.2416i + 0.2354j + 0.1649k + 0.1389l$	5.4746	III	[−0.5616, 0.7222]	[V, II]	$0.4757 + 0.5065i + 0.588ij + 0.5428k$	↑	$0.4843 + 0.4627i + 0.5200j$	↑	$0.5114 + 0.4708i$	→	0.5207	↓
T	$0.2372 + 0.2435i + 0.2591j + 0.1527k + 0.1075l$	5.7004	III	[−0.5253, 0.7850]	[V, II]	$0.4934 + 0.4845i + 0.6292j + 0.5869k$	↑	$0.5046 + 0.4350i + 0.5652j$	↑	$0.5370 + 0.4349i$	→	0.5525	↓

(Adapted with permission from Xuecai et al. 2018 Copyright © Elsevier)

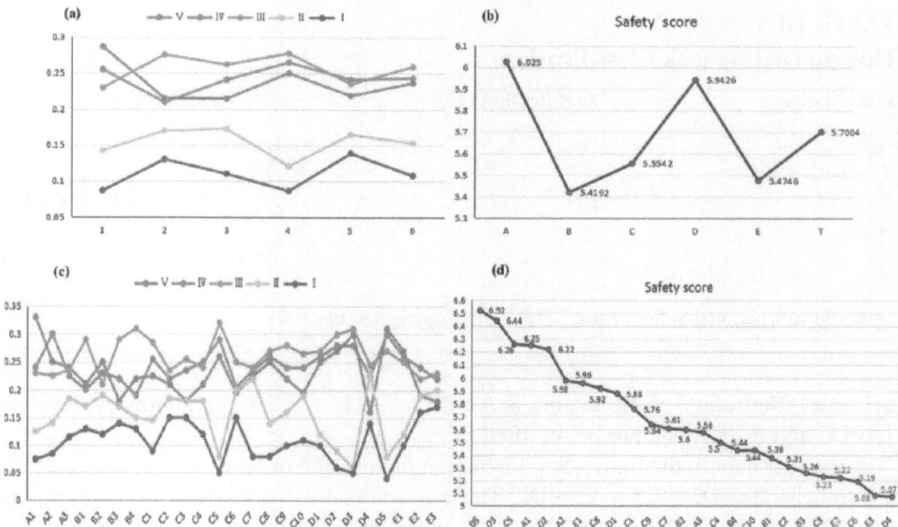

FIGURE 18.5 Human factors data comparison chart (Reproduced with permission from Xuecai & Deyong 2018 Copyright © Elsevier)

interval for each factor as depicted in Table 18.2, and drew the connection degree interval of each of the human factors shown in Figure 18.6.

From the point of view of the safety score, the safety order of each level is A > D > T > C > E > B. The risk of each level and the entire system is between Level III and Level IV. This indicates that the safety situation is within the acceptable range. From risk level V (absolutely safe), the safety order is A > D > T > E > B > C. From the risk level I (absolutely dangerous), the safety order is D > A > T > C > B > E. For the whole system and 5 levels, the maximum level of safety that can be achieved is D > A > T > C > B > E. From the point of view of the most dangerous degree that can be

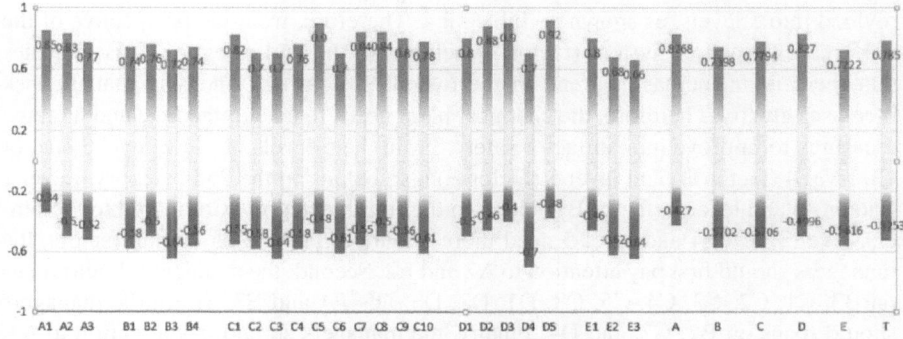

FIGURE 18.6 Human factors risk interval (Reproduced with permission from Xuecai & Deyong 2018 Copyright © Elsevier)

TABLE 18.3

Human Factors Risk Classification

Risk Change	Rectification Order	Human Factors
↑ increase	1st	B1, B3, B4, C2, C3, C6, C10, D4, E2, E3
	2nd	A2, A3, B2, C1, C4, C7, C8, C9, D1, E1
	3rd	A1, C5, D2, D3, D5

(Adapted with permission from Xuecai et al. 2018 Copyright © Elsevier)

achieved, the safety of the factors is A > D > T > E > B > C. In summary, level B, level C and level E need to be rectified first.

From the data in the figure, for the human factors, the order of safety scores above 6 points is D5 > D3 > C5 > A1 > D2. These 5 factors are the best factors for the safety situation. There are 10 factors that score between 5.5 and 6 points, A2 > E1 >C8 > D1 > C1 > C9 > C7 > B2 > A3 > C4. There are 10 factors that score between 5 and 5.5 points, B4 > C10 > B1 > C2 > B3 >C3 > E2 > C6 > E3 > D4. For risk level I (absolutely dangerous) to occur, the safety order is D5 > C5 = D3 > D2 > A1 > C7 = C8 >A2 > A1 > C9 = D1 = E1 > C10 > A3 > B2 = C4 > B1 = B4 > B3 = D4 >C2 = C3 = C6 > E2 > E3. From the maximum risk level that can be achieved, the safety rating of the factors is A1 > D5 > D3 >D2 > E1 > C5 > A2 > B2 > C8 > D1 > A3 > C1 > C7 > B4 > C9 >B1 > C2 > C4 > C6 > C10 > E2 > B3 > C3 > E3 > D4. In summary, human factors were roughly divided into 3 categories, as shown in Table 18.3.

18.6.4 RISK TREND ANALYSIS

In the process of the risk analysis, people are concerned about the current safety factor situation, but they also care about trends in risk factors. In the set pair analysis, the partial connection number method was used to predict the risk trend. Therefore, using the formulae (18.6–18.8), the partial connection number of the factors was calculated and shown in Table 18.2. According to the risk trends, the factors can be divided into 5 levels, as shown in Table 18.4. Therefore, from the perspective of the risk trend, the whole system (factor T) belongs to the 2nd category. This indicates that there is an increasing trend in the overall system risk and also that the risk increases greatly. Therefore, the chemical plant needs to take immediate and targeted measures to improve human factor safety. From the 5 levels of view, level 3 (factor C), level 4 (factor D) and level 5 (factor E) also belong to the 2nd category. Level 1 (factor A) and level 2 (factor B) belong to the fourth category. Compared to the others, level 3, level 4 and level 5 need more attention. For the 25 human factors, the managers should first pay attention to A2 and E2. Second, the managers should focus on B3, C1, C2, C3, C4, C5, C8, D1, D2, D3, D5, E1 and E3. Third, the managers should focus on B2, C7, and D4. Finally, the managers should pay attention to A1, A4, B1, B4, C6, C9, and C10. Since B4 shows a decreasing trend, it can be ignored temporarily.

TABLE 18.4

Risk Change Classification

Risk Tend Change	Order	Human Factors	Partial Connection Number			
			1st Order	2nd Order	3rd Order	4th Order
	1st	A2, E2	↑	↑	↑	↑
	2nd	B3, C1, C2, C3, C4, C5, C8, D1, D2, D3, D5, E1, E3, C, D, E, T	↑	↑	↓	↓
	3rd	B2, C7, D4	↑	↓	↑	↑
	4th	A1, A3, B1, C6, C9, C10, A, B	↑	↓	↓	↓
	5th	B4	↓	↓	↓	↓

increase

origin

decrease

(Adapted with permission from Xuecai et al. 2018 Copyright © Elsevier)

18.6.5 RISK PREDICTION

In this study, 12 safety management research cycles were performed. In the first six cycles, the chemical plant still uses the original safety management methods. Due to the lack of people, the managed data are discretized. Therefore, in the case of the discrete data, the SPA–Markov method proposed in this case study can be used to predict the risk.

First, it is required to test the validity of the SPA–Markov chain risk prediction method. In this case study, the fifth and sixth cycles were predicted. By calculating the error between the predicted value and the actual value, the practicability of the method is verified. According to Formulas (18.9–18.10), the state transition probability matrices for each period are calculated as follows:

$$P_{12} = \begin{bmatrix} 1 & 0 & 0 & 0 & 0 \\ 0.108 & 0.887 & 0.005 & 0 & 0 \\ 0 & 0 & 0.886 & 0.114 & 0 \\ 0 & 0 & 0 & 0.756 & 0.244 \\ 0 & 0 & 0 & 0 & 1 \end{bmatrix}$$

$$P_{23} = \begin{bmatrix} 0.923 & 0.077 & 0 & 0 & 0 \\ 0 & 1 & 0 & 0 & 0 \\ 0 & 0.019 & 0.981 & 0 & 0 \\ 0 & 0 & 0.254 & 0.746 & 0 \\ 0 & 0 & 0 & 0.2 & 0.8 \end{bmatrix}$$

$$P_{34} = \begin{bmatrix} 1 & 0 & 0 & 0 & 0 \\ 0.034 & 0.901 & 0.065 & 0 & 0 \\ 0 & 0 & 0.946 & 0.012 & 0.042 \\ 0 & 0 & 0 & 1 & 0 \\ 0 & 0 & 0 & 0 & 1 \end{bmatrix}$$

$$P_{45} = \begin{bmatrix} 0.958 & 0.042 & 0 & 0 & 0 \\ 0 & 1 & 0 & 0 & 0 \\ 0 & 0.098 & 0.902 & 0 & 0 \\ 0 & 0 & 0.902 & 0.908 & 0 \\ 0 & 0 & 0 & 0.049 & 0.951 \end{bmatrix}$$

$$P_{56} = \begin{bmatrix} 1 & 0 & 0 & 0 & 0 \\ 0.035 & 0.965 & 0 & 0 & 0 \\ 0 & 0.051 & 0.916 & 0.033 & 0 \\ 0 & 0 & 0 & 1 & 0 \\ 0 & 0 & 0 & 0.123 & 0.877 \end{bmatrix}$$

Since the original safety management method has been adopted, it can be approximated that the weights of the state transition probability matrices of the respective periods are the same. The average state transition probability matrices were calculated as:

TABLE 18.5

The Predicted Connection Number and Safety Score

No	Data Category	The Five-Element Connection Number	Safety Score	Relative Error
Cycle 5	Predictive date	$0.2533 + 0.2210i + 0.2561j + 0.1389k + 0.1307l$	5.6546	1.14%
	Actual date	$0.2408 + 0.2533i + 0.2516j + 0.1335k + 0.1208l$	5.7196	
Cycle 6	Predictive date	$0.2482 + 0.2365i + 0.2521j + 0.1362k + 0.1270l$	5.6854	1.56%
	Actual date	$0.2496 + 0.2573i + 0.2304j + 0.1567k + 0.1060l$	5.7756	

(Adapted with permission from Xuecai et al. 2018 Copyright © Elsevier)

$$P_{1-4} = \begin{bmatrix} 0.974 & 0.026 & 0 & 0 & 0 \\ 0.048 & 0.929 & 0.023 & 0 & 0 \\ 0 & 0.006 & 0.938 & 0.042 & 0.014 \\ 0 & 0 & 0.085 & 0.834 & 0.081 \\ 0 & 0 & 0 & 0.067 & 0.933 \end{bmatrix}$$

$$P_{1-5} = \begin{bmatrix} 0.970 & 0.030 & 0 & 0 & 0 \\ 0.036 & 0.947 & 0.017 & 0 & 0 \\ 0 & 0.029 & 0.929 & 0.031 & 0.011 \\ 0 & 0 & 0.087 & 0852 & 0.061 \\ 0 & 0 & 0 & 0.062 & 0.938 \end{bmatrix}$$

The average connection numbers are calculated as $\mu_{1-4} = 0.2488 + 0.2293i + 0.2544j + 0.1437k + 0.1238l$, $\mu_{1-5} = 0.2472 + 0.2341i + 0.2538j + 0.1417k + 0.1232l$. The predicted connection numbers and safety scores of the fifth and sixth cycles were predicted and are shown in Table 18.5.

The two predictions are 1.14% and 1.56%, respectively and are less than 2%, so they can be used to make a rough prediction of the future safety score. This indicates that the proposed SPA–Markov risk assessment method is effective.

Therefore, the dates of the following six cycles were predicted by using the SPA–Markov risk assessment method. The average connection number of the first six cycles is calculated as $\mu_{1-6} = 0.2476 + 0.2380i + 0.2499j + 0.1442k + 0.1203l$, and the average transfer matrix is:

$$P_{1-6} = \begin{bmatrix} 0.9762 & 0.0238 & 0 & 0 & 0 \\ 0.0354 & 0.9506 & 0.0140 & 0 & 0 \\ 0 & 0.0336 & 0.9262 & 0.0318 & 0.0084 \\ 0 & 0 & 0.0692 & 0.8820 & 0.0488 \\ 0 & 0 & 0 & 0.0744 & 0.9256 \end{bmatrix}$$

From the data analysis point of view, if the original management methods are applied, the risk level of the human factors is III. The safety score is between 5.6172 and 5.7868. Ultimately, the stable risk level of the human factor is IV, and the safety score is 6.9809 (approximately 500 cycles). This indicates that the current safety management measures can only control the risk at a medium level, which is not an effective way to improve the human factor safety level.

18.6.6 RISK MANAGEMENT

In the next six cycles, the chemical plant adopts the safety management method, proposed in this chapter to manage the safety assessment. The evaluation data are shown in Table 18.6. As the data are not discrete, if the SPA–Markov method is used to predict the need for weight repair, it will lead to a greater prediction bias. Therefore, this study uses the data fitting method shown in Table 18.7 for eliminating such biases. Figure 18.7 shows the following six cycles are predicted. From the forecast data, applying the management method proposed in this chapter, the safety level of human factors is V in the 12th cycle, and the safety score is 7.4109, while the safety score using the original management method is 5.7868. This shows that the safety management method proposed in this chapter is effective. To make a clearer comparison of the two management methods, this investigation makes a comparative study of the data, analysing 12 cycles elaborated in Figure 18.8.

When applying the original management method, the data were discrete, and the safety situation was relatively stable. When using the safety management method proposed here, the safety status is obviously improved. In total, there are 12 unsafe

TABLE 18.6
Using the Safety Management Proposed in this Article

Cycle	The Five-Element Connection Number	Safe Score	Risk Level
1	$0.2876 + 0.2662i + 0.2476j + 0.1132k + 0.0854l$	6.1148	II
2	$0.3030 + 0.2904i + 0.2527j + 0.0856k + 0.0683l$	6.3484	II
3	$0.3302 + 0.3061i + 0.2432j + 0.0684k + 0.0521l$	6.5878	II
4	$0.3687 + 0.3104i + 0.2257j + 0.0520k + 0.0432l$	6.8188	II
5	$0.3871 + 0.3261i + 0.2033j + 0.0479k + 0.0356l$	6.9624	II
6	$0.4068 + 0.3309i + 0.1884j + 0.0416k + 0.0323l$	7.0766	II

(Adapted with permission from Xuecai et al. 2018 Copyright © Elsevier)

TABLE 18.7
Data Fitting Table

Category	Fitting Formula	R2
Level V	$y = 0.2759x^{0.2025}$	0.9314
Level IV	$y = 0.2665x^{0.1207}$	0.9901
Level III	$y = 0.278e^{-0.06x}$	0.8884
Level II	$y = 0.1195x^{-0.57}$	0.9803
Level I	$y = 0.092x^{-0.562}$	0.9713

(Adapted with permission from Xuecai et al. 2018 Copyright © Elsevier)

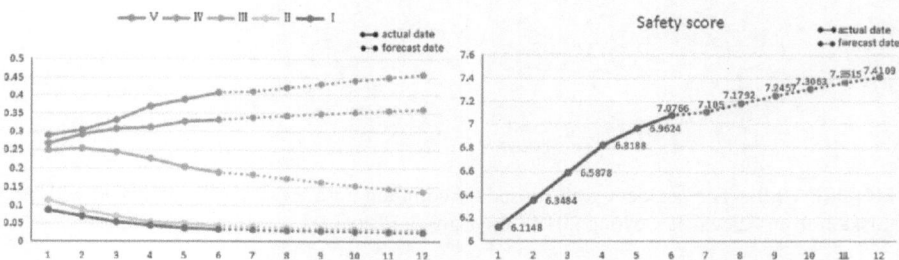

FIGURE 18.7 The data obtained by applying the management method proposed in this chapter (Reproduced with permission from Xuecai & Deyong 2018 Copyright © Elsevier)

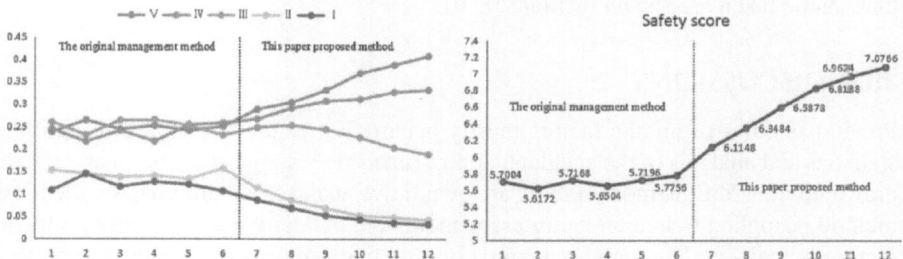

FIGURE 18.8 Comparison of two safety management methods (Reproduced with permission from Xuecai & Deyong 2018 Copyright © Elsevier)

human factors in common. The relationship between the factors is shown in Figure 18.9. Table 18.8 shows that an accident has the following nine types of paths The most effective safety "stimulus" measures are shown in Table 18.9 for the 12 most common unsafe human factors. From the point of view of the risk category, factor A is clearly an organizational factor, and factor D is an individual level factor. However, factors B, C and D include both organizational and individual factors. In the organizational level, unsafe factors are mainly caused by poor safety awareness or the limited safety management ability of the safety management personnel. The individual unsafe factors are mainly caused by the lack of staff safety skills and routine

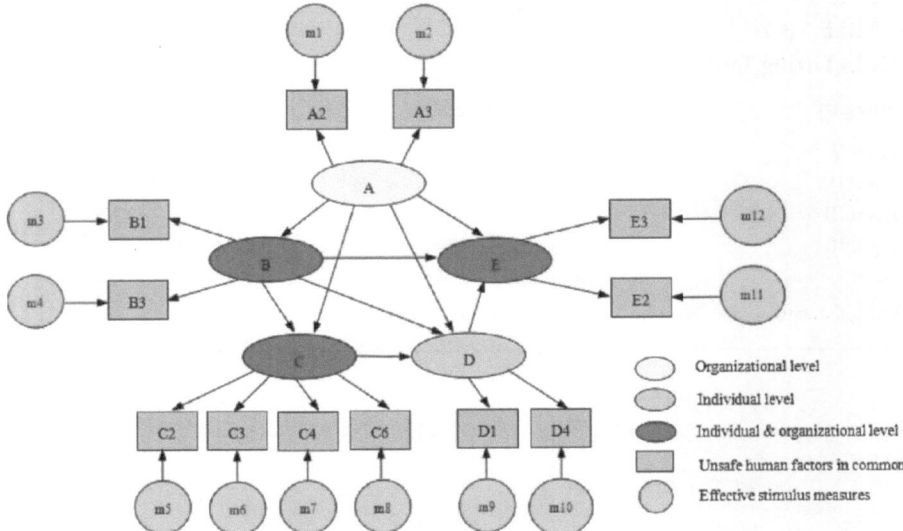

FIGURE 18.9 The accident path caused by common unsafe factors (Reproduced with permission from Xuecai & Deyong 2018 Copyright © Elsevier)

violations (Tabibzadeh, 2014). For the different types of human factors, the measures that can be taken are shown in Table 18.10.

18.7 DISCUSSIONS

Previous studies of human factors mostly occurred after the accident. Through the statistics and analysis of the accident, appropriate safety suggestions or measures are drawn up. In addition, most of them are qualitative analyses. In this chapter, the SPA method is applied to human factor assessment, and the daily management of human factors is realized. Through the evaluation and management of human factors, the safety level of human beings has been greatly improved, thereby effectively reducing accidents. This is more important than the analysis of human factors after an accident. The human factors risk assessment model established is an improvement on the HFACS framework. Combining the model with the SPA, the dynamic assessment of human factors can be achieved. The calculated number of connections consists of five data, which gives us a clear idea of the proportion of the five risk levels. The evaluation results are made up of multiple components, and the errors caused by a single result are avoided. The safety score is the embodiment of the safety situation, and it allows us to clearly see the risk level of each factor and level. Another advantage of SPA is the ability to calculate the volatility range of risk. This gives us an idea of the extent to which the risk of factors and systems varies. The meaning of the partial coefficient is similar to the concept of acceleration in physics. It mainly shows the trend of factors or systematic risk change. Similarly, the risk of future changes is also a matter of great concern to us. For the lack of human factors management, the

TABLE 18.8
Common Path of the Accident

No	Accident Path	Interpretation
1	A-B-C-D-E	This indicates that there are problems in all aspects, and the hidden dangers exist for a long time. The overall human factor management is poor.
2	A-B-D-E	Similar to path 1. Because there is no factor C, so the accident is mostly caused by improper organization or supervision.
3	A-B-E	Accidents are mainly caused by improper organization or supervision. People cause accidents in normal operation.
4	A-C-D-E	The accident is mainly caused by improper organization and unsafe behaviour. Has nothing to do with supervision. Hidden latency is longer.
5	A-D-E	The accident is mainly caused by improper organization and unsafe behaviour. Has nothing to do with supervision. Hidden latency is shorter.
6	B-C-D-E	The organizational influence is safety. The accident occurred mainly due to unsafe supervision and unsafe arts. Hidden latency is longer.
7	B-D-E	The organizational influence is safety. The accident occurred mainly due to unsafe supervision and unsafe arts. Hidden latency is shorter.
8	C-D-E	The whole safety process, organization and supervision are safe. Staff mental or physical problems cause accidents to occur due to long work hours.
9	D-E	This type of accident is mainly due to staff skill-based errors or routine violations.

(Adapted with permission from Xuecai et al. 2018 Copyright © Elsevier)

enterprise's security score will show a discrete distribution. Therefore, the application of the SPA–Markov chain prediction model was proposed to predict the risk. The error rate is less than 2%, so the method can be applied to practice. Due to the strengthening of human management, its safety features are non-discrete. When the SPA–Markov chain prediction model is applied to predict the risk, the coefficients need to be corrected. This creates errors. Therefore, in this chapter the method of fitting to predict was adopted. The case study involves ABC analysis and the S-O-R model were used for human risk management. By applying the ABC method, we can obtain the antecedents of the risk to human factors, and estimate the possible consequences. This is important for us to analyse the emergence of risk factors and to propose measures to help avoid accidents. Through use of the "S-O-R" method, we can obtain more effective safety "stimulus" means so that the corresponding safety measures can be put into place. This chapter also analyses the path of the accident, which is also important for the prevention of accidents. The analysis of common unsafe factors can rapidly improve the safety level of human factors. Finally, two different types of human factors are discussed. The leadership of the enterprise is the builder of a culture of safety, the maker of the security system and the implementer

TABLE 18.9

The Major Unsafe Human Factors and their Effective Stimulus Measures

No	Human Factor	Effective Stimulus Measure
1	A2	m1: foster an enterprise safety culture. Put the concept of safety production into every employee's mind. Provide staff with accident avoidance skills.
	A3	m2: establish a sound enterprise safety management system and safety system. Professional safety personnel participate in safety management.
2	B1	m3: establish a supervision and management system. An object supervised by multiple people. Have penalties and incentives.
	B3	m4: involve relevant experts and scholars to participate in the resolution of the problem. For effective safety measures or safety methods have rewards.
3	C2	m5: strengthen the training of job skills and safety emergency skills. Reasonable arrangements for staff to work time. Staff training qualified before induction.
	C3	m6: strengthen communication with employees and solve the psychological problems of employees. Adjust work intensity. Reasonable holidays.
	C4	m7: sick, injured and drunk to work is strictly prohibited. A reasonable arrangement of breaks.
	C6	m8: strengthen moral and safety skills training, to prevent fluke psychology, provincial psychology, impetuous psychology and other unhealthy psychologies.
4	D1	m9: strengthen pre-job training. Simplify operation flow. Try to be mechanized. Staff regular skills learning and safety learning. Penalties for violations
	D4	m10: standardize operation flow and operation procedure. Strengthen staff skills training. Strong supervision. Penalties for violations
5	E2	m11: strengthen the construction of enterprise's own rescue team. Equipped with rescue supplies and rescue vehicles as required. Set up a rescue alliance with the perimeter rescue team and the hospital
	E3	m12: equipped with rescue supplies and rescue vehicles as required. There are detailed rescue methods and command systems. Request rescue teams and hospitals and other professional rescue personnel to participate in the rescue.

(Adapted with permission from Xuecai et al. 2018 Copyright © Elsevier)

of safety management Therefore, the human factors in the organization should be managed from the perspective of safety awareness and safety skills. However, the individual-level human factors should be managed from the perspectives of safety skills, work attitude and personal health status. It is a long-term and complicated process to improve the safety of people working in a chemical plant. The author will continue to use the method to solve the existing problems in the plant through the comprehensive, reasonable and accurate assessment and management of human factors.

18.8 SUMMARY

A new method is investigated in this chapter to assess and manage human factors. Based on the HFACS framework, this chapter establishes the human factor risk assessment model. Through employing the model in common with set pair analysis, an assessment of the safety of human factors was realized. The calculated number of

TABLE 18.10

Different Types of Human Factors to Improve Measures

Human Factors Category	Measure
Organization level of human factors	• Strengthen the management of safety awareness and safety management ability • Establish a good corporate safety culture • Establish a sound enterprise safety management system and safety rules and regulations • Implement the safety supervision system and the safety responsibility system • Strengthen staff operation skills and safety skills training • The establishment of enterprise emergency rescue teams, equipped with adequate safety substances and equipment • Reasonable arrangement of staff hours and breaks • Organize different types of recreational activities to improve the mental state of staff
Individual level of human factors	• Strengthen operational skills and safety skills to learn and reduce skill-based errors • Familiar with the enterprise safety management system and safety supervision system to reduce the decision error • Be familiar with your own work scenes and mechanical equipment to reduce perceptual errors • Safety management and safety supervision in accordance with the correct requirements • Follow the correct process to reduce routine errors • Ensure adequate rest time, adequate preparation before work • Participate in recreational activities and adjust the mental state • Establish a correct working attitude to prevent the occurrence of a poor mental state. • Master the rescue skills, learn to save and save each other

(Adapted with permission from Xuecai et al. 2018 Copyright © Elsevier)

connections consists of 5 data points, which give us a clear idea of the proportion of the 5 risk levels. Therefore, we can determine the safety score of the object, risk level, and risk development interval. By calculating the partial connection number, we obtain the risk trend of the factors. According to the evaluation results, we can divide the human factors into different categories and adopt different security policies. For the lack of human factors management, the security score of the enterprise will show a discrete distribution. This chapter investigates about establishing a SPA–Markov chain risk assessment model to predict human factor risks. The error rate was found to be less than 2%, so the method can be applied in practice. This chapter employs ABC analysis and the S-O-R model were used for human risk management. From an application point of view, this method is effective in improving the safety of human factors. At the same time, 12 common unsafe factors and their effective safety "stimulus" measures were summarized and the accident path was researched. Various suggestions are also proposed according to the different levels of the human factors.

APPENDIX 18A

TABLE 18A.1

Factor Weights and Questionnaire Statistics

Factor	Expert 1	Expert 2	Expert 3	Expert 4	Expert 5	Expert 6	Expert 7	Expert 9	Expert 10	W1	W2	W3	W	Questionnaire data V	IV	III	II	I	Normalized data (assessment matrix R) V	IV	III	II	I
A1	0.40	0.38	0.39	0.41	0.40	0.40	0.38	0.38	0.39	0.391	0.575	0.483	0.083559	66	48	46	25	15	0.330	0.240	0.230	0.125	0.075
A2	0.35	0.34	0.35	0.36	0.38	0.40	0.36	0.35	0.33	0.360	0.248	0.304	0.052592	50	60	45	28	17	0.250	0.300	0.225	0.140	0.085
A3	0.25	0.28	0.26	0.23	0.22	0.20	0.26	0.27	0.28	0.249	0.177	0.213	0.036849	48	45	47	37	23	0.240	0.225	0.235	0.185	0.115
A	0.20	0.19	0.18	0.19	0.17	0.17	0.20	0.18	0.18	0.182	0.164	0.173											
B1	0.26	0.28	0.27	0.28	0.29	0.27	0.31	0.30	0.28	0.283	0.217	0.250	0.047750	42	40	58	34	26	0.210	0.200	0.290	0.170	0.130
B2	0.24	0.20	0.21	0.20	0.20	0.22	0.19	0.21	0.23	0.210	0.278	0.244	0.046604	50	46	42	38	24	0.250	0.230	0.210	0.190	0.120
B3	0.28	0.28	0.29	0.30	0.27	0.30	0.31	0.30	0.29	0.290	0.208	0.249	0.047559	36	44	58	34	28	0.180	0.220	0.290	0.170	0.140
B4	0.22	0.24	0.23	0.22	0.24	0.21	0.19	0.21	0.20	0.217	0.297	0.257	0.049087	44	38	62	30	26	0.220	0.190	0.310	0.150	0.130
B	0.10	0.12	0.12	0.11	0.13	0.11	0.11	0.11	0.13	0.114	0.267	0.191											
C1	0.14	0.15	0.13	0.14	0.12	0.13	0.15	0.16	0.16	0.142	0.062	0.102	0.024888	45	51	57	29	18	0.225	0.255	0.285	0.145	0.090
C2	0.10	0.09	0.09	0.08	0.10	0.11	0.10	0.10	0.08	0.093	0.139	0.116	0.028304	42	44	47	37	30	0.210	0.220	0.235	0.185	0.150
C3	0.11	0.10	0.09	0.09	0.08	0.09	0.11	0.12	0.08	0.097	0.105	0.101	0.024644	36	47	51	36	30	0.180	0.235	0.255	0.180	0.150
C4	0.09	0.10	0.11	0.09	0.08	0.08	0.08	0.07	0.08	0.086	0.078	0.082	0.020008	42	50	48	36	24	0.210	0.250	0.240	0.180	0.120
C5	0.07	0.08	0.08	0.09	0.10	0.13	0.14	0.06	0.10	0.082	0.054	0.068	0.016592	52	58	64	16	10	0.260	0.290	0.320	0.080	0.050
C6	0.14	0.15	0.14	0.14	0.16	0.13	0.15	0.15	0.16	0.147	0.055	0.101	0.024644	39	41	50	40	30	0.195	0.205	0.250	0.200	0.150
C7	0.08	0.09	0.09	0.08	0.08	0.08	0.11	0.07	0.08	0.080	0.142	0.111	0.027084	45	47	48	44	16	0.225	0.235	0.240	0.220	0.080
C8	0.10	0.11	0.10	0.09	0.07	0.10	0.11	0.08	0.07	0.091	0.101	0.096	0.023424	50	52	54	28	16	0.250	0.260	0.270	0.140	0.080
C9	0.09	0.08	0.09	0.10	0.09	0.10	0.09	0.11	0.07	0.101	0.073	0.087	0.021228	44	48	56	32	20	0.220	0.240	0.280	0.160	0.100
C10	0.08	0.07	0.08	0.09	0.10	0.07	0.07	0.08	0.12	0.081	0.191	0.136	0.033184	39	48	53	38	22	0.195	0.240	0.265	0.190	0.110
C	0.30	0.30	0.30	0.28	0.29	0.31	0.29	0.32	0.29	0.300	0.188	0.244											
D1	0.27	0.28	0.25	0.25	0.25	0.26	0.28	0.29	0.27	0.267	0.353	0.310	0.077810	50	52	54	24	20	0.250	0.260	0.270	0.120	0.100
D2	0.19	0.20	0.18	0.18	0.19	0.18	0.20	0.18	0.19	0.185	0.111	0.148	0.037148	54	56	60	18	12	0.270	0.280	0.300	0.090	0.060
D3	0.15	0.13	0.17	0.16	0.18	0.15	0.13	0.14	0.14	0.150	0.260	0.205	0.051455	60	56	62	12	10	0.300	0.280	0.310	0.060	0.050
D4	0.31	0.32	0.30	0.32	0.28	0.32	0.33	0.32	0.32	0.313	0.103	0.208	0.052208	32	45	49	46	28	0.160	0.225	0.245	0.230	0.140
D5	0.08	0.07	0.10	0.09	0.10	0.09	0.06	0.08	0.08	0.085	0.173	0.129	0.032379	62	60	54	16	8	0.310	0.300	0.270	0.080	0.040
D	0.32	0.29	0.30	0.33	0.28	0.30	0.30	0.29	0.35	0.304	0.199	0.251											
E1	0.36	0.38	0.34	0.34	0.42	0.38	0.41	0.42	0.35	0.380	0.430	0.405	0.057105	54	52	50	24	20	0.270	0.260	0.250	0.120	0.100
E2	0.34	0.32	0.34	0.33	0.30	0.30	0.31	0.30	0.33	0.315	0.227	0.271	0.038211	38	48	44	38	32	0.190	0.240	0.220	0.190	0.160
E3	0.30	0.30	0.32	0.33	0.28	0.32	0.28	0.30	0.32	0.305	0.343	0.324	0.045684	36	44	46	40	34	0.180	0.220	0.230	0.200	0.170
E	0.08	0.10	0.10	0.09	0.13	0.11	0.09	0.10	0.11	0.100	0.182	0.141											

REFERENCES

Cai, M., Gong, S., Li, X., 2008. Technique of human error failure analysis based on analytic hierarchy process. *J. Saf. Sci. Technol.* 4 (2), 74–77.

Celik, M., Cebi, S., 2009. Analytical HFACS for investigating human errors in shipping accident. *Accid. Anal. Prev.* 41 (1), 66–75.

Chauvin, C., Lardjane, S., Morel, G., Clostermann, J.-P., Langard, B., 2013. Human and organisational factors in maritime accidents: analysis of collisions at sea using the HFACS. *Accid. Anal. Prev.* 59 (5), 26–37.

Chen, S.-T., Wall, A., Davies, P., Yang, Z., Wang, J., Chou, Y.-H., 2013. A Human and Organisational Factors (HOFs) analysis method for marine casualties using HFACS-Maritime Accidents (HFACS-MA). *Saf. Sci.* 60 (12), 105–114.

Chong, T., Yi, S., Heng, C., 2017. Application of set pair analysis method on occupational hazard of coal mining. *Saf. Sci.* 92, 10–16.

Dai, W.Z., 2000. A new kind of model of the dynamic multiple attribute decision making based on new effective function and its application. *Control Decis.* 15 (2), 197–200.

Daramola, A.-Y., 2014. An investigation of air accident in Nigeria using the Human Factors Analysis and Classification System (HFACS) framework. *J. Air Transp. Manag.* 35 (4), 39–50.

Fu, G., 2013. *Safety Management—A Behavior-based Approach to Accident Prevention.* Science Press, Beijing.

Harvey, B., 2016. The Oaks Colliery disaster of 1866: a case study in responsibility. *Bus. Hist.* 58 (4), 501–531.

Heinrich, W.H., Peterson, D., Roos, N., 1980. *Industrial Accident Prevention.* McGraw-Hill Book Company, New York.

Konstantin, P., 2010. The effects of error management climate and safety communication on safety: a multilevel study. *Accid. Anal. Prev.* 42, 1498–1506.

Kumar, P., 2016. Categorization and standardization of accidental risk-criticality levels of human error to develop risk and safety management policy. *Saf. Sci.* 85 (1), 88–98.

Lenne, M.-G., Salmon, P.-M., Liu, C.-C., Trotter, M., 2012. A systems approach to accident causation in mining: an application of the HFACS method. *Accid. Anal. Prev.* 48 (3),111–117.

Li, C., Sun, L., Jia, J., Cai, Y., Wang, X., 2016. Risk assessment of water pollution sources based on an integrated k-means clustering and set pair analysis method in the region of Shiyan. *China. Sci. Total Environ.* 557–558, 307–316.

Madigan, R., Golightly, D., Madders, R., 2016. Application of Human Factors Analysis and Classification System (HFACS) to UK rail safety of line incidents. *Accid. Anal. Prev.* 97, 122–131.

Mehrabian, A., Russell, J.A., 1974. *An Approach to Environmental Psychology.* MIT Press, Cambridge.

Nie, B., Xin, H., Xin, S., Li, A., 2016. Experimental study on physiological changes of people trapped in coal mine accidents. *Saf. Sci.* 88, 33–43.

Patterson, J.M., Shappell, S.A., 2010. Operator error and system deficiencies: analysis of 508 mining incidents and accidents from Queensland, Australia using HFACS. *Accid. Anal. Prev.* 42(4), 1379–1385.

Reason, J., 1990. *Human Error.* Cambridge University Press, New York.

Shappell, S.A., Wiegmann, D.A., 2001. Applying reason: the Human Factors Analysis and Classification System (HFACS). *Hum. Factors Aerosp. Saf.* 1, 59–86.

Shappell, S.A., Wiegmann, D.A., 2004. HFACS analysis of military and civilian aviation accidents: a North American comparison. In: Proceedings of International Society of Air Safety Investigators, Australia, Queensland, November 2–8.

Soner, O., Asan, U., Celik, M., 2015. Use of HFACS-FCM in fire prevention modelling on board ships. *Saf. Sci.* 77, 25–41.

Stewart, J.M., 2002. *Managing for Word Class Safety*. A Wiley Interscience Publication, New York, pp. 1–31.

Stewart, J.M., 2011. The turnaround in safety at the Kenora pulp paper mill. *Prof. Saf.*, 34–44.

Su, M.R., Yang, Z.F., Chen, B., et al., 2009. Urban ecosystem health assessment based on energy and set pair analysis—a comparative study of typical Chinese cities. *Ecol. Modell.* 220 (18), 2341–2348.

Tabibzadeh, M., 2014. A Risk Analysis Methodology to Address Human and Organizational Factors in Offshore Drilling Safety: With an Emphasis on Negative Pressure Test. University of Southern California, Los Angeles.

Tao, J., Fu, M., Sun, J., Zheng, X., Zhang, J., Zhang, D., 2014. Multifunctional assessment and zoning of crop production system based on set pair analysis-a comparative study of 31 provincial regions in mainland China. *Commun. Nonlinear Sci. Numer. Simul.* 19 (5), 1400–1416.

Tarlor, G., Hegney, R., Easter, K., 2001. *Enhancing Safety*, 3rd ed. West one, West Australia.

Wei, C., Dai, X., Ye, S., Guo, Z., Wu, J., 2016. Prediction analysis model of integrated carrying capacity using set pair analysis. *Ocean Coast. Manag.* 120, 39–48.

Wiegmann, D.A., Shappell, S.A., 1997. Human factors analysis of post accident date: applying theoretical taxonomies of human error. *Int. J. Aviat. Psychol.* 7, 67–81.

Wiegmann, D.A., Shappell, S.A., 2001a. Human error analysis of commercial aviation accidents: application of the Human Factors Analysis and Classification System. *Aviat. Space Environ. Med.* 72, 1006–1016.

Wiegmann, D.A., Shappell, S.A., 2001b. Applying the Human Factors Analysis and Classification System to the analysis of commercial aviation accident date. In: Proceedings of 11th International Symposium on Aviation Psychology, Ohio State University, Columbus, OH.

Wiegmann, D.A., Shappell, S.A., 2001c. Human error perspectives in aviation. *Int. J. Aviat. Psychol.* 11, 341–357.

Wiegmann, D.A., Shappell, S.A., 2003. *A Human Error Approach to Aviation Accident Analysis: The Human Factors Analysis and Classification System*. Ashgate, Aldershot, UK.

Wu, T., 2009. Application on the analysis of developmental trend of the student mark with five-element partial connection number. *Math. Pract. Theory* 39 (5), 53–59.

Xuecai, X., Deyong, G., 2018. Human factors risk assessment and management: Process safety in engineering. *Proc. Safety Environ. Prot.* 113, 467–482.

Zhan, Q., Zheng, W., Zhao, B., 2017. A hybrid human and organizational analysis method for railway accidents based on HFACS-Railway Accident (HFACS-RAs). *Saf. Sci.* 91, 232–250.

Zhao, K.-Q., 1989. *Set pair and set pair analysis a new concept and systematic analysis method*. In: *Proceeding of the National Conference on System Theory and Regional Planning*, Baotou, China, pp. 87–91.

Zhao, K.-Q., 2000a. *Set Pair Analysis and its Preliminary Application*. Hangzhou Science and Technology Press, Hangzhou, pp. 20–21.

Zhao, K.-Q., 2000b. *Set Pair Analysis and its Application*. Zhejiang Science and Technology Press, Hangzhou.

Zhao, K.-Q., 2005. Partial connection number. In: *Progress of Artificial Intelligence in China*. Beijing University of Posts and Telecommunications Press, Beijing, pp. 884–885.

Zhou, X.-H., Zhang, J.-J., 2013. Risk comprehensive assessment method and its application based on the five-element connection number. *Syst. Eng. Theory Pract.* 33(8), 2169–2175.

19 Analysis of Off-Site Emergency Procedures and Reciprocation for Nuclear Accidents

19.1 INTRODUCTION

Since the 1950s, as a result of the development of civil nuclear power, relatively few accidents have taken place with off-site consequences. Because of the efforts made in providing the necessary preventive and protective measures to mitigate such hazards for all types of nuclear facilities has reduced the chances of occurring. Such preventive measures include the structures, systems and components (SSC) of the nuclear facility building, and also involving the management of the facility. Due to the adoption of such measures, there is a reduction in the probability of potential risk from high-consequence accidents, which eventually motivates the high financial incurring market to cover safety. However, with the risk from off-site consequences cannot be eliminated entirely. It was found that with each accident, and looking over the overall impacts associated with the accident, there is a great challenge ahead related to the policies and practices undertaken as a preventive measure to mitigate the consequences of the accident. For example, the permanent relocation of people within the evacuation zones of Chernobyl has been shown to be less optimal from the economic perspective, whereas a short-term evacuation policy followed by the later return of the displaced could have been adopted, coupled with aggressive remediation (Waddington et al., 2017a).

From the Fukushima Daiichi incident in 2011, two such examples of challenges are as follows: (a) during a prolonged release of radionuclides it is generally suggested that if a later evacuation is required, then short-term sheltering may be detrimental, since the potential risk of being exposed to increased radiation occurs during evacuation (Gering et al., 2013); (b) there were reported to have been no radiation-induced deaths following the accident at Fukushima Daiichi. By contrast, during the subsequent evacuation and relocation it was observed that almost 1800 had lost their lives (The Reconstruction Agency, 2014), and many others had experienced detrimental health effects (Yabe et al., 2014). The five levels of the defence-in-depth philosophy is usually accessed in most countries for adopting nuclear safety (IAEA, 2012a). The first four levels can be viewed as on-site SSC. Due to the high cost of SSC, a decision is required about its cost effectiveness while implementing it during unlikely events (Health and Safety Executive, 2001). The main mitigating safety measures in Level 5 generally refer to off-site emergency preparedness. However, by following the international guidelines, such arrangements are generally benchmarked (e.g. IAEA, 2002a). Between different national policies, clear variations are observed that depend upon the factors, including the national approach to dealing with civil emergencies, their political recognition, public trust in the government, and public perception and aversion to radiological risk, as explained in IAEA (2012b).

19.2 STUDY ON NUCLEAR ACCIDENTS

This study provides a comparison of UK's approach with international guidance on emergency preparedness and responses to accidents. The effect of the events at Fukushima Daiichi and the current emergency preparedness and response procedures were explained in Section 19.3. The UK's approach to performing the economic assessments and of nuclear accidents and factors that affect the severity of a nuclear accident (from the health and economic perspectives) were discussed in sections 19.4 and 19.5 (Heffron et al., 2016). Considering the economic factors (concern to the licensee and/or government), this study reviews the post-accident remediation measures to reduce radiation doses, the risks from ionizing radiation considering the health and safety of people and the environment (cf. the Safety Objective in IAEA, 2006). In this study, the likely effects of a major nuclear reactor accident, risks from conventional hazards, and COCO-2 economic costing model (Charnock et al., 2013; Higgins et al., 2008) was applied in order to assess the economic and health costs in the UK realistic demography (Ashley et al., 2017a,b).

19.3 DIFFERENT PHASES OF A NUCLEAR ACCIDENT

The three chronological phases of emergency preparedness for a nuclear accident are the planning, response, and recovery phases, although the distinctions between them are not entirely clear-cut. The boundaries of the phases should not be viewed as definitive as there may be some overlap.

19.3.1 PLANNING PHASE

19.3.1.1 Requirements

Countries with nuclear facilities mandates for emergency preparedness and planning which is generally supported in the form of legal representation (plans, evaluating their basis, exercising, and implementing if needed) that varies between countries in case of an accident. Article 16.1 of the Convention on Nuclear Safety (IAEA, 1994) states that "Each contracting party shall take the appropriate steps to ensure that there are on-site and off-site emergency plans that are routinely tested for nuclear installations and cover the activities to be carried out in the event of an emergency. For any new nuclear installation, such plans shall be prepared and tested before it commences operation above a low power level agreed by the regulatory body." The Convention of Nuclear Safety has 77 contracting parties with 65 signatories, including the United Kingdom and the European Union (under the auspices of EURATOM) (IAEA, 2014a). In the UK, the Nuclear Installations Act 1965 (HM Government, 1965), as amended by the Energy Act 2013 (HM Government, 2013), refers to emergency preparedness within Section 4.32: "Conditions that may be attached to a license by virtue of subsection (1) may in particular include provision-"clause (c):"with respect to preparations for dealing with, and measures to be taken on the happening of, any accident or other emergency on the site;" The Office of Nuclear Regulation (ONR) is responsible for administrating these acts and the requirements regarding emergency preparedness are covered in standard License Condition 11 (Office of Nuclear Regulation, 2013a, p. 11). The ONR, in its guidance document "Licensing Nuclear Installations" (Office of Nuclear Regulation, 2014), states as part of the supporting evidence required when applying for a Nuclear Site License the applicant should include: "details of appropriate emergency arrangements and a suitable emergency plan (this may be limited in extent for the period before nuclear fuel is brought onto the site)." The ONR also requires that before the start of active commissioning, adequate emergency arrangements should be in place and exercised as appropriate. The Health and Safety at Work Act 1974 (HM Government, 1974) covers thre protection of employees and the public from work activities; under it, specific regulations are made that deal with radiation and nuclear emergencies such as the Radiation (Emergency Preparedness and Public Information) Regulations 2001 (Health and Safety Executive, 2002).

A separate act, the Civil Contingencies Act 2004 (HM Government, 2004), centers on the roles for responders to all kinds of emergencies. Article 9 of REPPIR 2001 covers off-site emergency plans, and Item #1 requires that the local authority should draw up guidelines to restrict exposure from a reasonably foreseeable off-site radiation emergency (REPPIR defines a radiation emergency as an event which is likely to result in a member of the public receiving an effective dose in excess of 5 mSv over the first year). It is ONR's duty to determine whether the local authority and operator has met the requirements of REPPIR. As stated by the Office of Nuclear Regulation (2013b): "REPPIR presents the legal framework for protection of the public through emergency preparedness for all radiation accidents" and "REPPIR addresses the need for both on-site and off-site emergency planning. However, for operators of nuclear licensed sites the requirement for an on-site emergency plan is already covered by the existing nuclear site license conditions (LC11 and LC9). For operators of nuclear licensed sites, compliance with the LCs should satisfy equivalent provisions in REPPIR. For operators of nuclear licensed sites REPPIR mandates additional legal requirements for off-site emergency planning and the provision of information to the public." Also, ONR is empowered under Regulations 9 (1) and 16 (1) to determine the areas of the Detailed Emergency Planning Zone (DEPZ) and the Public Information Zone. REPPIR is the UK Regulation that puts into force the EC Basic Safety Standards Directive 96/29/EURATOM (European Commission, 1996) and Public Information Directive 89/618/EURATOM (European Commission, 1989). Since the accident at Fukushima Daiichi, these directives have been repealed and replaced by 2013/59/EURATOM with a deadline of February 6, 2018 for Member States to ensure their legislation is in compliance (European Commission, 2013, p. 59). The revision of the EC Basic Safety Standards Directive was in part due to the IAEA's Basic Safety Standards being revised as part of changes suggested by the International Commission for Radiological Protection (ICRP) as detailed in IAEA (2014b) and also due to the events at Fukushima Daiichi. National-level emergency planning for civil nuclear installations is coordinated in the United Kingdom by the Department for Business, Energy and Industrial Strategy (BEIS), and is facilitated through various fora. In 2015, the Department's predecessor (the Department for Energy and Climate Change, DECC) published revised guidance, the "National Nuclear Emergency Planning and Response Guidance" which was compiled with input from those with expertise in and responsibilities for nuclear emergency planning (HM Government, 2015).

19.3.1.2 Contents of An Emergency Plan

An off-site emergency plan necessities countermeasures (evacuation, sheltering, administering stable iodine, decontamination, and food banning) to minimize the risks arising from exposure of public to radiation in the surrounding localities that can be implemented over short and long time frames. The countermeasures include procedures for establishing the maintenance of essential infrastructure, transport, communication, command and control centers. The detailed policies explaining the emergency planning procedures of different zones was presented in OECD (2003). The areas covering the precautionary countermeasures extend radially outwards by 6–7 km, which includes communities that might be affected by the accidents as detailed by the Office of Nuclear Regulation (2016). The concept of "extendibility" in emergency planning was introduced in the Hinkley Point C inquiry 1990 due to the formation of an irregularly shaped emergency area and an "Extended Emergency Planning Zone" (EEPZ) was developed that could extend up to a 15 km radius (Health and Safety Executive, 1990). The countermeasures also include restricting food and drinking water beyond EEPZ due to long-term ingestion and contamination for years. However, the accident at Fukushima Daiichi advised a site-by-site basis extending up to 30 km of the EEPZ and is influenced by characteristics and the local geographical surroundings (HM Government, 2015). This enables the operator and local authorities to work together to determine the most effective countermeasures and public communication strategies. The predistribution of iodine, namely "door-to-door" distribution, schools, hospitals, evacuation reception and collection centers and the pre-accident information supplied to the residents were determined in the plan (REPPIR 2001).

The detailed description of off-site emergency plans are publicly available for Sizewell (Suffolk Resilience, 2017), Hinkley Point (Somerset County Council, 2008), Dungeness (Kent County Council, 2015), Wylfa (Isle of Anglesey County Council, 2011), Hunterston (Ayrshire Civil Contingencies Team, 2015), and Torness (East Lothian Council, 2016). However, some details on off-site arrangements produced by the operator are provided in the Emergency Plan for each nuclear power plant, Heysham (British Energy Generation Ltd., 2007a) and Hartlepool (British Energy Generation Ltd., 2007b). The location of emergency control, reception centers, and the need for primary health services, the requirements for specific equipment and the capability of monitoring the treatment of contaminated people has to be identified as a part of the off-site emergency plan. As shown in Table 19.1, in the UK, the countermeasures are called "Emergency Reference Levels" (ERLs) (Morrey, 1997). The use of Operational Intervention Levels (OILs), which are prescriptive limits based on dose rates, as outlined in Table 19.2, were advised by the recent IAEA guidance on nuclear emergency planning (IAEA, 2015). Furthermore, ICRP uses the concept of reference levels for existing and emergency exposure situations (ICRP, 2009a). The reference level represents the level of residual dose, or risk and for emergency exposure was in the range of 20-100 mSv effective dose per year. Protective measures against all exposure is suggested by ICPR if the dose increases to 100 mSv. Euratom's Basic Safety Standards Directive also details the use of reference levels (European Commission, 2013). In the case of countermeasures for foods, the UK uses Community Food Intervention Levels (CFILs) established by the EU under Council Regulation (Euratom) (European Commission, 2016). The intervention levels of food are based on the food intake of representative persons and may vary for different foodstuffs consumed locally or worldwide. The food intervention levels are often very conservative, as the ingestion doses are limited to all potential consumers (Waddington et al., 2017). The UK ERLs are presently being reviewed and are compared with the use of OILs. OILs are straightforward to implement, comprise single values (cut-offs between an interventions), and rely on assumptions and require a large number of accurate dose measurements. The use of both ERLs and OILs in future emergency management plans is advantageous.

TABLE 19.1

UK Emergency Reference Levels (ERLs) that Provide Guidance for when Specific Countermeasures Should be Adopted

Countermeasure	Body Organ	Dose Equivalent Level[a] (mSv)	
		Lower	Upper
Sheltering	Whole body[b]	3	30
	Thyroid, lung, skin[c]	30	300
Evacuation	Whole body[b]	30	300
	Thyroid, lung, skin[c]	300	3000
Administration of stable iodine	Thyroid	30	300

a These values should be interpreted as approximate figures.

b The numerical values for whole body ERLs may also be used for comparison with the quantities of effective dose and effective dose equivalent.

c These single organ ERLs were specified prior to the definition of effective dose by the ICRP. With exception to stable iodine ERLs, their use now would not normally be expected

(Adapted from Ashley et al. 2017 Copyright © Elsevier)

TABLE 19.2

Dose Rate Limits as Suggested by the IAEA (2013) and Operational Intervention Levels (OILs) Contained in IAEA (2015). OILs are Intended as Prescriptive Limits, as Opposed to the Indicative ERLs in Table 19.1

Time After Release	Dose Rate Limit	Intervention	Effective dose to Representative Person
<1 day (OIL1)	≥1000 µSv/h	OIL1: immediate safe evacuation (plus associated countermeasures)	100 mSv over 7-day exposure period
<10 days (OIL2)	≥100 µSv/h	OIL2: preparation for relocation (plus associated countermeasures) to be done within a week to a month thereafter	100 mSv over 1-year exposure period
>10 days–1 month (OIL2) <7 days (OIL3)	≥25 µSv/h ≥1 µSv/h	OIL3: stop distribution and consumption of non-essential food and water that is potentially at risk. Assess thereafter using OIL7	10 mSv over 1 year of consumption
>2 days (OIL7)	1000 Bq/kg ^{131}I or 200 Bq/kg ^{137}Cs	OIL7: stop consumption if non-essential. Replace essential foods or relocate public if replacements are not available	10 mSv over 1 year of consumption

(Adapted from Ashley et al. 2017 Copyright © Elsevier)

19.3.2 RESPONSE PHASE

The response to events that involve, or have the potential to involve, a significant release of radioactivity in a nuclear facility depends on its magnitude and its potential to cause harm. In some cases (delay initial incident and the release), countermeasures may be unnecessary as the event is terminated by on-site actions without the occurrence of an off-site release. The response phase comprises two sub-phases: pre-release and post-release actions.

19.3.2.1 Pre-Release Phase

The response phase before an actual release and the event has the potential to lead to a significant radioactive release is pre-release stage. The site's nuclear emergency arrangements are triggered as soon as the operator realizes the potential for an off-site release and this would involve notifying the off-site groups about restoring the facility and mitigate any releases from the site. The different technologies and designs of nuclear power plants adopt different accident sequences and the time period between the initial alarm and a subsequent release can vary significantly. The containment is predicted to stay intact for at least 24 hours, as per Level-2 PSA of the European Pressurized Reactor (EPR). However, in a small number of cases the release phase is predicted to start less than two hours after the initiating event (AREVA and EDF Energy, 2012). Headed by a senior police commander, the off-site plan in the UK is effectively run by the local police emergency organization who is advised by representatives of many agencies. The radiation emergency management was provided by the government's specialist scientific and technical advisor, the senior member of ONR on temporary attachment to DECC. Considering ONR, Environment Agency and Food Standards Agency resources, the local responders were advised by the site operators and since 2013 a Science and Technical Advice Cell (STAC) has provided independent science and technical advice.

During an emergency, Lead Government Departments (LGD) are responsible for accessing technical and scientific advice in a timely manner at national level. The roles of the Science Advisory Group for Emergencies (SAGE), STAC, and the response structure are detailed in the National Nuclear Emergency Planning and Response Guidance (HM Government, 2015). The precautionary countermeasures (sheltering and evacuation) may be initiated by alerts from the emergency siren and local radio/TV for guidance to mitigate radiation exposure and are based on potential for the incident to escalate, wind speed/direction and changes in meteorological conditions. Emergency management software such as RODOS (Ehrhardt and Weis, 2000) can assist in identifying the source term, the time available, predictions of the prospective releases, and the precautionary countermeasures to be adopted. No precautionary measures were instigated during the accidents at Three Mile Island 2 (TMI-2) (Moss and Sills, 1981) and Chernobyl (Medvedev, 1990). At Chernobyl, during the low-power experiments, the unstable reaction conditions and the overridden safety systems resulted in an explosion occurring within one minute. At TMI-2, the fuel melt and hydrogen release persisted even after the incident and progressed contaminating the building. However, the reports projected that there was no release of activity to the environment, restricting the time taken for the precautionary measures close at the site (Smith and Beresford, 2005) and the local population had not received any prior information of the hazards of radiation during an accident (International Chernobyl Project and IAEA, 1991). Table 19.3 details the off-site emergency countermeasures at Fukushima Daiichi on March 11, 2011. The on-site systems worked adequately even after the initial seismic event cut the off-site power. However, the associated tsunami flooded the emergency power systems after 45 min, disengaging on-site power. The reactor pressure vessel was damaged after $\sim$11 h and the fuel damage in Unit #1 started $\sim$4 h after the tsunami. Fukushima Daini, a nearby nuclear power station, located $\sim$16 km from Fukushima Daiichi lost its off-site power due to the earthquake and residents within 3 km were ordered to evacuate those within 10 km were ordered to shelter. Whilst precautionary measures for nuclear-related incidents may be able to limit the radiological risk, there is the potential for such actions to induce widespread panic. The precautionary measures were able to limit the radiological risk. However, such actions may result in panic as at TMI, though pregnant women and children were suggested to evacuate, many residents left the area, increasing the risks from such panic than the radiological risks.

19.3.2.2 Post-Release Phase

During the environmental release, the aim of the emergency plans is to manage and include wider consequences (the health of the public) by effectively implementing the countermeasures that depend on the amount and isotopic composition of the radioactive material and the different parts of society that are affected. To limit the

TABLE 19.3

Brief Timeline for Countermeasures at Fukushima Dai-ichi

Friday 11 March

14.46: Earthquake occurred: off-site electrical power lost

15.27: TEPCO (the utility) reported the site struck by three tsunami waves, on-site emergency diesel back-up generators affected and became inoperative

15.42: TEPCO first emergency report to government

19.03: Nuclear emergency declared

20:50: Evacuation within 2 km of plant by Fukushima Prefecture

21:23: Prime Minister extended evacuation to 3 km and sheltering for 3–10 km

Saturday 12 March

05:44: Evacuation extended to 10 km

18:25: Evacuation extended to 20 km

Tuesday 15 March

11.01: Government ordered sheltering 20–30 km

(Adapted from Ashley et al. 2017 Copyright © Elsevier)

exposure of short-lived radiation, the administration of stable iodine is a key intervention that can be executed for people living close to the site and the community susceptible to the largest off-site radiation doses. The different methods employed for the predistribution of stable iodine prophylaxis is detailed by the World Health Organization (1999). The distribution efficacy of stable iodine at different sites ranges from 3.5 to 70% (National Research Council, 2004) and on the site in Barrow-in-Furness, UK, for example, it was recommended that the employed scheme was 60% effective after two years of adoption (Astbury et al., 1999). The detailed discussion on administration and iodine distribution in Europe was presented in Johnson (2003) and European Commission Directorate-General for Energy et al. (2010). For small releases, sheltering and iodine prophylaxis may be adequate. Evacuation may be necessary when the projected dose exceeds the ERLs and this is based on demographics and the age of the potential evacuees. Young children have a higher radiological risk and evacuating elderly may lead to medical complications. The emergency plan must be adaptive and robust to account for the release and withstand political and public pressure as the response phase evolves. In case of a major release, countermeasures were needed up to 40 km from the plant and the distance increases as the time progresses (Health and Safety Executive, 1994). The accident zone at Fukushima Daiichi tend to be wedge-shaped rather than circular and may with change in weather conditions. The information about the emergency response of the accident at TMI-2 was considered from Moss and Sills (1981).

An advisory note suggested that pregnant women and pre-school children within five miles of the station might wish to evacuate. However, two days after the initial accident (i.e., when initial difficulties were seen at the plant) a significant amount of individually-driven, voluntary self-evacuation was observed that was prompted by the potential release of material from the reactor's containment. From NRC estimates, voluntary self-evacuation comprised: ~21,000 people (60%) living within

5 miles of the plant; ~56,000 people (44%) living within a 5–10 mile radius of the plant; and~67,000 people (32%) living within a 10–15 mile radius of the plant. Most of these people left within two days of the event and the median time period before returning was seven days after the accident. The lead-up and response to the accident at Chernobyl is well documented (International Chernobyl Project and IAEA, 1991; Smith and Beresford, 2005), and is briefly summarized here. The steam explosion at 01:24 on April 26, 1986 destroyed part of the graphite core and the roof of Reactor #4. This led to fires throughout the turbine hall and also inside the building. Two on-site workers were killed by the explosion. Firefighters from Pripyat, the nearest town to Chernobyl and located ~3 km from the site, arrived at the site within minutes and within half an hour they were joined by other fire-fighters. It is noted that these fire-fighters did not have specialist training for events involving radioactive materials. The radiation levels in some of the accessible places to fight the fires exceeded 100 Gy/h[5] and due to the lack of radiation monitoring, personal dosimetry, and awareness, no measures were taken to limit the exposure and doses to the emergency personnel. 132 emergency workers were affected by acute radiation sickness in the first 12 hours following the accident and were initially hospitalized in Pripyat. Small squads of emergency personnel on site provided first aid, evacuated those who needed further medical assistance, and distributed potassium iodide tablets. After 12 hours, a specialized emergency team arrived and within 36 h examined more than 350 persons on-site. In the first three days of the accident, 299 people suspected of acute radiation sickness were sent to specialized treatment centers and hospitals; thereafter approximately 200 people were admitted to these centers and hospitals for monitoring of acute radiation sickness. By May 10, 1986, the fire at Reactor #4 was extinguished and efforts were in place to stabilize the site (as briefly described in the recovery section). In the months that followed, a total of 28 workers died from acute radiation sickness, as reported in Table 19.1 of the International Chernobyl Project and IAEA (1991). From an off-site perspective, no official information had been given to those living in Pripyat on the day of the accident. Around 44,000 people were evacuated ~36 h after the accident to Polesskoe. Once they arrived in Polesskoe, it took a further three days for doctors to perform blood tests and to refer those showing acute radiation sickness to nearby hospitals. The evacuation zone was expanded to a radial distance of 10 km on May 1, 1986 and the decision to evacuate radially out to 30 km on May 2, 1986 and took four days to complete. In total, ~116,000 people and 60,000 cattle were evacuated from an area of 3500 km², although a number either refused to leave or returned surreptitiously. From May 10, 1986, absorbed dose rate contours were produced that defined three separate areas: prohibited zones (200 μGy/h totaling 1100 km²); the initial evacuation zone (50 μGy/h totaling 3000 km²); and a strict-controlled area (30 μGy/h totaling 8000 km²). Further areas, where the dose rate exceeded 50 μGy/h, were evacuated after May 10, 1986. Since the accident at Chernobyl, ~350,000 people have been evacuated from the areas surrounding the plant. According to the IAEA (2005) report, no public deaths have been directly attributed to the accident although official estimates are that ~4000 deaths are likely to result from the doses received. Since that report was released, UNSCEAR (2008) concluded that the contamination of milk with 131I, for which prompt countermeasures were lacking, resulted in large doses to the thyroids of members of the general

public; this led to a substantial fraction of the more than 6000 thyroid cancers observed to date among people who were children or adolescents at the time of the accident (by 2005, 15 cases had proved fatal). Furthermore, current radiation-induced thyroid cancer risk models predict that the excess risk continues into later life as a proportion of the background risk of thyroid cancer, which increases with attained age (Wakeford, 2016). Aspects of the response phase from the accident at Fukushima Daiichi are contained in the "Report of the Japanese Government to the IAEA Ministerial Conference on Nuclear Safety" (Government of Japan, 2011) and are presented below. Upon the release caused by the hydrogen explosion at Unit #1 of Fukushima Daiichi, the evacuation zone was extended to a radius of 20 km from the nuclear power plant (at 18:25; March 12) – this was partly driven by the anticipation of the potential for further incidents to occur at the other units. This order was given at 18:25 on March 12. At 17:39 on March 12, the evacuation zone for Fukushima Daini was also extended to 10 km. The evacuation involved 78,200 people from both of these areas. On March 15 at 11:00, ~62,400 residents within the 20–30 km radius of Fukushima Daiichi were ordered to 'stay-in-house'. However, this order went well beyond the typical times associated with sheltering. It was observed that within the extended 'stay-in-house' zone that voluntary self-evacuation was evident, especially as time progressed, and that the standard of living for those within that area significantly decreased as time progressed. On March 25, the zone was reclassified and voluntary self-evacuation was supported. Given the damage caused by the earthquake and tsunami, roads and other infrastructures that would have assisted the emergency management plan were severely damaged, which hampered the evacuation. One observation from the incident at Fukushima Daiichi is that what may be an optimal response for one particular release (i.e. sheltering, which is typically limited to a maximum of two days), may become suboptimal if there are subsequent releases (i.e. people sheltering for too long, or evacuating through the plume of the second release) (Gering et al., 2013). This is highlighted (Figure 19.1) by the release pattern from Fukushima Daiichi, with the main releases from Units #2 and #3 occurring two days after the release at Unit #1. For the last set of evacuees within the 20 km radius additional stable iodine needed to be administered as they evacuated through the plume. In total, 164,218 people evacuated, including those who voluntarily evacuated (Yasumura, 2014). UNSCEAR have reported that "No radiation-related deaths or acute diseases have been observed among the workers and general public exposed to radiation from the accident" and that "The doses to the general public, both those incurred during the first year and estimated for their lifetimes, are generally low or very low. No discernible increased incidence of radiation-related health effects are expected among exposed members of the public or their descendants. The most important health effect is on mental and social well-being, related to the enormous impact of the earthquake, tsunami and nuclear accident, and the fear and stigma related to the perceived risk of exposure to ionizing radiation" (United Nations, 2013). The UN also note that an estimated 160 workers received doses in excess of 100 mSv, but the associated increased incidence of cancer is expected to be indiscernible in this cohort (United Nations, 2013). Yasumura (2014) points out that whilst there have been no prompt radiation-induced deaths, prompt deaths arising from the evacuation were observed, especially among the elderly and infirm. We posit that

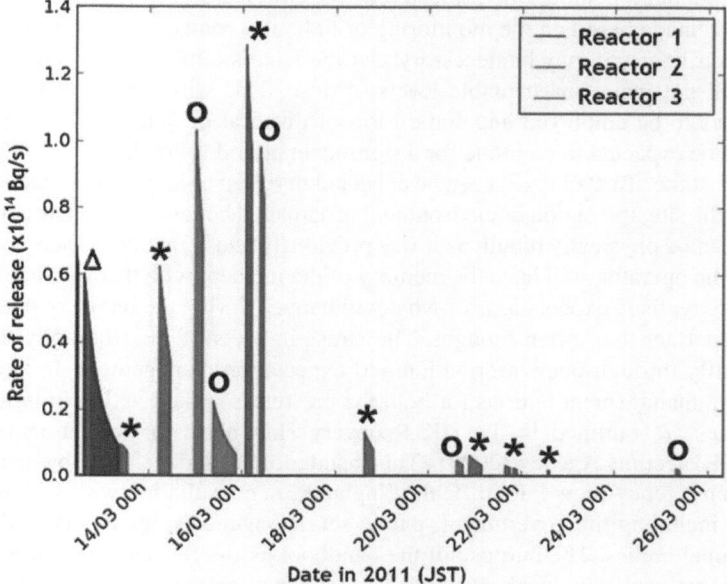

FIGURE 19.1 Rate of all radio nuclides released into the atmosphere from the accident at Fukushima Dai-ichi. (Δ) denotes unit #1, (o) denotes unit #2, and (*) denotes unit #3 (Reproduced from Ashley et al. 2017a,b Copyright © Elsevier)

radiation-induced deaths and deaths arising from implementing the evacuation policies should both be considered as part of the nuclear accident. By doing this, it can hopefully lead to more effective emergency management procedures. One way in which evacuation-related deaths can be gauged is by looking at the total distances that evacuees travel (in person-km or vehicle-km) and to compare this against associated deaths per person-km and/or vehicle-km associated with car travel. A compilation of fatalities per vehicle-km can be found in World Health Organization (2013). An initial study of non-radiological evacuation risks was performed by Aumonier and Morrey (1990) and suggested that within the UK an upper risk estimate of 1×10^{-8} per person-km should be assigned for both fatalities and injuries, although it is noted in that work that such risks are likely to be lower during an evacuation; and the risk from preparing to evacuate were difficult to assess, but presumably would be higher than the domestic daily accident rate of 3.3×10^{-7} per person for fatalities. It also noted that stress from the evacuation is inseparable from the stress of the accident itself; and that self-evacuation poses an issue in assessing the overall collective risk associated with evacuation.

19.3.3 RECOVERY PHASE

The recovery phase begins once the event is brought under control. Following certain immediate countermeasures, such as short-term evacuation, decisions on whether the affected population should be relocated or the timescale on which they can return need to be made based on the monitoring of radiation contamination. Before returning to their homes it may be necessary that the land is remediated so that the occupants will not incur unacceptable levels of dose. This will require one of various techniques to be employed and some form of disposal of materials affected. Food bans can be expected to continue for a significant period following the accident. It is seldom that the affected region can be expected to return to its previous state entirely. Perhaps the site, the region or environment is damaged beyond repair. Even when the system can be physically rebuilt as it was previously, those involved, their communities and the operator, will have the memory of the incident with them for a long time. Managing realistic expectations of what will happen during the recovery phase is far more important than often thought. The stress impacts of the Chernobyl accident arose partly through poor information and expectation management. In the United Kingdom, management options for both the pre-release phase and long-term recovery phases are outlined in the UK Recovery Handbook for Radiation Incidents (Health Protection Agency, 2009). This handbook was developed by the Health Protection Agency (now Public Health England), in consultation with various stakeholders, including the government, public service agencies, regulatory bodies, and professional bodies. The purpose of the handbook is to offer guidance in non-crisis times to national and local authorities, emergency services, radiation protection experts, agriculture and food production sectors, the water industry, and others who may be affected by a radiological instance in developing their recovery strategies. A useful extension of the handbook may involve coupling it to a Geographical Information Systems database to allow an immediate comparison between different management strategies throughout the UK. Further information on the underpinning science, the range and complexity of the issues responders will face, and pointers to delivering a recovery strategy are provided in Part 3 of the National Nuclear Emergency Planning and Response Guidance (HM Government, 2015). At the time of the accident at Fukushima Daiichi, international guidance on the protection of people in emergency exposure situations and for those living in long-term contaminated areas was provided by ICRP 109 (ICRP, 2009a, p. 109) and ICRP 111 (ICRP, 2009b, p. 111; ICRP, 2009c), respectively. This guidance uses the concept of reference levels for existing and emergency exposure situations, with recommended reference levels for emergency exposure situations in the band of 20–100 mSv effective dose (acute or per year). The reference level represents the level of residual dose or risk above which it is generally judged to be inappropriate to plan to allow exposures to occur. ICRP considers that a dose rising toward 100mSv will almost always justify protective measures; and that protection against all exposures, above or below the reference level, should be optimized. (For comparison and contrast, in ICRP 103 (ICRP, 2007), reference levels for existing exposure situations are recommended to be within 1–20 mSv of projected dose per year. Euratom's Basic Safety Standards Directive also details the use of reference levels (European Commission, 2013).) The accident at Fukushima Daiichi has led to unforeseen issues and concerns with this guidance; notably, the guidance had not covered the necessary remediation level explicitly, so it implicitly assumed that it should be performed down to the baseline level of 1 mSv per year over background, which is the dose limit to the public from normal operations. Whilst remediation to a low threshold may be sensible from a cost–benefit perspective in densely populated urban areas, it is not necessarily sensible from a cost–benefit perspective in areas where there are lower levels of permanent occupation. Another factor that requires further consideration in this guidance is the time required to remediate the contaminated area. It can be argued that in the majority of cases, evacuation followed by swift remediation and repopulation is more beneficial to those affected than protracted evacuation and relocation (Yumashev et al., 2017). Present guidance suggests that relocation is allowed if doses in contaminated areas do not exceed 20 mSv per year on the basis that further remediation will take place. However, the available workforce to perform the remediation can be significantly diminished by the scale of the work needed and hence this may inhibit the ability for those people who are relocated to return. In short, there are still unanswered questions that surround the transition from the response phase to the recovery phase, and whether there can be a transition from recovery to normality. A detailed timeline of the events of both the on-site and off-site recovery of TMI-2 is presented in IAEA and JRC (2012) and is detailed in the USNRC "Programmatic Environmental Impact Statement" NUREG-0683 (US Nuclear Regulatory Commission, 1981). As mentioned earlier, following the very limited release of radioactivity at TMI-2, the majority of those who self-evacuated returned within two weeks and there were no detectable health effects on plant workers or the public. Estimates suggest that ~52% of people living within 20 miles self-evacuated (including ~72% of mothers who had preschool children) (Dohrenwend, 1983). For those who were living in the affected areas at the time of the accident, a sharp rise in non-specific distress, "demoralization", was observed in April 1979 which appeared to reduce sharply toward background levels in follow-up measurements made in May 1979 and mid-July 1979; this contrasts with public distrust which remained high during this period (Dohrenwend, 1983). A significant amount of research has been performed and published on the recovery phase of Chernobyl and is encompassed in research presented in this special issue. Further information can be found in IAEA (2002b), IAEA (2002b), IAEA (1996), International Chernobyl Project and IAEA (1991), Medvedev (1990), and Smith and Beresford (2005). Details of the continuing off-site recovery at Fukushima Daiichi are best described in Chapter 36 of 2014 World Bank publication "Mega disasters" (Ranghieri and Ishiwatari, 2014). Following the incident ~160,000 left their hometowns for transition shelters. As of mid-2012, ~62,000 of those within the evacuation zone had evacuated away from the Fukushima prefecture, with this number decreasing to ~48,000 by mid-2014. The location of these people across Japan is shown in Figure 19.2. It is noted that the displacement of people from the Fukushima prefecture has placed strains on housing and other civic services within other prefectures hosting those evacuated and that in certain places tensions exist between the hosts and evacuees. 3194 premature deaths have been ascribed to "physical and mental fatigue" from the earthquake (Table 19.4). As of September 30, 2014, 1793 deaths have been recorded from those within the Fukushima prefecture which have been ascribed to physical and mental fatigue from the accident at Fukushima Daiichi (The Reconstruction Agency, 2014). Preliminary investigations into the psychological distress caused by those evacuated due to the Great East Japan Earthquake and accident at Fukushima Daiichi have recently been published (Yabe et al., 2014) and show

FIGURE 19.2   The number of evacuees from the Fukushima prefecture, as of February 13th 2014 (Reproduced from Ashley et al. 2017a,b Copyright © Elsevier)

protracted mental health trauma up to the end of the 2012 Japanese financial year. The percentage of those who were surveyed from the evacuation zones with serious mental illnesses are between 4–5 times higher than background (~12–15% cf. 3%). It would be of interest to see how these numbers, and those reported in Table 19.4, are related to other measures of morbidity (such as Quality Affected Life Years). Nearly four years on from the accident at Fukushima Daiichi the majority of the evacuated areas are still yet to be repopulated, as shown in Figure 19.3. Even once an area has been rehabilitated, only a fraction of those people may return due to the lack of public services. This has been observed with the lifting of evacuation orders for Hirono Town, where only ~25% of the town's inhabitants have returned within 15 months of the evacuation orders being lifted (Ranghieri and Ishiwatari, 2014). It is noted that this effect has also been observed following the evacuation of New Orleans due to Hurricane Katrina in August 2005. The population of New Orleans 11 months after the hurricane was only ~47% than that recorded in the 2000 US Census; by July 2014, this had risen to ~79% (Plyer, 2015). Experience from the world's previous major nuclear accidents provides us with two important lessons: (a) well-meaning countermeasures introduced purely on the basis of radiological protection "rules" can do harm as well as good; and (b) more generally, that the non-radiological health consequences of a nuclear accident may well be more significant than those caused directly by exposure to radiation (World Health Organization, 2005a,b, 2006; Waddington et al., 2017a). Given these important and now well-documented

TABLE 19.4

The Number of Earthquake-related Deaths in the Great East Japan Earthquake (by Both Prefecture and Time) up to September 30 2014

Prefecture	Number of Attributable Deaths in the Time Period Since the Great East Japan Earthquake										Prefecture Total
	<1 week	1 wk–1month	1–3 months	3–6 months	6–12 months	12–18 months	18–24 months	24–30 months	30–36 months	36–42 months	
Iwate	93	120	116	59	36	14	5	1	2	0	446
Miyagi	232	332	212	77	28	8	5	3	2	1	900
Yamagata	0	1	0	0	0	1	0	0	0	0	2
Fukushima	111	256	333	315	349	189	129	82	29	0	1793
Ibaraki	19	12	5	4	4	0	0	0	0	0	41
Saitama	1	0	0	0	0	0	0	0	0	0	1
Chiba	2	1	0	1	0	0	0	0	0	0	4
Tokyo Met.	1	0	0	0	0	0	0	0	0	0	1
Kanagawa	2	1	0	0	0	0	0	0	0	0	3
Nagano	1	1	1	0	0	0	0	0	0	0	3
Total	462	724	667	456	414	212	139	86	33	1	3194

(Adapted from Ashley et al. 2017 Copyright © Elsevier)

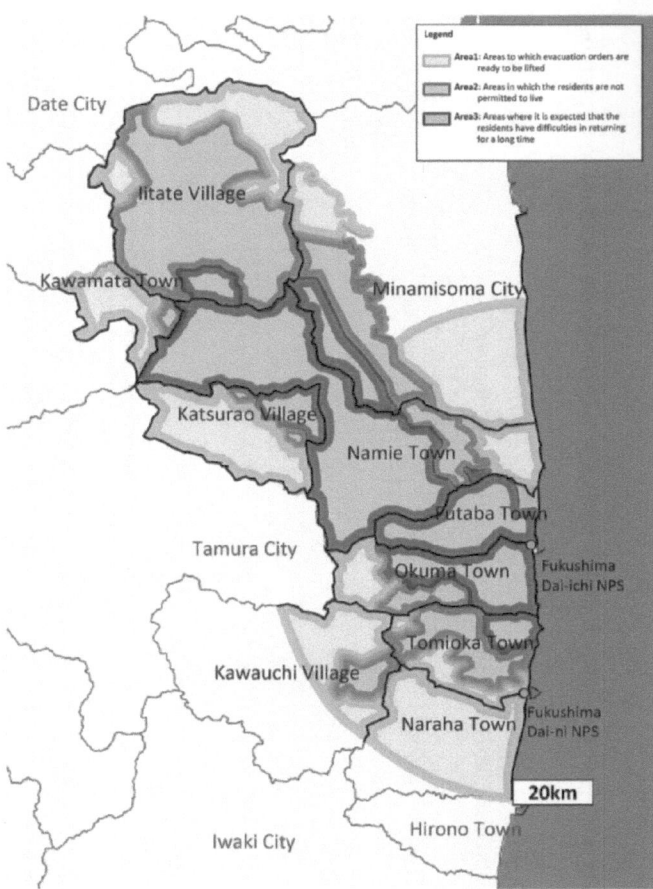

FIGURE 19.3   Areas where evacuation orders have been issued following the accident at Fukushima Dai-ichi (Reproduced from Ashley et al. 2017a,b Copyright © Elsevier)

findings, there is a need to guard against the danger that future international guidance on nuclear emergency planning will not be driven by radiation dose-related criteria to the exclusion of social and economic needs. It is sensible to bear in mind at all times the principles of radiological protection, as codified by the ICRP (2007), which require that "any decision that alters the radiation exposure situation should do more good than harm" (the "principle of justification"); and "the likelihood of incurring exposure, the number of people exposed, and the magnitude of their individual doses should all be kept as low as reasonably achievable, taking into account economic and societal factors" (the "principle of optimization").

19.4 ECONOMIC COSTS ANALYSIS

A balance needs to be struck between the costs of implementing safety measures to reduce the frequency and/or mitigate the consequences of an accident and the costs of the accident should it occur. This requires a consideration of what the costs of an accident would be. In general, the costs of developing and exercising an emergency plan, and the implementation costs if an accident occurs, are not specifically considered. The following paragraphs provide a brief synopsis of typical economic consequences associated with nuclear accidents. In the UK, the computer model COCO-2 has been developed to assess these economic consequences as outlined in Higgins et al. (2008).

19.4.1 FACTORS AFFECTING THE ECONOMIC COSTS OF A NUCLEAR ACCIDENT

Details on the basis of economic costs are taken from an overview into the economic implications of a nuclear accident that can be found in the Organization for Economic Development's report "Methodologies for Assessing the Economic Consequences of Nuclear Reactor Accidents" (OECD-NEA, 2000). From an economic perspective, the cost of a nuclear accident can be viewed as the cost of restoring those affected by such an accident to their pre-accident state, as far as possible. Some of these costs are tangible, in that these are quantifiable costs that relate to an identifiable asset or source, but there are also intangible losses such as the stress suffered by people who have had their lives disrupted. Costing intangible losses of this kind is extremely difficult a priori. These issues are discussed in Higgins et al. (2008).

Tangible losses can involve both direct and indirect costs. Direct costs are those that are directly attributable to the accident. Indirect costs can be defined as those that are secondary effects from the accident. Direct costs are typically easier to quantify than indirect costs. It is also noted that certain benefits may be observed by placing certain countermeasures or actions in place6and these have to also be factored in to the economic costing. Due to the immediate and potentially protracted effects from a radiological release, both "short-term" and "long term" costs need to be accounted for, ranging from the costs of prompt countermeasures to the costs associated with latent and hereditary health effects. Countermeasures thus have the important role in cost-effectively counterbalancing detrimental health and social effects. The costs associated with countermeasures span population movement, agricultural countermeasures and restrictions, and decontamination. Population movement covers evacuation (transportation, temporary accommodation, and food); managing the evacuation, the evacuees, and the evacuated area; loss of income, capital value and investment on land and property; and the health effects from the worry and upheaval of the accident. Costs associated with agricultural countermeasures and restrictions include: the loss of food (including the replacement cost of alternative supplies); the loss of capital value of land and stock during the period of restriction; and storage/ processing/disposal costs. Decontamination costs include: providing the necessary labour (and accounting for adverse health effects), equipment and materials, as well as transporting and disposing of the generated waste. Microeconomic approaches can be used to ascertain these costs if the accident only affects a single country (e.g. the accident at TMI-2), whereas macroeconomic approaches may also need to be used if an accident has ramifications beyond the directly affected country (e.g. in the instance of Chernobyl). The costs associated with radiation-induced health effects span: early effects, such as acute radiation sickness, see e.g. UNSCEAR (1962);

latent effects, such as radiation-induced cancers, detailed further in National Research Council (2006) and UNSCEAR (2008); and hereditary effects (UNSCEAR, 2001).

In addition, non-radiological health effects must be accounted for, including the health effects caused by applying countermeasures (such as physical injury caused by evacuation and psychological detriments caused by the upheaval). Thus, there are direct health costs, indirect health costs (e.g. loss of salary whilst recuperating), and non-market costs arising from the anguish. It appears that there is no overall consensus in how such health costs can be accounted for over both the short term and the long term. The human capital (HC) approach, although simple, is limited and outmoded as it is restricted to considering the direct and indirect costs of the lost output. Restorative actions can have an overall benefit by more than compensating for a particular loss due to the accident. However, it may not be the optimal use of the funds when contrasted with the situation if no accident had occurred (i.e. opportunity cost). It must be noted that there may be alternative uses for the land which may help preserve its value from the working production, whilst willingness to pay (WTP), which allows for the inclusion of non-market costs, is difficult to evaluate. However, recent research has raised severe doubts on the WTP valuations used in the UK for the finding the value of a prevented fatality (Thomas and Vaughan, 2015a,b,c) as well as the validity of the concept in the context of radiological protection (Thomas and Vaughan, 2013; Thomas and Waddington, 2017a; Thomas, 2017a) .

A more modern and objective technique, the Judgement-or J-value method (Thomas et al., 2006a,b; Thomas et al., 2010), has now been developed, based on the life-quality index (Nathwani and Lind, 1997; Nathwani et al., 2009). The J-value is a revealed preference method that is able to place an objective value on the increase in life expectancy that the safety measure brings about by balancing this against the utility that the person being protected loses by his notional payment of the cost of protection in line with the Kaldor–Hicks compensation principle (Kaldor, 1939; Hicks, 1939). Rather than being reliant on the subjective opinions of a small group of people, the J-value is instead grounded in objective actuarial and economic statistics that characterize the lives and behaviors of millions of citizens. Recently validated (Thomas, 2017b; Thomas and Waddington, 2017b) the J-value allows immediate fatalities and loss of life in the longer term (e.g. after exposure, either of workers or of the general public, to nuclear radiation) to be differentiated but measured on the same scale. The J-value methodology has been applied as part of the NREFS study (NREFS, 2017) to assess the relocation measures after Chernobyl and Fukushima Daiichi (Waddington et al., 2017a), the remediation measures taken after those two accidents (Waddington et al., 2017c), and the sheep meat restrictions in the UK after Chernobyl (Waddington et al., 2017).

19.4.2 ECONOMIC MODELLING IN THE UK

To assess these costs in a UK perspective, the COCO-2 model developed by the Health Protection Agency (now Public Health England) (Higgins et al., 2008), is included within the Level-3 Probabilistic Safety Assessment code "PACE" (Charnock et al., 2013). COCO-2 uses an input–output methodology to represent the effect of lost production. The model accounts for various economic activities in terms of

TABLE 19.5
Countermeasures and Associated Breakdown of Costs that are Considered in the COCO-2 Methodology, as Described Further with Formulae to Calculate Gross Value-Added (GVA)

Action	Losses
Sheltering	Cost of lost GVA[a]
Evacuation	Cost of population movement[b]
	Cost of temp. accommodation and temporary lost use of domestic assets
	Cost of lost GVA[a]
Relocation	Cost of population movement
	Cost of temp. accommodation and temporary lost use of domestic assets
	Permanent loss of housing and domestic assets Cost of lost GVA[a]
	Loss of industrial buildings and assets[c]

[a] Both direct and indirect business losses, and direct and indirect tourism losses are accounted for.
[b] Cost of population movement for evacuation is not yet implemented in PACE.
[c] Including capital value of industrial, commercial, and retail buildings.
(Adapted from Ashley et al. 2017 Copyright © Elsevier)

Gross Value Added (GVA), representing the benefit that businesses provide on a square-km basis which are lost as a result of the accident. It also accounts for losses in the supply chain caused by the primary loss of GVA. Furthermore, direct and indirect health costs are accounted for by including net output losses, the costs of medical treatments, and WTP costs for both fatalities and morbidities. The types of short-term and long-term countermeasures that are included in COCO-2 are presented in Table 19.5. COCO-2 does not account for costs of producing and distributing iodine prophylactics, as such costs are typically low when compared to the costs of sheltering, evacuation, and relocation; such costs would also need to account for the shelf-life of such prophylactics, which is typically given as five years. From the 2004 US study (National Research Council, 2004), a range of costs for producing and distributing stable iodine are seen. This variation depends on how stable iodine is distributed (e.g. door-to-door delivery cf. pick-up from a distribution centre); in certain cases costs are approximately centred around $0.50 per distributed tablet. Switzerland has recently brought in a new emergency preparedness plan that involves pre-distributing stable iodine to everyone living within 50 km of a nuclear power plant (4.6 million out of a total population of 8.1 million) at a reported cost of $31 million (Bosley and Bennett, 2014). Agricultural costs in COCO-2 account for various foodstuffs with modifiers that attempt to incorporate any seasonality of product into the GVA and output estimates generated. COCO-2 also includes indirect losses, and the loss of capital stock. It is noted that in an emergency situation, a wide range of countermeasures are available, as outlined in the UK Recovery Handbook (Health Protection Agency, 2009) and for such a range of countermeasures, a wide range of costs estimate scan be derived (e.g. placing ferrocyn in animal feed that limits137Cs transfer to milk which would be a cheaper counter measure compared to slaughtering and disposing of livestock). Further details of modelling the economic effects of a large nuclear accident in the UK are provided in a companion paper (Ashley et al., 2017a,b).

19.5 FACTORS IMPACTING HEALTH AND ECONOMIC COST

This section is concerned with a number of factors that will influence significantly the costs, in terms of both health and economics, involved with a big nuclear accident. The first of these is the fundamental initial condition particular to the nuclear plant, namely where it is sited, and how close it lies to population centers. A further initial condition, this time specific to the accident, is the "source term", that comprises the quantity and isotopic constitution of the nuclear material released, the time delay before the release occurs and the duration of the release. The weather at the time of the accident (wind speed and direction, dry or wet weather) is an exogenous variable that will then determine where the fallout is deposited. The final factor is one that is to a large degree under the control of the authorities after a big nuclear accident, namely the extent of the harvesting and sale of food from areas subject to some degree of radioactive contamination.

19.5.1 SITING AND DEMOGRAPHY

A complete account of the history on the siting of nuclear power stations in the United Kingdom is provided in Grimston et al. (2014). In summary, past differences in attitudes to the siting nuclear power stations in the UK have occurred in four phases. In the first phase (roughly 1945–1965, that included the Magnox reactor programme) siting decisions were made with considerable caution, with distance from populations being a key driver of siting policy. In the second phase (roughly 1965–1985, which included the AGR programme), siting, with regard to the density of local population, was more relaxed due to the belief that the safety and reliability of the reactors had been improved significantly. The third phase (roughly 1985–2005, which included the decision to build Sizewell B) reverted back to conservative siting plans. The main reason was the PWR technology was new to the UK, but other factors played a part such as the accidents at TMI-2 and Chernobyl, In the fourth phase (2005 onwards) a more positive attitude toward siting of future nuclear plants is observed, due to greater public and political support even after the accident at Fukushima Daiichi, although all the approved sites (with one exception) are those already used for nuclear power plants. One of the earliest tools to account for the impact on the public located near a nuclear power station was developed by Farmer, who developed a methodology that considered the trade-off between the frequency and consequence of an accident occurring (Farmer, 1967). This chapter indicated that the relative risk for a remote site compared to a hypothetical 'town' is approximately a factor of 10 lower; and the relative risk for a remote site compared to a hypothetical 'city' is approximately a factor of 100 lower. Subsequent work has noted that certain risk factors may be increased at a remote site (compared to a 'town' or 'city' site), such as less reliable grid interconnections due to longer transmission lines. Since this report, probabilistic safety assessments (Fullwood, 1999), notably Level-3 PSAs as outlined in OECD-NEA (2000), have been developed further to both ascertain the probabilities and corresponding consequences of various accident sequences for various reactor technologies. Present UK policy has involved the use of strategic siting assessments to look at potential sites for new nuclear reactors in the UK (DECC,

2011a,b). The strategic siting assessment includes 12 factors that were used in aiding the decision to grant development consent, including: (1) demographics; (2) proximity to military activities; (3) flooding, tsunami and storm surge; (4) coastal processes; (5) proximity to hazardous industrial facilities and operations; (6) proximity to civil aircraft movements; (7) internationally designated sites of ecological importance; (8) nationally designated sites of ecological importance; (9) areas of amenity, cultural heritage and landscape value; (10) size of site to accommodate operation; (11) access to suitable sources of cooling; (12) capability of the site to store spent fuel and intermediate-level waste. It should be noted that the population around the nuclear facility is not uniform and varies throughout the year; also, certain demographics will be affected and will behave in different ways. For instance, for a specific incident, young families may be encouraged to evacuate whereas the elderly and infirm may be advised against evacuating. Self-evacuation is almost impossible to gauge, but is almost certainly going to happen in societies where access to personal transport is high, e.g. self-evacuation following the accident at TMI-2 whereas the evacuation of Pripyat following the accident at Chernobyl was organized by the authorities. Additional variability will depend on the nature of the surroundings, whether there are hospitals, schools, homes, etc.: and depend on what people are doing at the time of the incident, i.e. inside or outside buildings, in home or away from home.

19.5.2 Source Terms

As alluded to earlier, the consequence of an event occurring is not just the number of people who may be affected but how they are affected. From a radiological standpoint, this centers on the quantities of radionuclides released over the time frame of the accident, referred to as the 'source term'. As there are various different accidents that can occur at a nuclear power plant, there are different modes in which nuclear material may be released that affect: (1) the quantities of radionuclides being released (generally described in terms of its activity); (2) the duration of pre-release (i.e. the time between the initial alarm and the release occurring); and (3) the duration of the release. In the accident at Chernobyl 15300 PBq of activity (excluding noble gases) was released, whereas for Fukushima Daiichi 520 (340–780) PBq of activity was released (Steinhauser et al., 2014). A summary of source terms for other nuclear incidents can be found in Sanderson et al. (1997). It is worth noting here that there is no direct relationship between the overall amount of activity released and the dose received – the isotopic composition, the physical form and its exposure pathway (e.g. external "shine" from the ground or the air, ingestion, and inhalation) are all significant factors. Such conversion factors for each radionuclide are provided by the International Commission for Radiological Protection (Eckerman et al., 2013). Other factors that can affect the severity of the source term are: whether the release is airborne or liquid in nature; the height of the release; and the energy (or heat) in the release.

19.5.3 Weather and Dispersion

For airborne releases, weather plays a significant role in how radionuclides are dispersed. There are several factors here that affect how radionuclides disperse including: wind speed, atmospheric turbulence (typically, in the past, described by Pasquill atmospheric stability classes (Pasquill, 1961)), and precipitation, which may vary over the release time. In general statistics on these aspects is available so as for the source term a statistical confidence level can be considered (though in this case on the basis of real statistics). A factor that influences weather is the time of year and the time of day, but it should be possible to include these in the statistical analysis. The earliest dispersion models that have been used focused on two-dimensional Gaussian techniques to characterize plumes; with two dispersion parameters used to approximate cross-winds and vertical winds, essentially fixed at the point of release, though some variation was possible by restarting calculations after some time. The outputs of such a code are given in the form of a series of graphs in the reportN-RPB-R91, (now Public Health England) (Clarke, 1979). More sophisticated codes have since been developed, such as the UK Met Office NAME III program (Jones et al., 2007) that uses Lagrangian modelling in three dimensions to calculate the behavior of plumes, which allows actual meteorological data to be used, though these require more computing power and are typically much more computationally expensive. A report by OECD-NEA (2000) discussed some aspects of emergency management codes as they were at that time. It is important, however, to recognize that a pre-event calculation can only give a probabilistic view by compiling a set of calculations for different weather conditions derived from various meteorological data, which may be used in planning. For example, it will indicate the likely extent of areas for specific countermeasures, which as the figures show are not uniformly distributed about the site. This will also indicate the likely costs which can be included in calculations of whether safety measures need to be improved. During an event, the weather type and direction may change and it is unlikely that it will match any of the pre-event scenarios. Thus, emergency preparedness plans must be flexible enough to deal with real-time changes. The use of a code such as PACE may be used at this stage to predict behavior, but this will require some way of postulating the source term also.

19.5.4 FOOD

The final major aspect concerns food. As the radioactive material is deposited it will move into the food cycle (and can affect sources of water). Besides what is grown in people's gardens, there is likely to be large areas which, even if for only a short time, will have bans on the sale (and hence distribution of) food both animal and vegetable. Two examples from UK experience are the banning of milk in Cumbria (then Cumberland) in 1957 for about a month in the vicinity of the plant, following the Windscale accident (Arnold, 1992) and the long-term restrictions on sheep following the Chernobyl accident which continued for some UK farms for over 25 years (Waddington et al., 2017). Clearly, as above, the time of year and, to a more limited extent the time of day, are both important factors. The levels of radiation in food which cause them to be banned also need to be established. Further details on how radionuclides enter the food-chain and how this is modelled can be found in Till and Grogan (2008).

19.6 SUMMARY

The accident at Fukushima Daiichi has highlighted and reinforced the need for emergency preparedness and response plans to be prepared prior to the operation of any nuclear facilities that have the potential to release material to the environment. This chapter has reviewed, from a UK perspective, the role of emergency preparedness and response plans, drawing on information from both Chernobyl and Fukushima Daiichi. The phases of a nuclear accident leading to an off-site release have been considered, and a review has been provided of the measures that can be adopted at each phase to mitigate effects on the public. The current UK approach has been examined, and a comparison has been made of UK and international stances. It is clear that a second major reactor accident has spurred efforts nationally and internationally to improve ways of coping with a big nuclear accident, should it happen. New analysis methods are emerging, and, in parallel, international committees of experts are attempting to refine the advice offered to governments who might face a large reactor accident in the future. One potential concern is the desire for too great a level of prescription, which might conflict with the degree of flexibility required by the fundamental principles of radiological protection as enunciated by the ICRP, specifically those of "justification" and "optimization". We observe also that it is necessary to consider societal effects, economic effects and non-radiological health effects in addition to radiological harm when drawing up plans and taking decisions after a major nuclear accident. The lesson from both Chernobyl and Fukushima Daiichi is that while great resources were put in place to keep radiation harm to the public to a very low level, this might have been at the expense of psychological and general health.

REFERENCES

AREVA, EDF Energy, 2012. Pre-Construction Safety Report. Chapter 15.4: Level-2 PSA (No. UKEPR-0002-154 Issue 06).

Arnold, L., 1992. *Windscale, 1957: Anatomy of a Nuclear Accident*. St. Martin's Press, New York.

Ashley, S.F., Vaughan, G.J., Nuttall, W.J., Thomas, P.J., 2017b. Considerations in relation to off-site emergency procedures and response for nuclear accidents. *Process Saf. Environ. Prot.* 112, 77–95.

Ashley, S.F., Vaughan, G.J., Nuttall, W.J., Thomas, P.J., Higgins, N.A., 2017a. Predicting the cost of the consequences of a large nuclear accident in the UK. *Process Saf. Environ. Prot.* 112, 96–113.

Astbury, J., Horsley, S., Gent, N., 1999. Evaluation of a scheme for the pre-distribution of stable iodine (potassium iodate) to the civilian population residing within the immediate countermeasures zone of a nuclear submarine construction facility. *J. Public Health* 21, 412–414, doi:10.1093/pubmed/21.4.412.

Aumonier, S., Morrey, M., 1990. Non-radiological risks of evacuation. *J. Radiol. Prot.* 10, 287, doi:10.1088/0952-4746/10/4/004.

Ayrshire Civil Contingencies Team, 2015. *Hunterston B Nuclear Power Station and Hunterston. A Decommissioning Site Off-Site Contingency Plan Redacted Version*. https://www.east-ayrshire.gov.uk/Resources/PDF/H/Hunterston-off-site-emergency-plan-redacted-version.pdf. (Accessed August12 2017).

Bosley, C., Bennett, S., 2014. *Switzerland Hands Out Iodine in Case of Nuclear Disaster.* Bloomberg, http://www.bloomberg.com/news/2014-11-05/switzerland-hands-out-iodine-in-case-of-nuclear-disaster.html. (Accessed August12 2017).

British Energy Generation Ltd, 2007a. *Heysham Power Stations: Emergency Plan.* British Energy Generation Ltd.

British Energy Generation Ltd, 2007b. *Hartle pool Power Station: Emergency Plan.* British Energy Generation Ltd.

Charnock, T.W., Bexon, A.P., Sherwood, J., Higgins, N.A., Field, S.J., 2013. PACE: a geographic information system based level 3 probabilistic accident consequence evaluation program. In: Presented at the ANS PSA 2013 International Topical Meeting on Probabilistic Safety Assessment and Analysis, Columbia, SC.

Clarke, R.H., 1979. A Model for Short and Medium Range Dispersion of Radionuclides Released to the Atmosphere. NRPB R-91.

DECC, 2011a. National Policy Statement for Nuclear Power GENERATION (EN-6). Volume I of II.

DECC, 2011b. National Policy Statement for Nuclear Power Generation (EN-6). Volume II of II-Annexes.

Dohrenwend, B.P., 1983. Psychological implications of nuclear accidents: the case of Three Mile Island. *Bull. N. Y. Acad. Med.* 59, 1060–1076.

East Lothian Council, 2016. *Torness Off-Site Emergency Plan.* http://www.eastlothian.gov.uk/downloads/file/2821/tornessoff-siteplan-publicversion. (Accessed August 12 2017).

Eckerman, K., Harrison, J., Menzel, H.-G., Clement, C.H., 2013. *ICRP Publication 119: Compendium of Dose Coefficients based on ICRP Publication 60. Ann. ICRP, ICRP PUBLICATION 123: Assessment of Radiation Exposure of Astronauts in Space,* 42, e1–e130. doi:10.1016/j.icrp.2013.05.003.

Ehrhardt, J., Weis, A., 2000. RODOS: Decision support system for off-site nuclear emergency management in Europe (EUR19144).

European Commission, 1989. Council Directive of 27 November 1989 on informing the general public about health protection measures to be applied and steps to be taken in the event of a radiological emergency, 89/618/EURATOM.

European Commission, 1996. Council Directive 96/29/Euratom of 13 May 1996 laying down basic safety standards for the protection of the health of workers and the general public against the dangers arising from ionizing radiation, 96/29/EURATOM.

European Commission, 2013. Council Directive 2013/59/Euratom of 5 December 2013 laying down basic safety standards for protection against the dangers arising from exposure to ionising radiation, and repealing Directives 89/618/Euratom,90/641/Euratom, 96/29/Euratom, 97/43/Euratom and 2003/122/Euratom, 2013/59/EURATOM.

European Commission, 2016. Council Regulation (Euratom) 2016/52 of 15 January 2016 laying down maximum permitted levels of radioactive contamination of food and feed following a nuclear accident or any other case of radiological emergency, and repealing Regulation (Euratom) No 3954/87and Commission Regulations (Euratom) No 944/89 and (Euratom) No 770/90.

European Commission Directorate-General for Energy, Jourdain, J.R., Herviou, K., 2010. Medical effectiveness of iodine prophylaxis in a nuclear reactor emergency situation and overview of European practices.

Farmer, F.R., 1967. *Siting criteria-a new approach.* In: *Containment and Siting of Nuclear Power Plants. Presented at the Symposium on the Containment and Siting of Nuclear Power Plants,* IAEA, Vienna, pp. 303–323.

Fullwood, R.R., 1999. *Probabilistic Safety Assessment in the Chemical and Nuclear Industries.* Butterworth-Heinemann, Oxford, United Kingdom.

Gering, F., Gerich, B., Wirth, E., Kirchner, G., 2013. Potential consequences of the Fukushima accident for off-site nuclear emergency management: a case study for Germany. *Radiat. Prot. Dosimetry* 155, 146–154, doi:10.1093/rpd/ncs323.

Government of Japan, 2011. Report of the Japanese Government to the IAEA Ministerial Conference on Nuclear Safety –Accident at TEPCO's Fukushima Nuclear Power Stations.

Grimston, M., Nuttall, W.J., Vaughan, G., 2014. The siting of UK nuclear reactors. *J. Radiol. Prot.* 34, R1, doi:10.1088/0952-4746/34/2/R1.

Health and Safety Executive, 1990. Outline Emergency Planning for Licensed Nuclear Power Stations.

Health and Safety Executive, 1994. Arrangements for responding to nuclear emergencies.

Health and Safety Executive, 2001. *Principles and guidelines to assist HSE in its judgements that duty-holders have reduced risk as low as reasonably practicable.* http://www.hse.gov.uk/risk/theory/alarp1.htm. (Accessed August 12 2017).

Health and Safety Executive, 2002. A guide to the Radiation (Emergency Preparedness and Public Information) Regulations 2001.

Health Protection Agency, 2009. HPA-RPD-064-UK Recovery Handbooks for Radiation Incidents: 2009 (Version 3).

Heffron, R.J., Ashley, S.F., Nuttall, W.J., 2016. Reform and issues in the global nuclear liability regime post Fukushima. *Prog. Nucl. Energy* 90, 1, doi:10.1016/j.pnucene.2016.02.019.

Hicks, J.R., 1939. *Value and Capital.* Oxford University Press, Oxford, UK.

Higgins, N.A., Jones, C., Munday, M., Balmforth, H., Holmes, W., Pfuderer, S., Mountford, L., Harvey, M., Charnock, T., 2008. COCO-2: a model to assess the economic impact of an accident. ISBN: 978-0-85951-628-0 (No. HPA-RPD-046).

HM Government, 1965. Nuclear Installations Act 1965.

HM Government, 1974. Health and Safety at Work Act 1974.

HM Government, 2004. Civil Contingencies Act 204.

HM Government, 2013. *Energy Act 2013.* UK Parliament: London.

HM Government, 2015. *Nuclear emergency planning: consolidated guidance.* https://www.gov.uk/government/publications/nuclear-emergency-planning-consolidated-guidance. (Accessed August 12 2017).

IAEA, 1994. Convention on Nuclear Safety. IAEA (Ed.), 1996. *One decade after Chernobyl: Summing Up the Consequences of the Accident proceedings of an International Conference on One Decade After Chernobyl: Summing Up the Consequences of the Accident, International Conference on One Decade After Chernobyl: Summing Up the Consequences of the Accident.* IAEA, Vienna.

IAEA (Ed.), 1996. *One decade after Chernobyl: Summing Up the Consequences of the Accident proceedings of an International Conference on One Decade After Chernobyl: Summing Up the Consequences of the Accident, International Conference on One Decade After Chernobyl: Summing Up the Consequences of the Accident.* IAEA, Vienna.

IAEA, 2002a. Preparedness and Response for a Nuclear or Radiological Emergency, GS-R-2.

IAEA, 2002b. *The human consequences of the Chernobyl nuclear accident: a strategy for recovery: a report.* IAEA Vienna, Austria.

IAEA, 2006. Fundamental Safety Principles SF-1 (No.STI/PUB/1273).

IAEA, 2012a. Safety of Nuclear Power Plants: Design, IAEA Safety Standards, Specific Safety Requirements. Vienna, Austria.

IAEA, 2012b. Lessons learned from the response to radiation emergencies (1945–2010).

IAEA, 2013. Actions to protect the public in an emergency due to severe conditions at a light water reactor (No. EPR-NPP-PPA). Vienna, Austria.

IAEA, 2014a. Status of Convention on Nuclear Safety.

IAEA, 2014b. Radiation Protection and Safety of Radiation Sources: International Basic Safety Standards (No. GSR Part 3).

IAEA, 2015. Preparedness and Response for a Nuclear or Radiological Emergency. IAEA Safety Standards Series (No.GSR Part 7).

IAEA, JRC, 2012. Chronology of Key Milestones and NRC Actions Taken During the Three Mile Island Unit 2 Recovery and Decontamination, Major Nuclear Accidents.

ICRP, 2007. The 2007 recommendations of the International Commission on Radiological Protection. *Ann. ICRP* 103, 81–123. doi:10.1016/j.icrp.2007.10.006.

ICRP, 2009a. ICRP 109: application of the commission's recommendations for the protection of people in emergency exposure situations. *Ann. ICRP* 39, 11–74. doi: 10.1016/j.icrp.2009.09.006.

ICRP, 2009b. ICRP 111: application of the commission's recommendations to the protection of people living in long-term contaminated areas after a nuclear accident or a radiation emergency. *Ann. ICRP* 39, 15–33, doi:10.1016/j.icrp.2009.09.005.

ICRP, 2009c. ICRP 111: application of the commission's recommendations to the protection of people living in long-term contaminated areas after a nuclear accident or a radiation emergency. *Ann. ICRP* 39, 35–46, doi:10.1016/j.icrp.2009.09.006.

International Chernobyl Project, IAEA, 1991. *The International Chernobyl Project: Technical Report: Assessment of Radiological Consequences and Evaluation of Protective Measures*. IAEA, Vienna.

Isle of Anglesey County Council, 2011. *Wylfa Nuclear Power Station Off-site Emergency Plan*. http://www.onr.org.uk/foi/2013/2013030177.pdf. (Accessed August12 2017).

Johnson, J.R., 2003. Guest Editorial – on the distribution of potassium iodide to members of the public in anticipation of an accidental release of radioiodine. *Radiat. Prot. Dosimetry* 104, 195–197.

Jones, A., Thomson, D., Hort, M., Devenish, B., 2007. The U.K. met office's next-generation atmospheric dispersion model, NAMEIII. In: Borrego, C., Norman, A.-L. (Eds.), *Air Pollution Modeling and Its Application XVII*. Springer, US, pp. 580–589.

Kaldor, N., 1939. Welfare propositions and interpersonal comparisons of utility. *Econ J. XLIX*, 549–552.

Kent County Council, 2015. *Dungeness B Nuclear Power Station Off Site Emergency Plan*. https://www.kent.gov.uk/data/assets/pdffile/0017/11339/Dungeness-off-site-emergency-plan.pdf. (Accessed August 12 2017).

Medvedev, Z.A., 1990. *The Legacy of Chernobyl*. W.W. Norton, New York, ISBN: 0393308146.

Morrey, M., 1997. Application of Emergency Reference Levels of Dose in Emergency Planning and Response. *Documents of the NRPB* 8:1, 21–34.

Moss, T.H., Sills, D.L. (Eds.), 1981. The Three Mile Island nuclear accident: lessons and implications, Annals of the New York Academy of Sciences. ISBN: 9780897661164.

Nathwani, J.S., Lind, N.C., 1997. Affordable Safety by Choice: the Life Quality Method. Institute for Risk Research, University of Waterloo, Waterloo, Ontario, Canada, ISBN: 9780969674795.

Nathwani, J.S., Pandey, M.D., Lind, N.C., 2009. *Engineering Decisions for Life Quality: How Safe is Safe Enough?* Springer, London, ISBN: 9781848826021.

National Research Council, 2004. *Distribution and Administration of Potassium Iodide in the Event of a Nuclear Incident*. National Academies Press.

National Research Council, 2006. Health Risks from Exposure to Low Levels of Ionizing Radiation: BEIR VII Phase 2. National Academy of Sciences, Washington, United States.

NREFS, 2017. *Managing Nuclear Risk: Environmental, Financial and Safety*. http://www.nrefs.org. (Accessed August 12 2017).

OECD, 2003. Short-term Countermeasures in Case of a Nuclear or Radiological Emergency. NEA.

OECD-NEA, 2000. Methodologies for Assessing the Economic Consequences of Nuclear Reactor Accidents.

Office of Nuclear Regulation, 2013a. LC 11 – Emergency Arrangement (No. NS-INSP-GD-011 (Rev 2)), Office for Nuclear Regulation (ONR) Compliance inspection – Technical inspection guides.

Office of Nuclear Regulation, 2013b. The Technical Assessment of REPPIR Submissions and the Determination of Detailed Emergency Planning Zones (No. NS-TAST-GD-082 Revision 2), Office for Nuclear Regulation (ONR) Compliance inspection –Technical assessment guides.

Office of Nuclear Regulation, 2014. Licensing Nuclear Installations (No. 3rd Edition).

Office of Nuclear Regulation, 2016. *Emergency planning areas around UK nuclear installations*. http://www.onr.org.uk/depz.htm. (Accessed August 12 2017).

Pasquill, F., 1961. The estimation of the dispersion of wind borne material. *Meteorol. Mag.* 90, 33–49.

Plyer, A., 2015. *Facts for Features: Katrina Impact. The Data Center*. https://s3.amazonaws.com/gnocdc/reports/TheDataCenterFactsforFeatures.pdf. (Accessed August 12 2017).

Ranghieri, F., Ishiwatari, M., 2014. Learning from Megadisasters: Lessons from the Great East Japan Earthquake (No. 89069). World Bank.

Sanderson, D.C.W., Cresswell, A., Allyson, J.D., McConville, P., 1997. *Review of Past Nuclear Accidents: Source Terms and Recorded Gamma-Ray Spectra*. http://eprints.gla.ac.uk/58967/. (Accessed August 12 2017).

Smith, J.T., Beresford, N.A., 2005. *Chernobyl: catastrophe and consequences*. Springer; Published in association with Praxis Pub., Berlin; New York; Chichester, UK.

Somerset County Council, 2008. Hinkley Point Essential Services Off-Site Plan for Hinkley Point A & B Nuclear Licenced Sites.

Steinhauser, G., Brandl, A., Johnson, T.E., 2014. Comparison of the Chernobyl and Fukushima nuclear accidents: a review of the environmental impacts. *Sci. Total Environ.* 470, 800–817, doi:10.1016/j.scitotenv.2013.10.029.

Suffolk Resilience, 2017. *Sizewell Off Site Emergency Plan. Issue3.5*. http://www.suffolkresilience.com/assets/PDF-plans/Sizewell/NPM-Sizewell-Off-Site-Plan-Issue-3.5-dated-28-Feb-17.pdf.(Accessed August 12 2017).

The Reconstruction Agency, 2014. *The death toll of the earthquake-related deaths in the Great East Japan Earthquake* (September 30, 2014). http://www.reconstruction.go.jp/topics/main-cat2/sub-cat2-1/20141226kanrenshi.pdf. (Accessed August 12 2017).

Thomas, P.J., 2017a. Age at death from a radiation-induced cancer based on the Marshall model for mortality period. *Process Saf. Environ. Prot.* 112, 143–178.

Thomas, P., 2017b. Corroboration of the J-value model for life-expectancy growth in industrialised countries. *Nanotechnol. Percept.* 13 (1), 31–44.

Thomas, P.J., Jones, R.D., Kearns, J.O., 2010. The trade-offs embodied in J-value analysis. *Process Saf. Environ. Prot.* 88 (3), 147–167, doi:10.1016/j.psep.2010.02.001.

Thomas, P.J., Stupples, D.W., Alghaffar, M.A., 2006a. The extent of regulatory consensus on health and safety expenditure. Part1: development of the J-value technique and evaluation of the regulators' recommendations. *Process Saf. Environ. Prot.* 84(5), 1–8, doi:10.1205/psep05005.

Thomas, P.J., Stupples, D.W., Alghaffar, M.A., 2006b. The extent of regulatory consensus on health and safety expenditure. Part2: applying the J-value technique to case studies across industries. *Process Saf. Environ. Prot.* 84 (5), 9–15, doi:10.1205/psep05006.

Thomas, P., Vaughan, G., 2013. All in the balance: assessing schemes to protect humans and the environment. *Nucl. Future* 9 (3), 41–51.

Thomas, P.J., Vaughan, G.J., 2015a. Testing the validity of the value of a prevented fatality (VPF) used to assess UK safety measures. *Process Saf. Environ. Prot.* 94, 239–261, doi:10.1016/j.psep.2014.07.001.

Thomas, P.J., Vaughan, G.J., 2015b. 'Testing the validity of the value of a prevented fatality (VPF) used to assess UK safety measures': reply to the comments of Chilton, Covey, Jones-Lee, Loomes, Pidgeon and Spencer. *Process Saf. Environ. Prot.* 93, 299–306, doi:10.1016/j.psep.2014.11.003.

Thomas, P.J., Vaughan, G.J., 2015c. Pitfalls in the application of utility functions to the valuation of human life. *Process Saf. Environ. Prot.* 98, 148–169, doi:10.1016/j. psep.2015.07.002.

Thomas, P., Waddington, I., 2017a. What is the value of life? A review of the value of a prevented fatality used by regulators and others in the UK. *Nucl. Future* 13 (1), 32–39.

Thomas, P., Waddington, I., 2017b. Validating the J-value safety assessment tool against pannational data. *Process Saf. Environ. Prot.* 112, 179–197.

Till, J.E., Grogan, H.A., 2008. *Radiological Risk Assessment and Environmental Analysis.* Oxford University Press, Oxford, New York.

U.S. Nuclear Regulatory Commission, 1981. Final programmatic environmental impact statement related to decontamination and disposal of radioactive wastes resulting from March 28, 1979, accident Three Mile Island Nuclear Station, Unit 2, Docket no. 50-320, Metropolitan Edison Company, Jersey Central Power and Light Company, Pennsylvania Electric Company. U.S. Nuclear Regulatory Commission, Office of Nuclear Reactor Regulation: Available from GPO Sales Program, Division of Technical Information and Document Control, U.S. Nuclear Regulatory Commission; National Technical Information Service, Washington, D.C.: Springfield, VA.

United Nations, 2013. Report of the United Nations Scientific Committee on the Effects of Atomic Radiation: Sixtieth Session (No. A/68/46).

UNSCEAR, 1962. UNSCEAR 1962 Report. Annex D: Somatic Effects of Radiation. United Nations, New York.

UNSCEAR, 2001. Hereditary Effects of Radiation UNSCEAR 2001 Report to the General Assembly, with Scientific Annex. United Nations, New York.

UNSCEAR, 2008. Effects of Ionizing Radiation United Nations Scientific Committee on the Effects of Atomic Radiation: UNSCEAR 2006. United Nations, New York.

Waddington, I., Thomas, P.J., Taylor, R.H., Jones, R.D., Thomas, P.J., 2017. J-value assessment of the cost effectiveness of UK sheep meat restrictions after the 1986 Chernobyl accident. *Process Saf. Environ. Prot.* 112, 114–130.

Waddington, I., Thomas, P.J., Taylor, R.H., Vaughan, G.J., 2017a. J-value assessment of relocation measures following the nuclear power plant accidents at Chernobyl and Fukushima Daiichi. *Process Saf. Environ. Prot.* 112, 16–49.

Waddington, I., Thomas, P.J., Taylor, R.H., Vaughan, G.J., 2017c.J-value assessment of remediation measures following the nuclear power plant accidents at Chernobyl and Fukushima Daiichi. *Process Saf. Environ. Prot.* 112, 50–62.

Wakeford, R., 2016. Chernobyl and Fukushima—where are we now? *J. Radiol. Prot.* 36, E1, doi:10.1088/0952-4746/36/2/E1.

World Health Organization, 1999. Guidelines for Iodine Prophylaxis following Nuclear Accidents – 1999 update (No. WHO/SDE/PHE/99.6). Geneva, Switzerland.

World Health Organization, 2005a. Chernobyl: the true scale of the accident. 20 Years Later a UN Report Provides Definitive Answers and Ways to Repair Lives, Joint News Release WHO/IAEA/UNDP,   http://www.who.int/mediacentre/news/releases/2005/pr38/en/. (Accessed August 12 2017).

World Health Organization, 2005b. Chernobyl: the true scale of the accident. 20 Years Later a UN Report Provides Definitive Answers and Ways to Repair Lives. Answers to Longstanding Questions. http://www.who.int/mediacentre/news/releases/2005/pr38/en/index1.html. (Accessed August 12 2017).

World Health Organization, 2006. *Health effects of the Chernobyl accident: an overview.* http://www.who.int/ionizing radiation/chernobyl/backgrounder/en/. (Accessed August 12 2017).

World Health Organization, 2013. Global status report on road safety 2013: supporting a decade of action.

Yabe, H., Suzuki, Y., Mashiko, H., Nakayama, Y., Hisata, M., Niwa, S.-I., Yasumura, S., Yamashita, S., Kamiya, K., Abe, M., Mental Health Group of the Fukushima Health Management Survey, 2014. Psychological distress after the Great East Japan Earthquake and Fukushima Daiichi nuclear power plant accident: results of a mental health and lifestyle survey through the Fukushima Health Management Survey in FY2011 and FY2012. *Fukushima J. Med. Sci.* 60, 57–67, doi:10.5387/fms.2014-1.

Yasumura, S., 2014. Evacuation effect on excess mortality among institutionalized elderly after the Fukushima Daiichi nuclear power plant accident. *Fukushima J. Med. Sci.* 60, 192–195, doi:10.5387/fms.2014-13.

Yumashev, D., Johnson, P., Thomas, P.J., 2017. Economically optimal strategies for medium-term recovery after a major nuclear reactor accident. *Process Saf. Environ. Prot.* 112, 63–76.

Index